Novel Challenges and Therapeutic Options for Liver Diseases

Novel Challenges and Therapeutic Options for Liver Diseases

Editor

Guido Gerken

MDPI

Basel • Beijing • Wuhan • Barcelona • Belgrade • Novi Sad • Cluj • Manchester

Editor
Guido Gerken
Gastroenterology and Hepatology
University Hospital of Essen
University Duisburg-Essen
Essen
Germany

Editorial Office
MDPI
St. Alban-Anlage 66
4052 Basel, Switzerland

This is a reprint of articles from the Special Issue published online in the open access journal *Journal of Personalized Medicine* (ISSN 2075-4426) (available at: www.mdpi.com/journal/jpm/special_issues/7XL1GUOPU1).

For citation purposes, cite each article independently as indicated on the article page online and as indicated below:

Lastname, A.A.; Lastname, B.B. Article Title. *Journal Name* **Year**, *Volume Number*, Page Range.

ISBN 978-3-7258-1528-9 (Hbk)
ISBN 978-3-7258-1527-2 (PDF)
doi.org/10.3390/books978-3-7258-1527-2

Contents

About the Editor

Guido Gerken

Prof. Guido Guenther Heinrich Gerken is an MD Professor of Medicine. He is director emeritus of the Department of Gastroenterology and Hepatology Universitaetsklinikum, Essen, Germany. He is a senior Consultant at the Helios Klinikum Niederberg in Velbert, Germany.

His clinical interests include internal medicine, gastroenterology, hepatology, clinical virology, clinical immunology, endoscopy, liver transplantation, autoimmunity, infectiology, visceral medicine, acute and terminal liver failure, molecular pathomechanisms of liver failure, fibrosis progression of chronic liver diseases, the evaluation of mini-laparoscopy for the diagnosis of chronic liver diseases, Crohn's disease, ulcerative colitis, autoimmune hepatitis, viral hepatitis and HIV coinfection, hepatocellular carcinoma (HCC), and the establishment of biomarkers in gastroenterology.

Since 1989, Prof. Guenther has conducted more than 150 clinical trials according to GCP guidelines, phase I-IV.

Some of his scientific interests include virus–host interactions in viral hepatitis, cellular immune response in hepatitis, auto-antibodies, molecular pathomechanisms in NAFL and NASH, the role of apoptosis and fibrogenesis, cholangiocellular carcinoma (CCC), and therapies with new immunosuppressants (DNA, peptides, and T-cell vaccines).

He has been awarded the following awards during his career:

"Prix jeunes chercheurs" of the French Association for the Study of the Liver (L'AFEF) (1989);
Asche Fellowship of the German Association for Gastroenterology (1989);
Annual Award of the Dr. Carl Duisberg Foundation (1992);
Albert Knoll Award (1992).

Publications: original articles (peer reviewed): >750 (in >180 Journals); H-index (1985–2024): 84; citations (without self-citations): 28,865; Impact Factors, cumulated (1985–2023): 5,827.4.

Preface

Dear Colleagues,

Liver diseases represent one of the greatest health problems worldwide and are one of the leading causes of death. Acute and chronic liver diseases and their consequences are the second most common medical care problem. The importance of the early detection and timely treatment of various liver diseases and their primary and secondary prevention should therefore be of the highest priority worldwide.

Particularly in the last three decades, groundbreaking developments in the diagnosis and therapy of liver diseases have been made thanks to scientific advancements. In this Special Issue, research topics range from viral diseases and autoimmune, vascular, metabolic, and genetic diseases to liver cirrhosis and its subsequent complications, acute chronic liver failure, and the development of primary liver cell carcinomas. Topics such as therapeutic immunization, cell-based tumor therapy, liver–gut interaction based on the microbiome, and precision medicine using the example of the molecular profiling of tumors are also covered herewith.

Guido Gerken
Editor

Journal of
Personalized
Medicine

MDPI

Editorial

Liver Diseases: Science, Fiction and the Foreseeable Future

Robert K. Gieseler *, Theodor Baars, Mustafa K. Özçürümez and Ali Canbay *

Department of Medicine, University Hospital Knappschaftskrankenhaus, Ruhr University Bochum, 44892 Bochum, Germany; theodor.baars@t-online.de (T.B.); mustafa.oezcueruemez@kk-bochum.de (M.K.Ö.)
* Correspondence: rk.gieseler@gmx.de (R.K.G.); ali.canbay@rub.de (A.C.); Tel.: +49-234-299-3401

Abstract: This Editorial precedes the Special Issue entitled "Novel Challenges and Therapeutic Options for Liver Diseases". Following a historical outline of the roots of hepatology, we provide a brief insight into our colleagues' contributions in this issue on the current developments in this discipline related to the prevention of liver diseases, the metabolic dysfunction-associated steatotic liver disease (or non-alcoholic fatty liver disease, respectively), liver cirrhosis, chronic viral hepatitides, acute-on-chronic liver failure, liver transplantation, the liver–microbiome axis and microbiome transplantation, and telemedicine. We further add some topics not covered by the contributions herein that will likely impact future hepatology. Clinically, these comprise the predictive potential of organokine crosstalk and treatment options for liver fibrosis. With regard to promising developments in basic research, some current findings on the genetic basis of metabolism-associated chronic liver diseases, chronobiology, metabolic zonation of the liver, aspects of the aging liver against the background of demography, and liver regeneration will be presented. We expect machine learning to thrive as an overarching topic throughout hepatology. The largest study to date on the early detection of liver damage—which has been kicked off on 1 March 2024—is highlighted, too.

Keywords: history of hepatology; MASLD/NAFLD; liver fibrosis/cirrhosis; chronic viral hepatitides; liver transplantation; liver–microbiome axis; organokine crosstalk; chronobiology; liver aging

check for updates

Citation: Gieseler, R.K.; Baars, T.; Özçürümez, M.K.; Canbay, A. Liver Diseases: Science, Fiction and the Foreseeable Future. *J. Pers. Med.* **2024**, *14*, 492. https://doi.org/10.3390/jpm14050492

Received: 26 April 2024
Accepted: 30 April 2024
Published: 4 May 2024

1. Memorable Origins

Allow us to begin with a brief yet meaningful journey into the past before we embark on a voyage towards the future of personalized hepatology with this Special Issue. As a humble remark in advance, however, the space available here only allows for a few select brushstrokes taken from several wall-filling paintings of life.

Knowing our past is the prerequisite for being able to shape the future based on a well-aligned inner compass and a clear vision. So, who were the forefathers and foremothers of our discipline? While many faces and names may come to mind, five of them stand out clearly—and regrettably, it seems as if each new generation needs to be reminded time and again of the first and perhaps most trailblazing among them. There is an extremely malevolent reason behind this, which is must-tell history: just like so many other Jews, Ismar Isidor Boas (1858–1938) was inevitably exposed to the maelstrom of Nazi racial fanatism in 1930s' Germany. If they had had their way, both he as a human being and his seminal œuvre should have been erased from the face of the earth. And, they almost succeeded: already in his seventies, Boas fled from Berlin to Vienna in 1936 in the wake of the 'Nuremberg (Racial) Laws', only to find himself exposed again to the threat of persecution after the Austrian *Anschluss* (annexation) to the so-called 'Third Reich'. Three days after the Nazis entered Vienna in 1938, he took his own life [1]. The legacy of his work had already been taken away from Professor Boas by the professional society he had helped to shape so much: the *Deutsche Gesellschaft für Gastroenterologie, Verdauungs- und Stoffwechselkrankheiten* (i.e., the German Society for Gastroenterology, Digestive and Metabolic Diseases) (DGVS). Or so it seemed—but a crucial document had been preserved. As it says on the DGVS website, "*While doing research for the commemorative publication*

"100 Years DGVS" in 2013, the transcript of proceedings of the then Secretary General was discovered. It contained the society's 1932/33 membership list" [2]. There it was, a record revealing the names of its former Jewish members crossed out in red pencil and expelled—a third of all colleagues. Following this discovery, the DGVS initiated the project "Gegen das Vergessen" (literally "Against Forgetting"), entitled "We Remember" in its English-language version. It can be found on the society's website and commemorates all of those *"Jewish physicians who were excluded from the specialist society, who were disenfranchised, persecuted, forced to flee Germany, or deported to concentration camps (. . .)"* [2]. With reference to Hoenig and Boyle, we would like to invite you to take a closer look at the fate of Ismar Boas: as early as 1988, they delved into his life, his work, and the circumstances that drove him to his death [1]. As far as our topic is concerned, Boas, at an age of less than thirty, founded the field of gastroenterology in 1886, which he shaped for over half a decade, in addition to numerous other achievements. Separating the field of gastroenterology from internal medicine—which he had to defend against considerable opposition—can in fact be considered a key prerequisite for the subsequent establishment of hepatology as yet another distinct discipline [1,2].

However, we can go back even further to the period from 1858 to 1861 when Friedrich Theodor Frerichs (1819–1885) laid the foundation for science-based hepatology with his two-volume opus 'Clinic of Liver Diseases' [3–5]. Accordingly, his successor Ernst von Leyden stated that Frerichs had transferred medicine to the ranks of the exact natural sciences [6]. Still, Frerichs was committed to maintaining the unity of internal medicine [5], and it took several more decades for hepatology to blossom as a separate field. In this respect, Hans Philipp Popper (1903–1988) [7,8] and Dame Sheila Patricia Violet Sherlock (1918–2001) [9,10] are reasonably assumed to be household names to the hepatological community as they were the ones who, in the 1940s, 'surgically resected' the new discipline of hepatology from the corpus of gastroenterology [7–10].

Finally, one might arguably include Nancy Leslie Rutherford Bucher (1913–2017) as another founding mother in this illustrious circle [11]. In 1964, she was the first to show that the most striking effect of aging on the liver is the decline of the organ's remarkable capacity to 'regenerate' (a process correctly termed 'compensatory hyperplasia') in the event of tissue loss [12]. Not only epistemologically, but also in terms of practical medical relevance, this decline must be viewed against an evolutionary background spanning eons that helps to explain aging-related hepatic and systemic pathological effects, which may lead to novel options for their antagonization [11,13].

Mirroring the progress of knowledge, hepatology has thus repeatedly shed its skin. In more recent times, the liver has increasingly been shown to intersect with other organs and systems, typically manifesting as "axes"—such as the gut–liver axis, the brain–liver axis, the gut–brain–liver axis, and the liver–microbiome axis—or, for example, reflected in the form of organokine crosstalk. Still, hepatology has not continued to divide into further sub-disciplines, but today consists of various subspecialties, some of which are featured in this Special Issue. It is probably correct to say that hepatology would not be what it is today had it not been for the luminaries briefly sketched above. They created the discipline's cornerstones. Let us try to prove ourselves worthy successors to these pre-eminent colleagues and their groundbreaking achievements with the perspectives presented in this Special Issue and with our future endeavors in hepatology.

2. Back to the Future: Forward-Looking Topics in This Special Issue

We will now briefly outline the contributions to this Special Issue and discuss some previous publications in the respective contexts.

2.1. Prevention of Liver Diseases

In this Special Issue, Muñoz-Restrepo et al. highlight successfully implemented measures for preventing liver diseases, such as vaccination strategies, novel medications, lifestyle changes, and preventive surgeries. However, they also point to the parallel worldwide increase in chronic liver diseases—prominently including metabolic dysfunction-associated steatotic liver disease (or non-alcoholic fatty liver disease, respectively) (MASLD/NAFLD) (cf. Section 2.2.) as well as chronic hepatitis B and hepatitis C (cf. Section 2.4.)—which collectively call for better prevention strategies that may also result from big-data analyses [14]. This will likely create options for personalized medicine.

2.2. MASLD/NAFLD

The MASLD/NAFLD pandemic, which affects >25% of humanity, is the greatest hepatological challenge. It can still only be treated inadequately, so dietary and behavioral measures are currently the best options available [15]. Following the initial suggestion by Eslam and colleagues on renaming NAFLD to MAFLD [16,17], we are now trending towards the term (see above) and acronym MASLD. Ultimately, this renaming will result in the review and update of the nomenclature and subphenotypes of this condition, which could impact personalized medicine interventions for patients suffering from this disease entity.

In this Special Issue, Kreimeyer et al. show that the treatment of patients with the severe inflammatory form of MASLD/NAFLD (i.e., NASH) with bile acid transporter (BAT) gene polymorphisms in ABCB4 or ABCB11 with ursodeoxycholic acid for 12 months significantly reduced GGT (all patients) and ALT (homozygous patients). Patients with the TM6SF2 polymorphism showed a significant reduction in both GGT and ALT. Thus, NASH patients with elevated GGT should be screened for BAT gene polymorphisms prior to treatment [18].

Also, in this Special Issue, Schwertheim et al. examined selected protein expressions in liver tissue from MASLD/NAFLD patients to clarify their potential alteration in the progression from simple steatosis to NASH. The expression of pNRF2, SOCS3, and RIG1 (hepatocytes), and (bile ducts) was significantly higher in NASH than in steatosis; thus, these proteins might be assessed as potential therapeutic targets [19].

2.3. Liver Cirrhosis

Rotational thromboelastometry is a viscoelastic method that allows to quickly assess the state of induced hemostasis in whole blood samples and to effectively reduce the number of transfusions, healthcare costs, and complications [20]. The pressure of blood flowing through a cirrhotic liver can be reduced via the placement of a transjugular intrahepatic portosystemic shunt (TIPS) and is employed for treating refractory ascites and/or variceal bleeding [21].

In this Special Issue, Bedreli et al. evaluated alterations and differences in coagulation in the portal and peripheral circulation (PORC; PERC) via rotational thromboelastometry during TIPS. In blood samples from a cohort of cirrhotic patients (MELD Score: 12; median age: 60 years) undergoing TIPS, the authors detected no coagulation differences between PERC and PORC, which contrasts previous reports that suggested increased clotting activity in PORC vs. PERC in patients with liver cirrhosis [22].

2.4. Chronic Viral Hepatitides

The viral hepatitides remain a major public health problem, with five biologically unrelated hepatotropic viruses responsible for the majority of the global burden. The highest numbers of chronic infections and deaths alike are due to the hepatitis B and hepatitis C viruses (HBV; HCV), with approximately 257 million or 71 million infected people, respectively [23]. However, infection with the hepatitis D virus (HDV)—which occurs in association with HBV—also affects between 12 and 72 million people worldwide, and it is associated with a more rapid progression to cirrhosis and liver failure as well as higher rates of hepatocellular carcinoma than infections with HBV or HCV alone [24].

In this Special Issue, Schlaak gives an overview of the current treatments and developments, stating that (i) the introduction of direct-acting antivirals for treating HCV has been a boon by all accounts (at least, one might add, for those patients living in countries where these drugs are available); (ii) nowadays, HBV is generally well controllable, although a "functional cure" has not yet been achieved, which calls for novel therapeutic strategies that are currently being developed; and (iii) HDV remains the most challenging type of chronic viral hepatitis, for which therapeutic approaches with better response rates must be conceived [25].

2.5. *Acute-on-Chronic Liver Failure*

Acute-on-chronic liver failure (ACLF) is a condition in patients with known chronic liver disease and acute decompensation (AD) of liver cirrhosis. This syndrome is characterized by severe systemic inflammation and proinflammatory precipitating events (such as infections), and it is associated with single or multiple organ failure. Therefore, patients with ACLF are at a high risk of death within 28 days after hospital admission [26].

In this Special Issue, Kimmann et al. present the pathophysiological and clinical background to AD and ACLF as well as the current interventional treatment options—with liver transplantation (LTx) as the only curative treatment option currently available—and they further expand on future therapeutic options for ACLF management of AD as well as of ACLF [27].

2.6. *Liver Transplantation*

Apart from ACLF (cf. Section 2.5), another indication for LTx comprises neuroendocrine tumors (NETs) with liver metastases, although this rare indication consists of only <1% of all LTx activity. Specifically, favorable and acceptable transplant candidates are NET patients with metastases that are confined to the liver, are not poorly differentiated, are non-resectable, and are treatment-resistant [28].

In their narrative systematic review on this therapeutic approach, which is still controversially debated, Palaniappan et al. critically appraise the existing literature regarding this modality and thus provide an important basis for further discussing the role of LTx in the setting of patients suffering from NETs with liver metastases that meet the required criteria [29].

2.7. *Liver–Microbiome Axis and Gut Microbiome Transplantation*

An ever-increasing body of knowledge clearly shows that the gut microbiome, dietary habits, and metabolic health (or, conversely, metabolic diseases) form closely associated functionally interactive intersections. More recently, it has become apparent that the gut microbiota functions as a mediator of the dietary impact on the metabolic status. Against this background, causal relationships are increasingly being elucidated, which may let us therapeutically address metabolic diseases by personalized nutrition in the future [30].

Another approach for treating metabolic diseases is fecal microbiome transplantation (FMT), which is dealt with by Stadlbauer in this Special Issue. Although differing from the approach of personalized nutrition, its intention is identical in that it aims at altering a patient's microbiome composition. However, FMT is presently approved for proof-of-concept studies only and is not yet ready for a broad application. Against this background, Stadlbauer presents an overview of the current knowledge and describes the tasks ahead to be tackled for making FMT available to larger patient populations [31]. In this regard, methodological advances and standardization approaches, such as the one just published by Lederer and colleagues, will certainly be instrumental [32].

2.8. *Telemedicine for Remote Monitoring*

Telehealth or telemedicine—i.e., remote electronic patient monitoring as well as the provision of medical information and services via telecommunication—has considerably increased during the SARS-CoV-2 pandemic [33].

Accordingly, Akbar et al., in this Special Issue, assess the role of telemedicine in monitoring MASLD/NAFLD during the pandemic. Their systematic review and meta-analysis demonstrate that telemedicine via mobile applications during the SARS-CoV-2 pandemic proved to be an option for monitoring lifestyle modifications in MASLD/NAFLD patients [34].

In 2022, Greiwe surmised that telemedicine will remain an effective tool in the future, regardless of its use during the pandemic [33]. This assumption is, for instance, confirmed by the situation in Germany, where telemedicine networks are currently being expanded in all federal states against the backdrop of the country's continuously increasing demographic challenges with rising care and treatment needs, while the staffing levels are chronically low [35].

3. Further Forward-Looking Topics

We will now provide a brief overview of some other future-oriented topics in hepatology not covered by articles featured in this Special Issue. This subjective non-exhaustive selection reflects some of the aspects that we believe may become increasingly important with respect to clinical applications, clinical trials, and clinically oriented basic research.

3.1. Clinical Opportunities: The Foreseeable Future

3.1.1. Predictive Potential: Organokine Crosstalk

MASLD/NAFLD increases the risk for cardiovascular disease (CVD) [36]. However, despite the high numbers of MASLD/NAFLD cases, the medical need for a reliable scoring system to predict the CVD risk remains unmet. We recently presented cumulative evidence supporting the assumption that this may be achieved by establishing an algorithm based on the systemic release of organokines, whose healthy pattern changes in disease. Specifically focusing on adipokines, hepatokines, and cardiokines, we thus hypothesized that an algorithm predictive of the CVD risk in patients with MASLD/NAFLD can be established and improved continuously via machine learning. Once implemented, such a score might be used to estimate the risk for CVD for prevention and early stage life-saving interventions [15]. In general, we venture that computing the diverse and complex data through machine learning will ultimately yield excellent applications for personalized medicine, which will likely extend even beyond the fields of hepatology and cardiology.

3.1.2. Therapeutic Potential: Vitamin D in Liver Fibrosis

We previously showed that vitamin D (VD) may be a treatment option early in the onset of liver fibrosis in patients with certain VD receptor (VDR) genotypes or VDR polymorphisms. Specifically, targets within the TGF-β pathway might provide opportunities for patients with detrimental VDR single-nucleotide polymorphisms [37]. In a systematic review on the benefit of VD supplementation, Sharifi and Amani highlighted partly contradictory findings between different clinical trials and pointed out that influencing factors, such as gender, VD co-supplementation with calcium, and gene polymorphisms, should be considered in future clinical studies [38]. The clear VDR genotype dependencies we already emphasized [37] may explain some of those contradictory clinical trial results. We therefore maintain our view that VD supplementation could be a future treatment option early in the onset of liver fibrosis, provided that the treated patient cohort is strictly defined.

3.2. Basic Research: What Is on the Horizon?

3.2.1. Predictive Potential: Genetic Underpinnings of Major Liver Diseases

Only very few studies into the genetic underpinnings of major liver diseases go back in time as far as a most recently published evolutionary investigation. Herein, a common variant—i.e., the risk allele *PNPLA3* p.I148M (rs738409)—of the gene for patatin-like phospholipase domain-containing 3 (PNPLA3), which is prominently associated with an increased risk to develop steatotic liver disease, MASLD/NAFLD (including progressive inflammation), liver cirrhosis, and hepatocellular carcinoma [39], was studied. Contrary to non-human primates, this risk allele was identified in all Neanderthals and Denisovans, which indicates that the risk allele emerged prior to the split between the Neanderthals and modern humans. Interestingly, however, present-day humans exhibit a wide range (i.e., 8%

to 72%) of the presence of *PNPLA3* p.I148M, depending on their ethnicity [40]. The authors may thus have identified the earliest evolutionarily known risk gene for developing certain severe liver diseases, but it is too early to assess whether and how this knowledge may be used in the future.

3.2.2. Predictive Potential: Early Detection of Liver Damage

Particularly in this Special Issue, which has a focus on future hepatology, we should not forget to mention that the project "A Biomarker-Based Platform for Early Diagnosis of Chronic Liver Disease to Enable Personalized Therapy", funded by the European Commission with a volume of EUR 15 million and designed for five years, was launched on 1 March 2024 [41]. This consortium involves 23 academic and industrial partners from 11 European countries, including Croatia (1), Denmark (3), France (1), Germany (6), Ireland (1), Italy (2), Slovakia (1), Spain (4), Sweden (1), and Switzerland (1), as well as the transnational European Liver Patients' Association (ELPA) located in Brussels, Belgium, with overarching support from Innovation Acta headquartered in Rome, Italy. This is the world's largest study to date on the early detection of liver damage—an immensely important field of research, particularly in view of the fact that, unlike other organs, the damaged liver often "complains" about its deplorable condition too late. This program is designed to detect liver fibrosis at an early stage to prevent the development of cirrhosis and liver cancer. We believe that medicine, science, and, first and foremost, the patients can look forward to the results of this project.

3.2.3. Therapeutic Potential: Metabolic Syndrome and MASLD/NAFLD

As mentioned, treatment options for metabolic diseases—and MASLD/NAFLD in particular—are extremely limited. Therefore, this topic is being intensively researched. We recently presented a study on the application of L-ornithine–L-aspartate (LOLA) (the stable salt of L-ornithine and L-aspartic acid) in human in vitro models of steatosis, insulin resistance and metabolic syndrome [42]. In addition to the known effects of LOLA on NH_3 detoxification (e.g., [43]), this agent normalized fatty acid transport regulation, branched-chain amino acid catabolism, energy consumption, and mitochondrial energy balance. We thus suggested that this relatively inexpensive active agent may significantly contribute to a safer, more effective, and gentler management of metabolic diseases, including MASLD/NAFLD [42].

3.2.4. Hepatic Chronobiology

To our knowledge, the first paper on a chronobiological aspect of the liver was published in 1981 [44]. Yet, this comparatively young sub-discipline has great future potential. Regarding the role of chronobiology in metabolic disease, Amatobi and colleagues have summed up a piece of fundamental insight in a wonderful sentence: "*Modern lifestyle is often at odds with endogenously driven rhythmicity, which can lead to circadian disruption and metabolic syndrome*". In a trailblazing study performed in *Drosophila melanogaster* (a well-established model in chronobiology), the authors elucidated several important relationships between metabolite cycling and the metabolic status, the disruption of circadian rhythmicity, and the propensity for metabolic disease [45]. Thus, while fruit flies and humans admittedly are separated by a large evolutionary gap, such relationships have long been established in the living world, allowing us to learn more regarding hepatic metabolic diseases in humans. Although not immediately implementable in the clinic, studies like this one showcase the great potential of basic research for yielding new medical approaches in the more distant future.

3.2.5. Metabolic Zonation of the Liver

More than 45 years ago, Jungermann and Sasse introduced the model of metabolic zonation of the liver, on which today's understanding of a dynamically organized organ with functionally specialized hepatocyte variants is based [46]. With his integrative view,

Jungermann was also among the first to recognize the important role of neuronal and inflammatory signaling in the regulation of the liver's metabolic functions [47]. We presume that the spectrum of implications of these important aspects of hepatic versatility is far from being fully explored. For example, the integration of metabolic zonation with chronobiological, age-related, and pharmacogenomic aspects will add new dimensions of complexity that should best be analyzed by intelligently programmed machine learning to improve disease prevention measures and optimize the treatment of chronic and malignant liver diseases.

3.2.6. Liver Aging and the Aging Societies

As early as 1985, Hans Popper (see above) farsightedly predicted that *"the effect of age on the liver and of the liver on aging is full of promise if available methodologies are rigorously applied"* [48]. Such promise meets urgent societal demand: already covering an extended period of time and clearly extrapolated into the future, demographics show an increasing trend towards population aging, which is not limited to the Western societies' Baby-Boomer issue [49]. In fact, the World Health Organization has predicted that the global proportion of people >60 years of age will increase from 12% to 22% between 2015 and 2050 [50]. Therefore, as the body of knowledge about the liver and aging, as well as the available methodologies, have increased considerably in recent decades, they can—and definitely should—now be applied rigorously to minimize the negative health effects and their social and economic implications. This includes all hepatic aspects of aging covering genomics and epigenomics [48], age-related changes of the hepatic transcriptome [51], dietary habits in conjunction with medically based guidelines for the food industry vis-à-vis the continuous increase in metabolic disorders [52], and, last but not least, important evolutionary underpinnings of hepatic biology that go along with an increased likelihood for developing certain chronic diseases, while chronic diseases in turn accelerate biological aging [11]. We therefore have an impressive "toolbox" at our disposal. It is now up to healthcare policies to set the necessary course: given our immense challenges, it would be inexcusable to shirk this responsibility.

3.2.7. Liver Regeneration

The liver's ability to regenerate after tissue loss [11,13] will always remain a central topic in many hepatological contexts. We can expect further fascinating findings in this area that will help expand our medical use of this capacity. Here is a prime example: hepatocyte membrane-specific phospholipids (i.e., short-chain fatty acids) synthesized by gut microbiota and delivered via the gut–liver axis have been demonstrated to promote liver regeneration. The authors concluded that this process is pivotal for hepatic compensatory hyperplasia [53]—and in their Editorial Commentary on this article, Jian et al. stated that *"the revolutionary clinical value of postoperative interventions based on gut microbiota in patients undergoing liver surgery will undoubtedly propel gut microbial interventions to become a standard of care in the future"* [54].

3.3. Machine Learning—An Overarching Aspect in Hepatology (and Beyond)

We already repeatedly mentioned the use of machine learning (ML) in various hepatological contexts. ML [55] is defined as a subset of artificial intelligence [56]. It has experienced rapid growth in all fields of research and application, including the medical area, where speed, accuracy, and efficiency have steadily improved. Therefore, as a result of data collection, analysis, and classification, ML has continuously enhanced the prediction of various diseases [55]. Our own experience with the application of ML has been very encouraging [57–60]. We further believe that computation using suitable ML techniques will enable us to establish algorithms that will permanently learn from the acquired data via probabilistic modeling [61]. All of those interested may find an excellent review and critical appraisal on the application of ML in hepatology by Spann et al. to be instructive [62]. Overall, while some desirable applications may remain a fiction in the near future, ML will undoubtedly play a very important role in hepatology—not to mention its hard-to-overestimate potential for personalized medicine as a whole.

4. Epilogue

Working tirelessly and altruistically in the organism's engine room, the liver is constantly exposed to all our impositions—be it gluttony, the abuse of alcohol, illegal drugs or medications, and the challenge of polypharmacy. Trying to protect us from mischief, it stands solid as a rock. However, nobody takes notice—this organ does not complain; the liver just perishes quietly when it can take no more. Similar thoughts might have crossed the mind of Dr. Xaime Quintanilla Ulla, who, as a physician and mayor of the Spanish municipality of Ferrol, declared to the local council in 1987, "*Al hígado habría que hacerle un monumento*" ("The liver should be given a monument"). So, in the same year, a monolith was erected in honor of this organ (Figure 1); it remains the only liver monument in the world to this day [63].

Figure 1. Xaime Quintanilla Ulla, M.D., dedicates the liver monument in 1987. The translated inscription reads "The liver gives life" (photo credit: Rodrigo R. Arda [63]).

This man had definitely recognized the preciousness of this remarkable organ. Given his medical background, this project was a matter close to his heart—and the liver truly deserves its monument. Not only may the rapidly increasing options of personalized medicine enable us to improve the management of chronic liver diseases, but with the liver's central role being closely intertwined with all bodily aspects, it eventually could also be key to the implementation of life-prolonging measures.

We hope you find this Special Issue a pleasant and insightful read.

Author Contributions: Conceptualization, R.K.G. and A.C.; writing—original draft preparation, R.K.G.; writing—review and editing, A.C. and R.K.G.; resources, T.B. and M.K.Ö. All authors have read and agreed to the published version of the manuscript.

Acknowledgments: R.K.G.s' heartfelt thanks go to Harro Jenss of the German Society of Gastroenterology (DGVS) for a pleasant and fruitful exchange on the history of gastroenterology/hepatology as well as for valuable input.

Conflicts of Interest: The authors declare no conflicts of interest.

References

1. Hoenig, L.J.; Boyle, J.D. The life and death of Ismar Boas. *J. Clin. Gastroenterol.* **1988**, *10*, 16–24. [CrossRef] [PubMed]
2. Jenss, H. Gegen das Vergessen (We Remember). Deutsche Gesellschaft für Verdauungs- und Stoffwechselkrankheiten (German Society of Gastroenterology). Available online: https://www.dgvs-gegen-das-vergessen.de/en/ (accessed on 15 March 2024). (In English).
3. Frerichs, F.T. *Klinik der Leberkrankheiten (Clinic of Liver Diseases)*; Friedrich Vieweg und Sohn: Braunschweig, Germany, 1858; Volume 1.
4. Frerichs, F.T. *Klinik der Leberkrankheiten (Clinic of Liver Diseases)*; Friedrich Vieweg und Sohn: Braunschweig, Germany, 1861; Volume 2.
5. Jenss, H. Friedrich Theodor Frerichs—Begründer der modernen Hepatologie—200. Geburtstag (Friedrich Theodor Frerichs—founder of modern hepatology—200th birthday). *Z. Gastroenterol.* **2019**, *57*, 291–292. [CrossRef]
6. Franken, F.H. *Friedrich Theodor Frerichs (1819–1885)—Leben und Hepatologisches Werk*; Falk Foundation e.V.: Freiburg im Breisgau, Germany, 1994.
7. Schmid, R. *Hans Popper 1903–1988*; National Academy of Sciences: Washington, DC, USA, 1994. Available online: https://www.nasonline.org/publications/biographical-memoirs/memoir-pdfs/popper-hans.pdf (accessed on 18 March 2024).
8. Kaiser, S.; Sziranyi, J.; Groß, D. The hepatopathologist Hans Popper (1903-1988): An early victim of National Socialism in Austria. *Pathologe* **2020**, *41* (Suppl. 1), 30–38. [CrossRef] [PubMed]
9. Williams, R. Dame Sheila Sherlock: MCA Patron and Hepatologist Extraordinaire. *Alcohol. Alcohol.* **2002**, *37*, 622–623. [CrossRef]
10. Campollo, O. 50 years of Hepatology: The Royal Free Hospital School of Hepatology. *Ann. Hepatol.* **2020**, *19*, 113–116. [CrossRef] [PubMed]
11. Gieseler, R.K.; Schreiter, T.; Canbay, A. The aging human liver: The weal and woe of evolutionary legacy. *Z. Gastroenterol.* **2023**, *61*, 83–94. [CrossRef]
12. Bucher, N.L.R.; Swaffield, M.N.; DiTroia, J.F. The Influence of Age upon the Incorporation of Thymidine-2-C14 into the DNA of Regenerating Rat Liver. *Cancer Res.* **1964**, *24*, 509–512.
13. Delgado-Coello, B. Liver regeneration observed across the different classes of vertebrates from an evolutionary perspective. *Heliyon* **2021**, *7*, e06449. [CrossRef]
14. Muñoz-Restrepo, A.M.; Navas, M.C.; Daza, J.; Girala, M.; Ridruejo, E.; Gerken, G.; Teufel, A. Prevention in Hepatology. *J. Pers. Med.* **2024**, *14*, 132. [CrossRef]
15. Baars, T.; Gieseler, R.K.; Patsalis, P.C.; Canbay, A. Towards harnessing the value of organokine crosstalk to predict the risk for cardiovascular disease in non-alcoholic fatty liver disease. *Metabolism* **2022**, *130*, 155179. [CrossRef]
16. Eslam, M.; Sanyal, A.J.; George, J.; International Consensus Panel. MAFLD: A Consensus-Driven Proposed Nomenclature for Metabolic Associated Fatty Liver Disease. *Gastroenterology* **2020**, *158*, 1999–2014.e1. [CrossRef] [PubMed]
17. Méndez-Sánchez, N.; Bugianesi, E.; Gish, R.G.; Lammert, F.; Tilg, H.; Nguyen, M.H.; Sarin, S.K.; Fabrellas, N.; Zelber-Sagi, S.; Fan, J.; et al. Global multi-stakeholder endorsement of the MAFLD definition. *Lancet Gastroenterol. Hepatol.* **2022**, *7*, 388–390. [CrossRef] [PubMed]
18. Kreimeyer, H.; Vogt, K.; Götze, T.; Best, J.; Götze, O.; Weigt, J.; Kahraman, A.; Özçürümez, M.; Kälsch, J.; Syn, W.K.; et al. Influence of the Bile Acid Transporter Genes ABCB4, ABCB8, and ABCB11 and the Farnesoid X Receptor on the Response to Ursodeoxycholic Acid in Patients with Nonalcoholic Steatohepatitis. *J. Pers. Med.* **2023**, *13*, 1180. [CrossRef] [PubMed]
19. Schwertheim, S.; Alhardan, M.; Manka, P.P.; Sowa, J.P.; Canbay, A.; Schmidt, H.H.; Baba, H.A.; Kälsch, J. Higher pNRF2, SOCS3, IRF3, and RIG1 Tissue Protein Expression in NASH Patients versus NAFL Patients: pNRF2 Expression Is Concomitantly Associated with Elevated Fasting Glucose Levels. *J. Pers. Med.* **2023**, *13*, 1152. [CrossRef] [PubMed]
20. Drotarova, M.; Zolkova, J.; Belakova, K.M.; Brunclikova, M.; Skornova, I.; Stasko, J.; Simurda, T. Basic Principles of Rotational Thromboelastometry (ROTEM®) and the Role of ROTEM-Guided Fibrinogen Replacement Therapy in the Management of Coagulopathies. *Diagnostics* **2023**, *13*, 3219. [CrossRef] [PubMed]
21. Rajesh, S.; George, T.; Philips, C.A.; Ahamed, R.; Kumbar, S.; Mohan, N.; Mohanan, M.; Augustine, P. Transjugular intrahepatic portosystemic shunt in cirrhosis: An exhaustive critical update. *World J. Gastroenterol.* **2020**, *26*, 5561–5596. [CrossRef]
22. Bedreli, S.; Manka, P.; Buechter, M.; Jahn, M.; Theysohn, J.M.; Canbay, A.; Katsounas, A. No Differences in Rotational Thromboelastometry Measurements between Portal and Peripheral Circulation in Cirrhotic Patients Undergoing TIPS. *J. Pers. Med.* **2023**, *13*, 424. [CrossRef] [PubMed]
23. Lanini, S.; Ustianowski, A.; Pisapia, R.; Zumla, A.; Ippolito, G. Viral Hepatitis: Etiology, Epidemiology, Transmission, Diagnostics, Treatment, and Prevention. *Infect. Dis. Clin. N. Am.* **2019**, *33*, 1045–1062. [CrossRef] [PubMed]
24. Negro, F.; Lok, A.S. Hepatitis D: A Review. *JAMA* **2023**, *330*, 2376–2387. [CrossRef]
25. Schlaak, J.F. Current Therapy of Chronic Viral Hepatitis B, C and D. *J. Pers. Med.* **2023**, *13*, 964. [CrossRef]
26. Arroyo, V.; Moreau, R.; Jalan, R. Acute-on-Chronic Liver Failure. *N. Engl. J. Med.* **2020**, *382*, 2137–2145. [CrossRef] [PubMed]
27. Kimmann, M.; Trebicka, J. Acute-On-Chronic Liver Failure: Current Interventional Treatment Options and Future Challenges. *J. Pers. Med.* **2023**, *13*, 1052. [CrossRef] [PubMed]
28. Fan, S.T.; Le Treut, Y.P.; Mazzaferro, V.; Burroughs, A.K.; Olausson, M.; Breitenstein, S.; Frilling, A. Liver transplantation for neuroendocrine tumour liver metastases. *HPB* **2015**, *17*, 23–28. [CrossRef]

29. Palaniappan, V.; Li, C.H.; Frilling, A.; Clift, A.K. Long-Term Outcomes of Liver Transplantation for the Management of Neuroendocrine Neoplasms: A Systematic Review. *J. Pers. Med.* **2023**, *13*, 1428. [CrossRef]
30. Sonnenburg, J.L.; Bäckhed, F. Diet–microbiota interactions as moderators of human metabolism. *Nature* **2016**, *535*, 56–64. [CrossRef]
31. Stadlbauer, V. Liver-Gut-Interaction: Role of Microbiome Transplantation in the Future Treatment of Metabolic Disease. *J. Pers. Med.* **2023**, *13*, 220. [CrossRef]
32. Lederer, T.; Hipler, N.M.; Thon, C.; Kupcinskas, J.; Link, A. Comparison of Fecal MicroRNA Isolation Using Various Total RNA Isolation Kits. *Genes* **2024**, *15*, 498. [CrossRef] [PubMed]
33. Greiwe, J. Telemedicine Lessons Learned During the COVID-19 Pandemic. *Curr. Allergy Asthma Rep.* **2022**, *22*, 1–5. [CrossRef]
34. Akbar, F.N.; Choirida, S.R.; Muttaqin, A.Z.; Ekayanti, F.; Nisa, H.; Hendarto, H. Telemedicine as an Option for Monitoring Metabolic Dysfunction-Associated Fatty Liver Disease (MAFLD) Patients Facing the COVID-19 Pandemic: A Systematic Review and Meta-Analysis. *J. Pers. Med.* **2024**, *14*, 281. [CrossRef]
35. Kurz, C. Digitale Lösungen: Telemedizin Wird Ausgebaut. *Dtsch. Arztebl.* **2024**, *121*, A402. Available online: https://www.aerzteblatt.de/pdf.asp?id=238226 (accessed on 22 March 2024).
36. Younossi, Z.M. Non-alcoholic fatty liver disease—A global public health perspective. *J. Hepatol.* **2019**, *70*, 531–544. [CrossRef] [PubMed]
37. Beilfuss, A.; Sowa, J.-P.; Sydor, S.; Beste, M.; Bechmann, L.P.; Schlattjan, M.; Syn, W.-K.; Wedemeyer, I.; Mathé, Z.; Jochum, C.; et al. Vitamin D counteracts fibrogenic TGF-β signalling in human hepatic stellate cells both receptor-dependently and independently. *Gut* **2015**, *64*, 791–799. [CrossRef]
38. Sharifi, N.; Amani, R. Vitamin D supplementation and non-alcoholic fatty liver disease: A critical and systematic review of clinical trials. *Crit. Rev. Food Sci. Nutr.* **2019**, *59*, 693–703. [CrossRef] [PubMed]
39. Kodama, T.; Takehara, T. Molecular genealogy of metabolic-associated hepatocellular carcinoma. *Semin. Liver. Dis.* **2024**. [CrossRef] [PubMed]
40. Geier, A.; Trost, J.; Wang, K.; Schmid, C.; Krawczyk, M.; Schiffels, S. *PNPLA3* fatty liver allele was fixed in Neanderthals and segregates neutrally in humans. *Gut* **2024**. [CrossRef] [PubMed]
41. European Commission. A Biomarker-Based Platform for Early Diagnosis of Chronic Liver Disease to Enable Personalized Therapy (LIVERAIM). Project ID: 101132901; Call: HORIZON-JU-IHI-2022-03-single-stage; Programme: HORIZON; DG/Agency: IHI. EU Funding & Tenders Portal. Available online: https://ec.europa.eu/info/funding-tenders/opportunities/portal/screen/how-to-participate/org-details/895467917/project/101132901/program/43108390/details (accessed on 15 March 2024).
42. Canbay, A.; Götze, O.; Kucukoglu, O.; Weinert, S.; Croner, R.S.; Baars, T.; Özçürümez, M.K.; Gieseler, R.K. L-Ornithine-L-Aspartate (LOLA) Normalizes Metabolic Parameters in Models of Steatosis, Insulin Resistance and Metabolic Syndrome. *Pharmaceutics* **2024**, *16*, 506. [CrossRef] [PubMed]
43. Butterworth, R.F.; Canbay, A. Hepatoprotection by L-Ornithine L-Aspartate in Non-Alcoholic Fatty Liver Disease. *Dig. Dis.* **2019**, *37*, 63–68. [CrossRef] [PubMed]
44. Scalabrino, G.; Ferioli, M.E.; Puerari, M.; Modena, D.; Fraschini, F.; Majorino, G. Changes in the circadian rhythm of ornithine decarboxylase in rat liver during chemical hepatocarcinogenesis. *J. Natl. Cancer Inst.* **1981**, *66*, 697–702. [PubMed]
45. Amatobi, K.M.; Ozbek-Unal, A.G.; Schäbler, S.; Deppisch, P.; Helfrich-Förster, C.; Mueller, M.J.; Wegener, C.; Fekete, A. The circadian clock is required for rhythmic lipid transport in *Drosophila* in interaction with diet and photic condition. *J. Lipid Res.* **2023**, *64*, 100417. [CrossRef]
46. Jungermann, K.; Sasse, D. Heterogeneity of liver parenchymal cells. *Trends Biochem. Sci.* **1978**, *3*, 198–202. [CrossRef]
47. Häussinger, D. Kurt Jungermann: A pioneer in liver research. Anat. Rec. A Discov. *Mol. Cell. Evol. Biol.* **2004**, *280*, 807. [CrossRef] [PubMed]
48. Hunt, N.J.; Kang, S.W.S.; Lockwood, G.P.; Le Couteur, D.G.; Cogger, V.C. Hallmarks of Aging in the Liver. *Comput. Struct. Biotechnol. J.* **2019**, *17*, 1151–1161. [CrossRef] [PubMed]
49. Knickman, J.R.; Snell, E.K. The 2030 problem: Caring for aging baby boomers. *Health Serv. Res.* **2002**, *37*, 849–884. [CrossRef] [PubMed]
50. World Health Organization. Ageing and Health. 1 October 2022. Available online: https://www.who.int/news-room/fact-sheets/detail/ageing-and-health (accessed on 4 April 2024).
51. Schreiter, T.; Gieseler, R.K.; Vílchez-Vargas, R.; Jauregui, R.; Sowa, J.P.; Klein-Scory, S.; Broering, R.; Croner, R.S.; Treckmann, J.W.; Link, A.; et al. Transcriptome-Wide Analysis of Human Liver Reveals Age-Related Differences in the Expression of Select Functional Gene Clusters and Evidence for a *PPP1R10*-Governed 'Aging Cascade'. *Pharmaceutics* **2021**, *13*, 2009. [CrossRef] [PubMed]
52. Agodi, A.; Maugeri, A.; Kunzova, S.; Sochor, O.; Bauerova, H.; Kiacova, N.; Barchitta, M.; Vinciguerra, M. Association of Dietary Patterns with Metabolic Syndrome: Results from the Kardiovize Brno 2030 Study. *Nutrients* **2018**, *10*, 898. [CrossRef] [PubMed]
53. Yin, Y.; Sichler, A.; Ecker, J.; Laschinger, M.; Liebisch, G.; Höring, M.; Basic, M.; Bleich, A.; Zhang, X.J.; Kübelsbeck, L.; et al. Gut microbiota promote liver regeneration through hepatic membrane phospholipid biosynthesis. *J. Hepatol.* **2023**, *78*, 820–835. [CrossRef]
54. Jiang, N.; Wang, Z.; Yuan, K. New insights in the metabolic function related to gut microbiota in the process of liver regeneration. *Hepatobiliary Surg. Nutr.* **2023**, *12*, 933–935. [CrossRef] [PubMed]

55. Sindhu, V.; Nivedha, S.; Prakash, M. An empirical science research on bioinformatics in machine learning. *J. Mech. Continua Math. Sci.* **2020**, *2*, 86–94. [CrossRef]
56. Russell, S.J.; Norvig, P. *Artificial Intelligence: A Modern Approach*, 4th ed.; Pearson: Hoboken, NJ, USA, 2021.
57. Sowa, J.P.; Heider, D.; Bechmann, L.P.; Gerken, G.; Hoffmann, D.; Canbay, A. Novel algorithm for non-invasive assessment of fibrosis in NAFLD. *PLoS ONE* **2013**, *8*, e62439. [CrossRef]
58. Sowa, J.P.; Atmaca, Ö.; Kahraman, A.; Schlattjan, M.; Lindner, M.; Sydor, S.; Scherbaum, N.; Lackner, K.; Gerken, G.; Heider, D.; et al. Noninvasive separation of alcoholic and non-alcoholic liver disease with predictive modeling. *PLoS ONE* **2014**, *9*, e101444. [CrossRef]
59. Canbay, A.; Kälsch, J.; Neumann, U.; Rau, M.; Hohenester, S.; Baba, H.A.; Rust, C.; Geier, A.; Heider, D.; Sowa, J.P. Noninvasive assessment of NAFLD as systemic disease—A machine learning perspective. *PLoS ONE* **2019**, *14*, e0214436. [CrossRef] [PubMed]
60. Spänig, S.; Emberger-Klein, A.; Sowa, J.P.; Canbay, A.; Menrad, K.; Heider, D. The virtual doctor: An interactive clinical-decision-support system based on deep learning for non-invasive prediction of diabetes. *Artif. Intell. Med.* **2019**, *100*, 101706. [CrossRef] [PubMed]
61. Ghahramani, Z. Probabilistic machine learning and artificial intelligence. *Nature* **2015**, *521*, 452–459. [CrossRef] [PubMed]
62. Spann, A.; Yasodhara, A.; Kang, J.; Watt, K.; Wang, B.; Goldenberg, A.; Bhat, M. Applying Machine Learning in Liver Disease and Transplantation: A Comprehensive Review. *Hepatology* **2020**, *71*, 1093–1105. [CrossRef]
63. de la Barrera, E. Ferrol Ciudad: Valón Tiene un Hígado Monumental. La Voz de Galicia: Arteixo A Coruña, Spain. 14 April 2023. Available online: https://www.lavozdegalicia.es/noticia/ferrol/ferrol/2023/04/13/ferrol-higado-monumental/00031681378177497559543.htm (accessed on 29 March 2024).

Journal of *Personalized Medicine*

MDPI

Review

Current Therapy of Chronic Viral Hepatitis B, C and D

Jörg F. Schlaak

Department of Internal Medicine, Ameos Hospital Oberhausen, Wilhelmstr. 34, 46145 Oberhausen, Germany; jfs63@web.de

Abstract: The majority of chronic viral hepatitis cases are induced via infection with the hepatitis B virus (HBV), hepatitis C virus (HCV), or hepatitis D virus (HDV). These patients are at increased risk for progressive liver disease leading to cirrhosis as well as hepatocellular carcinoma (HCC). HBV infection is well controlled by the currently available nucleosides as well as nucleotides, and the development of cirrhosis can be prevented. Additionally, it has been shown that HBV-induced liver fibrosis can regress during successful antiviral treatment; however, a "functional cure", i.e., loss of HBsAg, is a rare event when these drugs are used. Therefore, novel therapeutic strategies are aiming at the selective suppression of HBsAg levels in combination with immunostimulation. The development of directly acting antivirals (DAAs) has revolutionized HCV therapy, as almost all patients can be cured via this treatment. Additionally, DAA therapy has few, if any, side effects, and is generally well tolerated by patients. HDV remains the most challenging type of chronic viral hepatitis. Although novel therapeutic options have recently been approved, response rates are still less favorable compared to HBV and HCV. This review discusses current and future options for the treatment of chronic HBV, HCV, and HDV infection.

Keywords: chronic hepatitis B; chronic hepatitis C; chronic hepatitis D; antiviral therapy; liver cirrhosis; DAA therapy; functional cure; HBsAg loss

check for updates

Citation: Schlaak, J.F. Current Therapy of Chronic Viral Hepatitis B, C and D. *J. Pers. Med.* **2023**, *13*, 964. https://doi.org/10.3390/jpm13060964

Academic Editors: Guido Gerken and Frank Vinholt Schiødt

Received: 25 March 2023
Revised: 22 May 2023
Accepted: 6 June 2023
Published: 7 June 2023

1. Introduction

Chronic viral hepatitis induced via infection with the hepatitis B virus (HBV), hepatitis C virus (HCV), or hepatitis D virus (HDV) is still a major cause for morbidity and mortality worldwide; HBV and HCV are among the 10 major causes of global mortality [1]. It is estimated that 300 million people are chronically infected with HBV, while 58 million are infected with HCV [2] and 20 million are infected with HDV [1,2]. These patients are at increased risk for progressive liver disease leading to cirrhosis as well as hepatocellular carcinoma (HCC).

HBV. In general, chronic HBV infection is well controlled by the currently available drugs, and the development of cirrhosis can be prevented. Additionally, it has been shown that HBV-induced liver fibrosis can regress during successful antiviral treatment [3,4]. This requires the careful selection of an antiviral drug (low resistance rate, etc.) and the adequate drug adherence of a patient. In some HBeAg (HBV envelope antigen)-positive patients, antiviral therapy can cause seroconversion into anti-HBe. The elimination of HBsAg (HBV surface antigen), with or without seroconversion into anti-HBs, however, occurs in only a small proportion of patients. Over a period of 5 years of therapy, only 10% of patients may experience HBsAg loss and seroconversion. This so-called "functional healing", however, is the goal of future treatment strategies against HBV as it leads to a significantly reduced induced risk of HCC and other complications of liver fibrosis.

HCV. Similarly to HBV, infection with HCV can lead to serious and potentially life-threatening complications, such as liver cirrhosis or HCC. In the past it was treated with type I interferons (IFNs), which was complicated by numerous side effects and was only effective in a small proportion of patients, largely depending upon the HCV genotype and underlying stage of liver disease. The development of directly acting antivirals (DAAs) has revolutionized HCV therapy, as almost all patients can be cured via this treatment independent of genotype, fibrosis stage, and other risk factors. In addition, DAA therapy has few, if any, side effects, and is generally well tolerated by patients.

HDV. HDV remains the most challenging type of chronic viral hepatitis when therapy options are considered [5]. An HDV prevalence of up to 1% of the world population has been suggested, although reliable data on the global HDV prevalence are still missing [6]. It can be assumed that HDV prevalence is even higher in different risk groups, such as patients infected with human immune deficiency virus (HIV), where it is a major cause of liver-related morbidity [7]. Early studies have demonstrated that HDV coinfection leads to liver cirrhosis, liver decompensation, and HCC in a significant proportion of patients [8,9], which could be confirmed in more recent monocentric series [10,11]. Although novel therapeutic options have recently been approved, response rates are still less favorable compared to those seen in the treatment of chronic HBV and HCV infection.

2. Current Therapy of Chronic Hepatitis B

2.1. Indication for Antiviral Therapy

The suppression of viral load as well as normalization of liver function tests define the indication for the antiviral therapy of chronic HBV infection. Treatment is clearly indicated when transaminases are elevated and the viral load exceeds 2000 IU/mL. Other causes of liver disease, such as nonalcoholic steatohepatitis (NASH), alcohol abuse, etc., or a coinfection with HDV should be excluded, however. Another clear indication for therapy is given in HBV patients with advanced liver disease (F3 fibrosis or liver cirrhosis). Here, treatment should be initiated in all individuals that tested positive for HBV DNA via PCR regardless of transaminase levels or viral load [12–14]. This is also true for patients with HBV-related HCC and positive HBV PCR, as the risk of tumor recurrence or progress can be reduced [15–17]. Further therapy indications include a reduction in maternal transmission in pregnancy, professional (e.g., medical staff) or social reasons to reduce the risk of transmission, extrahepatic manifestations of HBV infection, and the prevention of HBV reactivation via immunosuppression.

The constellation that was earlier described as being an "asymptomatic HBsAg carrier" (normal transaminase levels and an HBV viral load < 2000 IU/mL) is usually no indication for therapy as the risk of liver disease progression and HCC development risk, as well as infectivity, are very low in these patients. The indication for therapy in patients with normal transaminase levels and a very high viral load (earlier named as the "immunotolerant stage") is still controversial. It has been suggested that antiviral therapy should be initiated in patients >30 years or in individuals with transaminases in the upper range of normal [12] (Figure 1).

2.2. Therapeutic Options for the Treatment of Chronic Hepatitis B Infection

Currently, there are two therapeutic strategies available for the treatment of chronic HBV infection: pegylated interferon alpha (Peg-IFN) and nucleoside or nucleotide analogues (NUCs). Peg-IFN is given once a week subcutaneously for 48 weeks and leads to long-term clinical responses (reduction in viral load and normalization of liver function tests) in about one-third of patients. Due to the relatively low response rates that are accompanied by relevant side effects, this therapy regimen is used in only a few patients. Most patients are treated with NUCs such as tenofovir (TDF) or entecavir (ETV), as this is well tolerated and leads to the reliable suppression of viral load as well as the normalization of transaminases. The downside of this approach is that a viral rebound may occur once NUC therapy has been stopped; thus, NUCs have to be given permanently.

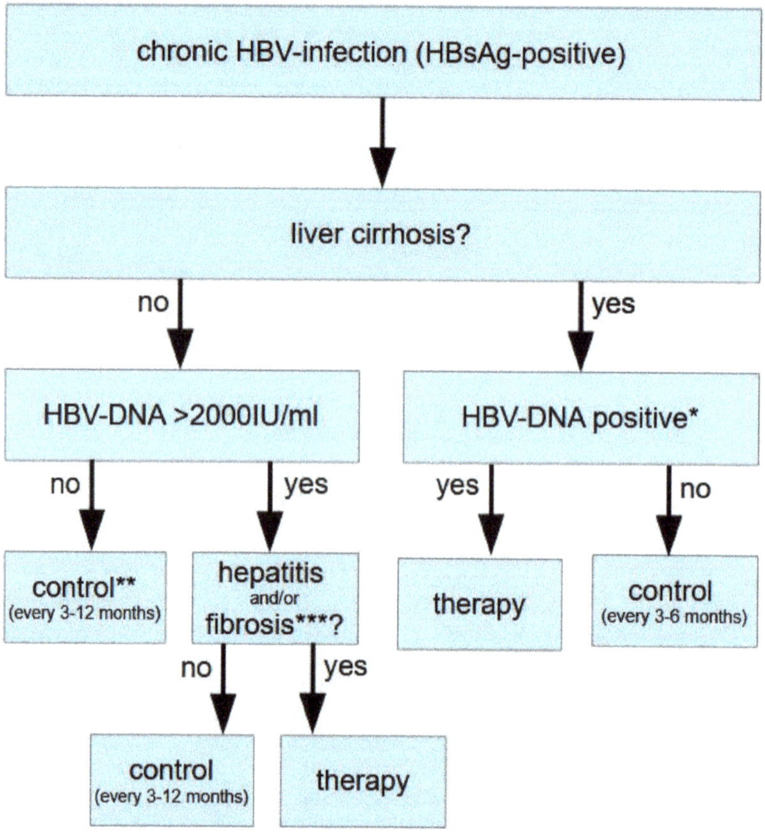

Figure 1. Algorithm for therapy indication in chronic HBV infection (modified from [18]). * Sensitive assay (<12 IU/mL), ** therapy can be indicated for other reasons (prophylaxis, extrahepatic manifestations, reduction in transmission, and HCC risk), and *** histology $\geq$ F2 fibrosis (Desmet).

a. PEG-IFN. IFN treatment may be considered in the presence of favorable baseline predictors, such as the absence of a high viral load (e.g., <1 million IU/mL), significantly increased transaminases, and HBV genotype A [12,13]. In Asian patients, genotype B/C age < 40 yrs., female gender, alanine aminotransferases (ALT) > 4× ULN, HBsAg levels < 25,000 U/mL and HBV DNA < 6 logIU/mL are predictive for a favorable response to PEG-IFN therapy [19]. Contraindications of PEG-IFN (e.g., previous psychiatric illnesses, Child cirrhosis stage B/C, autoimmune diseases, etc.) should be excluded before treatment is initiated. Successful IFN therapy is associated with a drop in HBsAg serum levels at therapy week 12 of more than 20% and below 20,000 IU/mL [18]. If this is not achieved, Peg-IFN therapy should be stopped and the patient should be switched to an NUC-based regimen. A de novo combination therapy with NUCs, an "add-on" of PEG-IFN to NUC therapy, or a switch from NUC- to PEG-IFN treatment is not recommended [18].

b. NUCs. Today, the most commonly used NUCs are tenofovir disoproxil (TDF; 245 mg/day) and entecavir (ETV; 0.5–1 mg/day). Lamivudine (LAM), adefovir (ADF), and telbivudine (TVD) are also licensed but are used less often, as high resistance rates have been observed. TDF and ETV are well tolerated, and resistance is only rarely seen. In the large majority of patients they lead to the suppression of the HBV viral load below the limit of detection and to the normalization of transaminases. If this is not achieved, a lack of compliance or therapy adherence of the patient should be considered. Long-term side effects include a slowly progressive reduction in bone density and the deterioration of renal function. Both side effects are more common under TDF as compared to ETV therapy. Therefore, a TDF analogue (tenofovir alafenamid (TAF)) has been developed with higher liver specificity that can be given at a lower dose (25 mg/day), therefore with fewer side effects [20,21]. It has been shown that long-term (5 years and more [22]) NUC therapy can lead to the regression of fibrosis or cirrhosis and may reduce the risk of HCC development. It is still a matter of debate, however, if TDF has stronger effects on the risk of HCC development than ETV [23–25].

c. Duration of NUC therapy. In (wild-type) HBeAg-positive patients, NUC therapy can be terminated 12 months after seroconversion into anti-HBe, with persistently negative HBV DNA thereafter. In (mutant) HBeAg-negative patients with detectable HBV DNA before therapy, the only definitive end point of NUC therapy is the loss of HBsAg with or without the detection of anti-HB antibodies. In patients with liver cirrhosis, a more cautious approach should be taken as it is recommended to prolong NUC therapy until anti-HB antibodies can be detected. This, however, is a rare event. Therefore, several studies have addressed the question as to whether it is possible to stop NUC therapy in the absence of these clear end points [26–28]. The current German and European guidelines consider ending therapy if HBV DNA has been undetectable for at least 3 years, advanced fibrosis is absent, and a close follow-up is guaranteed [12,18]. Further criteria include low HBsAg levels (<100 IU/mL) that have been shown to predict HBsAg loss in Asian patients [29]. Individual studies have identified further parameters, such as a low viral load before therapy (<200,000 IU/mL), low ALT, age < 40 yrs., female gender, and the absence of liver cirrhosis as favorable predictive parameters [30,31]. A relapse usually occurs 1–12 months after the cessation of NUC therapy. Therefore, liver function tests and HBV DNA should be controlled every 4 weeks in the first 6 months and every 12 weeks thereafter. In the case of a relapse, a rise in HBV DNA is initially observed, followed by an elevation in ALT elevation [30,31].

d. HBV and pregnancy. In rare cases, the activation of an HBV infection has been observed during pregnancy that can also lead to acute liver failure. In most patients, however, the attenuation of inflammatory activity is seen. After pregnancy, ALT flares can occur in the first 3–6 months after birth. Therefore, ALT and HBV DNA levels should be controlled in HBsAg-positive pregnant women every 3 months until 6 months after birth [18]. In the case of a patient becoming pregnant during antiviral treatment with an NUC or PEG-IFN, therapy with LAM, TVD, and TDF can be continued while PEG-IFN should be stopped, and ETV or ADF should be switched to TDF [18]. In therapy, naïve women initiation of antiviral treatment should be considered when active hepatitis has been diagnosed or HBV DNA > 200,000 IU/mL has been detected. It has been demonstrated that a high viral load increases the vertical mother-to-child transmission of HBV by up to 32%, while this risk is minimized when HBV DNA is suppressed < 200,000 IU/mL [32]. Therefore, antiviral treatment should not be initiated when HBV DNA is below this level. De novo antiviral treatment should be started as early as possible after the first trimester, preferably using TDF, and the patient should be informed about the possible risks and benefits [33]. NUC therapy can be stopped after birth in the absence of a medical indication (e.g., inactive hepatitis) or to prevent the transfer of toxic metabolites during breast feeding; however, it has been demonstrated that breast feeding is safe when TDF is used as an antiviral agent [34,35].

e. Prevention of HBV reactivation during immunosuppression. The reactivation of an inactive HBV infection is a potentially life-threatening complication of chemotherapy or immunosuppression. Therefore, HBsAg and anti-HBc should be tested before the initiation of chemotherapy or an immunosuppressive treatment [12,36,37]. In HBsAg-positive carriers, the reactivation of active hepatitis occurs in 15–50% patients after the start of chemotherapy, while this rate reaches 75% after bone marrow transplantation and fulminant hepatitis; fatal outcomes have also occurred. In HBsAg-negative/anti-HBc-positive individuals, the reactivation rate can reach 10%. As a result, it is recommended that HBsAg-positive patients with a high (>10%) risk of reactivation *must* be treated with

NUCs, while patients with a moderate (1–10%) risk *should* be treated with NUCs. HBsAg-positive patients with a low (<1%) risk of reactivation should be controlled every 8 weeks (Table 1). As mentioned, the reactivation risk in HBsAg-negative/anti-HBc-positive patients is much lower. Here, only patients with a high risk should be treated (Table 2). The risk of reactivation depends on the type of drug or immunosuppression used (Tables 1 and 2) [18]. For prophylaxis and preemptive therapy, highly potent NUCs (TDF and ETV) should be used [38,39] for 6–12 months [12,36]. After B-cell-depleting chemotherapy and high-risk constellations, prophylactic NUC therapy should be given for 18 months [40].

Table 1. Risk of HBV reactivation in HBsAg-positive, anti-HBc-positive patients (modified from [18]).

High risk (>10%)	B-cell depletion, anthracyclins, corticosteroids (>4 w, >10 mg/d), cyclophosphamide, stem cell transplantation, high-dose chemotherapy, and TACE/resection in HCC patients
Moderate risk (1–10%)	TNF inhibitors, cytokine/integrin inhibitors, tyrosine kinase inhibitors, mTOR inhibitors, JAK1/2 inhibitors, DAA therapy for HCV infection, and corticosteroids (>4 w, <10 mg/d)
Low risk (<1%)	Azathioprine, 6-mercaptopurine, methotrexate, intra-articular steroids, and corticosteroids < 1 w

Table 2. Risk of HBV reactivation in HBsAg-negative, anti-HBc-positive patients (modified from [18]).

High risk (>10%)	B-cell depletion, stem cell transplantation, and TACE/resection in HCC patients
Moderate to low risk (<10%)	Anthracyclins, corticosteroids (>4 w, >10 mg/d), TNF inhibitors, cytokine/integrin inhibitors, tyrosine kinase inhibitors, mTOR inhibitors, JAK1/2 inhibitors, corticosteroids (>4 w, <10 mg/d), DAA therapy for HCV infection, sorafenib for HCC, azathioprine, 6-mercaptopurine methotrexate, and intra-articular steroids

2.3. *Future Options in the Treatment of Chronic Hepatitis B Infection*

While viral replication can be controlled and the progression of liver disease can be prevented in most patients that are treated with the currently licensed antiviral drugs against HBV, the long-term control of an infection ("functional cure") is rarely seen, as HBsAg levels remain mostly elevated despite effective antiviral therapy. It has been demonstrated that HBsAg itself may suppress innate immune responses against HBV in particular, and thus promote the chronicity of an infection [41–44]. Therefore, several approaches are being developed that are aiming at HBsAg elimination. These can be divided into two groups that are acting directly on the virus (DAA) or that are aiming to improve the antiviral innate and/or adaptive immune response (IAS), respectively (Table 3, from [45]).

Table 3. Novel therapeutic strategies against chronic HBV infection (modified from [45]).

Mechanism of Action	Example Compounds
a. Directly Acting Antivirals (DAAs)	
Entry inhibitors	Bulevirtide
Capsid assembly inhibitors	JNJ-6379
HBsAg secretion inhibitors	REP-2139, REP-2165
Polymerase inhibitors	
Small interfering RNA (siRNA)	JNJ-3989
Antisense oligonucleotides (AO)	
b. Activation of Innate Immunity	
TLR-8 agonists	GS-9688
c. Activation of Adaptive Immunity	
Checkpoint inhibitors	ASC22
Therapeutic vaccination	GS-4774, TG-1050

a. Directly acting antivirals (DAAs). Bulevirtide (BLV) is an inhibitor of HBV entry and targets NTCP (sodium taurocholate cotransporting polypeptide), which functions as a receptor for HBV into a host cell. It is already licensed for the treatment of HBV/HDV coinfection (see below). Here, it has been shown to lead to HBsAg elimination in some cases when combined with PEG-IFN.

Capsid assembly modulators (CAMs) lead to the reduced formation of cccDNA and the introduction of HBV DNA into the nucleus by inhibiting the assembly of the HBV core protein. In clinical studies this has led to the suppression of HBV DNA without affecting HBsAg levels [46].

Very promising clinical results have been generated with HBsAg secretion inhibitors, such as REP-2139 or REP-2165, which are nucleic acid polymers (NAPs). In combination with PEG-IFN, they have demonstrated a high rate of a functional cure (e.g., sustained suppression of HBsAg after therapy) [47]. During therapy, ALT flares have been observed, suggesting the activation of the immune system.

Several small interfering RNAs (siRNAs) and antisense oligonucleotides (Aos) are currently in clinical development that are targeting the production of HbsAg and other viral proteins. In some studies a marked reduction in HbsAg levels was observed [48].

The data so far suggest that a combination of DAAs that lower HbsAg levels with IASs, such as PEG-IFN, may lead to a functional cure in a substantial proportion of patients. It will be of interest to see whether a combination of several DAAs may further enhance this effect.

b. Immune-activating strategies (IASs). Our own in vitro and in vivo data have suggested that the activation of the innate immune system may efficiently suppress HBV replication, while high levels of HBsAg may attenuate this immune activation [41–44]. Human data also suggest that HCV, which activates the innate immune system, can suppress HBV replication in HCV/HBV-coinfected patients. Therefore, toll-like receptor (TLR) agonists were developed to activate the innate immune system. Here, GS-9688, which activates TLR8, has shown some promising first clinical results in human studies and animal models [49,50].

Therapeutic vaccination may represent another approach that is aimed at activating the adaptive immune system. Vaccination with GS-4774 has been shown to induce HBV-specific T cells but lack the suppression HBsAg levels [51]. Similar results were obtained with TG-1050, which is based on an adenoviral system that encodes several HBV proteins [52].

Taken together, it is most likely that a "functional cure" will only be reached when the antiviral immune response against HBV is boosted via the activation of the innate and/or adaptive immune system. This immune activation in turn will only be achievable when HBsAg levels are suppressed by DAAs, as HBsAg may directly suppress the innate immune system [41].

3. Current Therapy of Chronic Hepatitis C

3.1. Indication for Antiviral Therapy

The therapeutic goal of antiviral therapy for chronic HCV infection is persistent virus suppression (SVR = "sustained virologic response"), which is defined by a lack of HCV RNA detection 12 weeks after the end of therapy. As the eradication of the virus does not lead to protective immunity, new infections are possible. Thus, a reinfection incidence of 6.4 per 100 patient years in patients with active intravenous drug use has been described [53]. Achieving an SVR is associated with a significant reduction in mortality, HCC development, and the need for a liver transplant [54,55]. While these positive effects are most obvious in patients with advanced fibrosis or compensated liver cirrhosis, they are less prevalent in patients with decompensated liver cirrhosis [56,57].

Every patient with a replicative HCV infection should be treated with antiviral therapy, provided that she or he will benefit from this treatment with respect to morbidity or mortality; when life expectancy is very limited, de novo therapy makes little, if any, sense, however. In the case of an initial diagnosis of HCV infection with the typical constellation of a chronic infection, antiviral therapy can be started immediately. Elevated transaminases and/or evidence of fibrosis are not necessary conditions. For patients with advanced fibrosis or cirrhosis, there is an urgent indication for treatment. Extrahepatic manifestations, professional reasons, the elimination of the risk of transmission, coinfections with HBV or HIV, and a patient's desire for treatment are also indications for treatment.

3.2. Therapeutic Options for the Treatment of Chronic Hepatitis C Infection

For different groups of patients, several interferon-free therapy options are available on the basis of HCV geno- and subtype, possible previous therapies, the presence as well as stage of liver cirrhosis, and kidney function. When choosing among the therapy options, the effectiveness in achieving an SVR, possible side effects or contraindications, concomitant diseases, drug interactions, the duration of therapy, and, if applicable, economic factors must be taken into account [58].

HCV therapy should be carried out with an interferon-free therapy regimen. In the case of known or foreseeable ribavirin side effects, a ribavirin-free therapy should preferably be used. Patients coinfected with the hepatitis B virus or HIV can be treated in a similar manner to HCV monoinfected patients. It should be noted that DAA therapy can rarely lead to HBV reactivation, as HCV activates the innate immune system, which in turn can suppress HBV replication [42,59,60].

1. Pangenotypic Regimes in DAA-Naïve Patients

a. Glecaprevir/pibrentasvir.

aa. DAA-naïve patients without cirrhosis. In patients without cirrhosis, the administration of glecaprevir (GPR) and pibrentasvir (PBR) leads to high SVR rates (8 weeks: 98%; 12 weeks: 99%) independent of numerous predictors, including HCV genotype, HIV coinfection, and non-DAA-based prior therapy [61–65]. Therefore, for all patients without cirrhosis, therapy with GPR and PBR for 8 weeks is recommended. For patients with HCV genotype 3 infection, this only applies to therapy-naïve patients, while therapy-experienced patients should be treated for 16 weeks.

ab. DAA-naïve patients with compensated cirrhosis. In patients with compensated cirrhosis, 8 (97.7% [66]) or 12 weeks (99% [67]) of therapy also lead to high SVR rates, independent of predictive factors. Therefore, treatment with GPR and PBR for all treatment-naïve patients with compensated cirrhosis is recommended for 8 weeks. For the retreatment of patients with HCV genotype 1, 2, and 4–6 infection, the duration of therapy is 12 weeks, whilst it is 16 weeks for previously treated patients with HCV genotype 3 infection.

b. Sofosbuvir/velpatasvir.

A large phase 3 study was performed in patients with HCV genotype non-3 infection using velpatasvir (VEL) in combination with sofosbuvir (SOF) for 12 weeks regardless of previous therapy, the presence of compensated cirrhosis, and numerous other predictors. In this study, an SVR rate of 99% was achieved [68]. In another phase 3 study, patients with HCV genotype 2 and 3 infection were included. Here, a 12-week therapy with VEL/SOF also led to an SVR rate of 99% [69]. As in a phase 2 study on patients with HCV genotype 1 or 2 infection, significantly lower SVR rates (77–90%) were observed when therapy was given for only 8 weeks [70]; VEL/SOF should be given to patients with HCV genotype 1, 2, and 4–6 infection for 12 weeks regardless of previous therapy and the presence of compensated cirrhosis. In HCV-genotype-3-infected patients without cirrhosis, the administration of VEL/SOF is recommended for 12 weeks. In patients with compensated cirrhosis, ribavirin can be added (Table 4).

Table 4. Pangenotypic regimens for DAA-naive patients without decompensated cirrhosis or advanced renal failure (modified from [58]).

Therapeutic Regimen	Duration (Weeks)	Patients without Cirrhosis		Patients with Compensated Cirrhosis		
		TN/TE	GT3 and TE	TN	TE	GT3 and TE
GPR/PBR	8	x		x		
GPR/PBR	12				x	
GPR/PBR	16		x			x
VEL/SOF	12	x	x	x	x	x

2. Genotype-Specific Regimens in DAA-Naïve Patients

Sofosbuvir/ledipasvir. The combination of sofosbuvir (SOF) with the NS5A inhibitor ledipasvir (LDV) was the first approved interferon-free, fixed coformulated regime. It is approved for the antiviral treatment of patients with HCV genotype 1, 4, or 6 with or without liver cirrhosis. The standard duration of therapy is 12 weeks and can be shortened to 8 weeks in therapy-naïve, noncirrhotic patients with HCV genotype 1 infection and a viral load of <6 million IU/mL HCV RNA. In patients with liver cirrhosis and/or negative predictors (e.g., failure of prior therapy, platelet counts of <75,000/nL), ribavirin can be added. In clinical practice, however, SOF/LDV is only rarely used due to the approval of pangenotypic regimens [58].

Grazoprevir/elbasvir. In this regimen, the NS3/4A protease inhibitor grazoprevir (GZR) is combined with the NS5A inhibitor elbasvir (EBR). It is licensed for the treatment of genotype 1 and 4 infections. Independent of the presence of compensated liver cirrhosis, the standard duration of therapy is 12 weeks. In patients infected with HCV genotype 1a, an extension of therapy to 16 weeks as well as the addition of ribavirin in patients with an initial viral load of over 800,000 IU/mL should be considered to reduce the risk of treatment failure. An extension of therapy to 16 weeks should also be considered in the presence of HCV genotype 4 with an initial HCV RNA of >800,000 IU/mL. Both SOF/LDV and GZR/EBR are only rarely used due to the approval of pangenotypic regimens, however [58].

3. Retherapy of DAA Failures

Patients that have failed therapy with an IFN-free DAA regimen should be retreated with a combination of voxilaprevir (VOX), VEL, and SOF for 12 weeks. This includes patients who have failed combination therapies consisting of SOF plus one NS3 protease inhibitor (e.g., simeprevir) or NS5A inhibitor (e.g., daclatasvir, ledipasvir, and velpatasvir) as well as nucleoside-free first-generation therapies (e.g., grazoprevir plus elbasvir or paritaprevir plus ombitasvir, with or without dasabuvir), each with or without the additional administration of ribavirin. This recommendation is based on two phase 3 studies that included patients with all HCV genotypes, various previous therapies, and patients with cirrhosis. The SVR rates ranged between 96% and 98% [71].

4. Treatment of Special Patient Populations

a. Patients with decompensated cirrhosis. Due to possible liver toxicity, NS3/4A protease inhibitors such as GZR, glecaprevir, and VOX are contraindicated in patients with decompensated cirrhosis. Consequently, therapy is limited to SOF in combination with NS5A inhibitors such as VEL and LDV in this population. The indication for antiviral therapy is given for all patients in whom a liver transplant can be avoided in the medium or long term, usually in patients with a MELD (model of end-stage liver disease) score of <20. In patients with a short-term need for a liver transplant, the indication for an antiviral therapy is much more difficult. The advantages of viral eradication before transplantation need to be weighed against serious potential side effects of the antiviral drugs in already very sick patients. Therefore, each case should be discussed with a liver transplantation center.

b. Patients with renal insufficiency. Patients with severe renal impairment (GFR < 30 mL/min or dialysis) should be treated in an equivalent manner to patients without renal insufficiency, with the following therapy options: GPR/PBR for 8, 12, or 16 weeks or GZR/EBR for HCV genotype 1 or 4 for 12 or 16 weeks. Here, previous therapies, comedications, and any comorbidities should be taken into account. Studies show high SVR rates of 98% across all HCV genotypes for GPR/PBR [64] and 99% in patients with HCV genotype 1 infection for GZR/EBR [72]. The additional administration of ribavirin should be avoided. Regimens including SOF should not be given in severe renal impairment.

4. Current Therapy of Chronic Hepatitis D

4.1. Indication for Antiviral Therapy

HDV/HBV-coinfected patients are at higher risk for the development of liver cirrhosis compared to HBV-monoinfected individuals [2,8,9,73]. Thus, about 20% of all cases of liver cirrhosis in HBsAg-positive patients are due to HDV coinfection, which is associated with significantly increased mortality [2,8,9]. HDV infection is also an independent risk factor for the development of hepatocellular carcinoma, with a relative risk of 1.3 in HBV/HDV- and 7.1 in HBV/HDV/HIV-coinfected patients compared to HBV-monoinfected patients [74]. Therefore, all HDV patients should be evaluated as possible candidates for antiviral therapy.

4.2. Therapeutic Options for the Treatment of Chronic Hepatitis D Infection

Pegylated IFN. PEG-IFN is approved for the treatment of hepatitis B and is also effective against HDV [75]. HDV RNA levels can be suppressed by up to 47% via standard IFN or PEG-IFN therapy. In the two large, controlled, and prospective HIDIT studies, the response rate at the end of therapy was 23–48%. Twenty-four weeks after the end of therapy, only about 25% of patients had negative HDV RNA [76,77]. During the long-term follow-up, however, about 50% of the patients had a late HDV RNA relapse [78]. Successful IFN therapy is associated with a more favorable long-term course, as the risk of developing clinical complications of liver cirrhosis was lower in these patients [11,79–81]. When a loss in HBsAg was achieved, a very favorable long-term course could be observed [79,82].

Nucleoside/nucleotide analogues. NUCs against HBV have no direct antiviral activity against HDV. There are negative studies for famciclovir [83], lamivudine [84], entecavir [85], and adefovir [76]. Likewise, TDF in combination with PEG-IFN showed no additional effect compared to PEG-IFN alone [77]. Nevertheless, it can be assumed that the therapy principles recommended for HBV monoinfection are also applicable for HDV/HBV coinfection. In the majority of cases, however, patients with hepatitis D have low HBV DNA levels [86,87], and will not benefit from HBV DNA suppression.

Bulevirtide. BLV is an entry inhibitor at the sodium taurocholate cotransporting polypeptide (NTCP) receptor, and has recently been approved for the treatment of HBV/HDV coinfection. HBV and HDV use NTCP as a receptor for virus entry [88]. BLV has been tested in several phase 2 studies. The results showed that monotherapy with this compound resulted in a dose-dependent decrease in HDV RNA levels; however, the studies also found evidence that combination therapy with PEG-IFN may be more effective than BLV monotherapy [89]. In a phase 2b dose-ranging study, patients received one dose of

BLV of 2 mg, 5 mg, or 10 mg in combination with TDF for 24 weeks. One patient group was treated exclusively with TDF monotherapy. At the end of the therapy, 46%, 47%, and 77% of patients, respectively, had a drop in HDV RNA of more than 2 logs compared to 3% in TDF monotherapy. While ALT levels also dropped, BLV had no effect on HBsAg levels [90].

BLV/PEG-IFN combination therapy for 48 weeks was also studied [91]. The median drop in HDV RNA was greater in the BLV/PEG-IFN group (−4.81 and −5.59 log for the combinations with 2 mg and 5 mg of BLV, respectively) compared to the PEG-IFN or BLV group (−1.30 and −2.84). A total of 53.3% of patients in the BLV 2 mg/PEG-IFN group and 26.7% of patients in the BLV 5 mg/PEG-IFN group were HDV-RNA-negative 24 weeks after the end of therapy. Additional data from a study with 10 mg BLV/PEG-IFNα-2a or 10 mg BLV/TDF for 48 weeks [91] showed that 86.7% of patients in the BLV/PEG-IFN group had undetectable HDV RNA at the end of therapy compared to 40% in the BLV/TDF group. Only combination therapy with PEG-IFN led to a decrease in HBsAg levels, while monotherapy with bulevirtide had no such effect. After the cessation of antiviral therapy, a rebound of HDV RNA was found in the majority of patients.

4.3. Future Options in the Treatment of Chronic Hepatitis D

Pegylated interferon λ. Pegylated interferon λ (PEG-IFNλ) is a type III interferon that stimulates cell-mediated immune responses through type III IFN receptors. In HDV patients treated with PEG-IFNλ, it was found that it is better tolerated than PEG-IFNα [92]. An SVR, as defined by a drop in HDV RNA levels by 2 logs 24 weeks after the cessation of antiviral therapy, was found in 36% of patients [93]. Further studies investigating IFNλ in patients with chronic hepatitis D are currently under way.

Lonafarnib. Lonafarnib is an orally active prenylation inhibitor that has demonstrated antiviral activity against HDV. Initial studies showed a significantly greater drop in HDV RNA levels when compared to a placebo, while no effects on HBsAg levels were seen. In phase II studies, lonafarnib was tested as a monotherapy, in combination with ritonavir, or in combination with PEG-IFN. The results showed that a combination therapy of low-dose lonafarnib with ritonavir or PEG-IFN was superior to a high-dose lonafarnib monotherapy with regard to antiviral activity and tolerability [94]. The LOWR-HDV-2 study [95] studied a triple regimen of 50 mg lonafarnib with ritonavir and PEG-IFN. A total of 63% of the patients in this study achieved the composite end point of a decline in HDV RNA of ≥2 logs and the normalization of ALT.

Nucleic acid polymers. Nucleic acid polymers (NAPs) have demonstrated very promising results when used for the treatment of HBV/HDV coinfection. In a phase 2 study, 12 patients with chronic hepatitis D and compensated liver disease received REP-2139 as an intravenous infusion once weekly, and after 15 weeks PEG-IFN was added for an additional 15 weeks. This was followed by PEG-IFN monotherapy for 33 weeks. HBsAg levels dropped in all patients during therapy, and 5 of 12 patients were negative for HBsAg at the end of therapy as well as positive for anti-HBs [96]. HDV RNA remained negative 18 months after the end of therapy in seven patients, and five patients were HBsAg-negative. During the long-term follow up, 7 of 11 patients were still HDV-RNA-negative 3.5 years after therapy [97].

5. Conclusions

In general, therapy for chronic viral hepatitis has improved significantly in the last two decades. The introduction of DAAs has revolutionized the treatment of chronic hepatitis C, as almost every patients can be cured from this infection with few, if any, side effects. Therefore, it is highly unlikely that new therapeutic developments will be seen for this entity in the future. The situation in hepatitis B is more complex, as chronic infection and disease progression can be controlled in most patients via the use of antivirals. A "functional cure", however, is achievable in only a minority of treated individuals with the currently licensed drugs. In this case, a number of promising new therapeutic approaches are currently under investigation that are aiming to reach this goal. Finally, chronic hepatitis

D remains the most challenging type of chronic viral hepatitis. While a novel compound has recently been licensed for the treatment of this disease, the overall results to control this infection and prevent disease progression are still not satisfactory. Here, further therapeutic approaches are desperately needed in the future.

Funding: This research received no external funding.

Institutional Review Board Statement: Not applicable.

Informed Consent Statement: Not applicable.

Data Availability Statement: Not applicable.

Conflicts of Interest: The author declares no conflict of interest.

References

1. Stanaway, J.D.; Flaxman, A.D.; Naghavi, M.; Fitzmaurice, C.; Vos, T.; Abubakar, I.; Abu-Raddad, L.J.; Assadi, R.; Bhala, N.; Cowie, B.; et al. The global burden of viral hepatitis from 1990 to 2013: Findings from the Global Burden of Disease Study 2013. *Lancet* **2016**, *388*, 1081–1088. [CrossRef] [PubMed]
2. Stockdale, A.J.; Kreuels, B.; Henrion, M.Y.R.; Giorgi, E.; Kyomuhangi, I.; de Martel, C.; Hutin, Y.; Geretti, A.M. The global prevalence of hepatitis D virus infection: Systematic review and meta-analysis. *J. Hepatol.* **2020**, *73*, 523–532. [CrossRef] [PubMed]
3. Marcellin, P.; Gane, E.; Buti, M.; Afdhal, N.; Sievert, W.; Jacobson, I.M.; Washington, M.K.; Germanidis, G.; Flaherty, J.F.; Aguilar Schall, R.; et al. Regression of cirrhosis during treatment with tenofovir disoproxil fumarate for chronic hepatitis B: A 5-year open-label follow-up study. *Lancet* **2013**, *381*, 468–475. [CrossRef] [PubMed]
4. Chang, T.T.; Liaw, Y.F.; Wu, S.S.; Schiff, E.; Han, K.H.; Lai, C.L.; Safadi, R.; Lee, S.S.; Halota, W.; Goodman, Z.; et al. Long-term entecavir therapy results in the reversal of fibrosis/cirrhosis and continued histological improvement in patients with chronic hepatitis B. *Hepatology* **2010**, *52*, 886–893. [CrossRef]
5. Hughes, S.A.; Wedemeyer, H.; Harrison, P.M. Hepatitis delta virus. *Lancet* **2011**, *378*, 73–85. [CrossRef] [PubMed]
6. Chen, H.Y.; Shen, D.T.; Ji, D.Z.; Han, P.C.; Zhang, W.M.; Ma, J.F.; Chen, W.S.; Goyal, H.; Pan, S.; Xu, H.G. Prevalence and burden of hepatitis D virus infection in the global population: A systematic review and meta-analysis. *Gut* **2019**, *68*, 512–521. [CrossRef]
7. Fernandez-Montero, J.V.; Vispo, E.; Barreiro, P.; Sierra-Enguita, R.; de Mendoza, C.; Labarga, P.; Soriano, V. Hepatitis delta is a major determinant of liver decompensation events and death in HIV-infected patients. *Clin. Infect. Dis.* **2014**, *58*, 1549–1553. [CrossRef]
8. Fattovich, G.; Boscaro, S.; Noventa, F.; Pornaro, E.; Stenico, D.; Alberti, A.; Ruol, A.; Realdi, G. Influence of hepatitis delta virus infection on progression to cirrhosis in chronic hepatitis type B. *J. Infect. Dis.* **1987**, *155*, 931–935. [CrossRef]
9. Fattovich, G.; Giustina, G.; Christensen, E.; Pantalena, M.; Zagni, I.; Realdi, G.; Schalm, S.W. Influence of hepatitis delta virus infection on morbidity and mortality in compensated cirrhosis type B. The European Concerted Action on Viral Hepatitis (Eurohep). *Gut* **2000**, *46*, 420–426. [CrossRef]
10. Niro, G.A.; Smedile, A.; Ippolito, A.M.; Ciancio, A.; Fontana, R.; Olivero, A.; Valvano, M.R.; Abate, M.L.; Gioffreda, D.; Caviglia, G.P.; et al. Outcome of chronic delta hepatitis in Italy: A long-term cohort study. *J. Hepatol.* **2010**, *53*, 834–840. [CrossRef]
11. Yurdaydin, C.; Keskin, O.; Kalkan, C.; Karakaya, F.; Caliskan, A.; Kabacam, G.; Onder, F.O.; Karatayli, S.; Karatayli, E.; Deda, X.; et al. Interferon Treatment Duration in Patients with Chronic Delta Hepatitis and its Effect on the Natural Course of the Disease. *J. Infect. Dis.* **2018**, *217*, 1184–1192. [CrossRef] [PubMed]
12. European Association for the Study of the Liver. EASL 2017 Clinical Practice Guidelines on the management of hepatitis B virus infection. *J. Hepatol.* **2017**, *67*, 370–398. [CrossRef] [PubMed]
13. Terrault, N.A.; Bzowej, N.H.; Chang, K.M.; Hwang, J.P.; Jonas, M.M.; Murad, M.H. AASLD guidelines for treatment of chronic hepatitis B. *Hepatology* **2016**, *63*, 261–283. [CrossRef] [PubMed]
14. Terrault, N.A.; Lok, A.S.F.; McMahon, B.J.; Chang, K.M.; Hwang, J.P.; Jonas, M.M.; Brown, R.S., Jr.; Bzowej, N.H.; Wong, J.B. Update on prevention, diagnosis, and treatment of chronic hepatitis B: AASLD 2018 hepatitis B guidance. *Hepatology* **2018**, *67*, 1560–1599. [CrossRef]
15. Huang, G.; Li, P.P.; Lau, W.Y.; Pan, Z.Y.; Zhao, L.H.; Wang, Z.G.; Wang, M.C.; Zhou, W.P. Antiviral Therapy Reduces Hepatocellular Carcinoma Recurrence in Patients with Low HBV-DNA Levels: A Randomized Controlled Trial. *Ann. Surg.* **2018**, *268*, 943–954. [CrossRef]
16. Chen, V.L.; Yeh, M.L.; Le, A.K.; Jun, M.; Saeed, W.K.; Yang, J.D.; Huang, C.F.; Lee, H.Y.; Tsai, P.C.; Lee, M.H.; et al. Anti-viral therapy is associated with improved survival but is underutilised in patients with hepatitis B virus-related hepatocellular carcinoma: Real-world east and west experience. *Aliment. Pharm. Ther.* **2018**, *48*, 44–54. [CrossRef]
17. He, L.; Liu, X.; Zhao, Y.; Zhang, S.; Jiang, Y.; Wang, X.; Yang, Z. Efficacy of Nucleot(s)ide Analogs Therapy in Patients with Unresectable HBV-Related Hepatocellular Carcinoma: A Systematic Review and Meta-Analysis. *Dis. Markers* **2017**, *2017*, 7075935. [CrossRef]

18. Cornberg, M.; Sandmann, L.; Protzer, U.; Niederau, C.; Tacke, F.; Berg, T.; Glebe, D.; Jilg, W.; Wedemeyer, H.; Wirth, S.; et al. S3-Leitlinie der Deutschen Gesellschaft für Gastroenterologie, Verdauungs- und Stoffwechselkrankheiten (DGVS) zur Prophylaxe, Diagnostik und Therapie der Hepatitis-B-Virusinfektion—(AWMF-Register-Nr. 021-11). *Z. Gastroenterol.* **2021**, *59*, 691–776. [CrossRef]
19. Sonneveld, M.J.; Hansen, B.E.; Piratvisuth, T.; Jia, J.D.; Zeuzem, S.; Gane, E.; Liaw, Y.F.; Xie, Q.; Heathcote, E.J.; Chan, H.L.; et al. Response-guided peginterferon therapy in hepatitis B e antigen-positive chronic hepatitis B using serum hepatitis B surface antigen levels. *Hepatology* **2013**, *58*, 872–880. [CrossRef]
20. Buti, M.; Gane, E.; Seto, W.K.; Chan, H.L.; Chuang, W.L.; Stepanova, T.; Hui, A.J.; Lim, Y.S.; Mehta, R.; Janssen, H.L.; et al. Tenofovir alafenamide versus tenofovir disoproxil fumarate for the treatment of patients with HBeAg-negative chronic hepatitis B virus infection: A randomised, double-blind, phase 3, non-inferiority trial. *Lancet Gastroenterol. Hepatol.* **2016**, *1*, 196–206. [CrossRef]
21. Toyoda, H.; Leong, J.; Landis, C.; Atsukawa, M.; Watanabe, T.; Huang, D.Q.; Liu, J.; Quek, S.X.Z.; Ishikawa, T.; Arai, T.; et al. Treatment and Renal Outcomes Up to 96 Weeks After Tenofovir Alafenamide Switch From Tenofovir Disoproxil Fumarate in Routine Practice. *Hepatology* **2021**, *74*, 656–666. [CrossRef] [PubMed]
22. Papatheodoridis, G.V.; Idilman, R.; Dalekos, G.N.; Buti, M.; Chi, H.; van Boemmel, F.; Calleja, J.L.; Sypsa, V.; Goulis, J.; Manolakopoulos, S.; et al. The risk of hepatocellular carcinoma decreases after the first 5 years of entecavir or tenofovir in Caucasians with chronic hepatitis B. *Hepatology* **2017**, *66*, 1444–1453. [CrossRef] [PubMed]
23. Choi, J.; Kim, H.J.; Lee, J.; Cho, S.; Ko, M.J.; Lim, Y.S. Risk of Hepatocellular Carcinoma in Patients Treated with Entecavir vs. Tenofovir for Chronic Hepatitis B: A Korean Nationwide Cohort Study. *JAMA Oncol.* **2019**, *5*, 30–36. [CrossRef] [PubMed]
24. Choi, W.M.; Choi, J.; Wong, G.L.; Han, S.; Lim, Y.S. Risk of hepatocellular carcinoma with tenofovir versus entecavir in chronic hepatitis B. *Lancet Gastroenterol. Hepatol.* **2021**, *6*, 87. [CrossRef]
25. Dave, S.; Park, S.; Murad, M.H.; Barnard, A.; Prokop, L.; Adams, L.A.; Singh, S.; Loomba, R. Comparative Effectiveness of Entecavir Versus Tenofovir for Preventing Hepatocellular Carcinoma in Patients with Chronic Hepatitis B: A Systematic Review and Meta-Analysis. *Hepatology* **2021**, *73*, 68–78. [CrossRef]
26. Tseng, C.H.; Chen, T.H.; Wu, J.L.; Lee, T.Y.; Borghi, J.A.; Lin, J.T.; Nguyen, M.H.; Hsu, Y.C. Serious adverse events after cessation of nucleos(t)ide analogues in individuals with chronic hepatitis B: A systematic review and meta-analysis. *JHEP Rep.* **2023**, *5*, 100617. [CrossRef]
27. Liaw, Y.F.; Jeng, W.J. Benefit of stopping finite nucleos(t)ide analogues therapy in chronic hepatitis B patients. *Gut* **2020**, *69*, 1898–1899. [CrossRef]
28. Liaw, Y.F. Finite nucleos(t)ide analog therapy in HBeAg-negative chronic hepatitis B: An emerging paradigm shift. *Hepatol. Int.* **2019**, *13*, 665–673. [CrossRef]
29. Cornberg, M.; Wong, V.W.; Locarnini, S.; Brunetto, M.; Janssen, H.L.A.; Chan, H.L. The role of quantitative hepatitis B surface antigen revisited. *J. Hepatol.* **2017**, *66*, 398–411. [CrossRef]
30. Chang, M.L.; Liaw, Y.F.; Hadziyannis, S.J. Systematic review: Cessation of long-term nucleos(t)ide analogue therapy in patients with hepatitis B e antigen-negative chronic hepatitis B. *Aliment. Pharm. Ther.* **2015**, *42*, 243–257. [CrossRef]
31. Papatheodoridis, G.; Vlachogiannakos, I.; Cholongitas, E.; Wursthorn, K.; Thomadakis, C.; Touloumi, G.; Petersen, J. Discontinuation of oral antivirals in chronic hepatitis B: A systematic review. *Hepatology* **2016**, *63*, 1481–1492. [CrossRef] [PubMed]
32. Brown, R.S., Jr.; McMahon, B.J.; Lok, A.S.; Wong, J.B.; Ahmed, A.T.; Mouchli, M.A.; Wang, Z.; Prokop, L.J.; Murad, M.H.; Mohammed, K. Antiviral therapy in chronic hepatitis B viral infection during pregnancy: A systematic review and meta-analysis. *Hepatology* **2016**, *63*, 319–333. [CrossRef] [PubMed]
33. Wang, M.; Bian, Q.; Zhu, Y.; Pang, Q.; Chang, L.; Li, R.; Tiongson, B.C.; Zhang, H.; Pan, C.Q. Real-world study of tenofovir disoproxil fumarate to prevent hepatitis B transmission in mothers with high viral load. *Aliment. Pharm. Ther.* **2019**, *49*, 211–217. [CrossRef]
34. Mofenson, L.M.; Baggaley, R.C.; Mameletzis, I. Tenofovir disoproxil fumarate safety for women and their infants during pregnancy and breastfeeding. *AIDS* **2017**, *31*, 213–232. [CrossRef]
35. Mugwanya, K.K.; John-Stewart, G.; Baeten, J. Safety of oral tenofovir disoproxil fumarate-based HIV pre-exposure prophylaxis use in lactating HIV-uninfected women. *Expert. Opin. Drug. Saf.* **2017**, *16*, 867–871. [CrossRef] [PubMed]
36. Reddy, K.R.; Beavers, K.L.; Hammond, S.P.; Lim, J.K.; Falck-Ytter, Y.T.; American Gastroenterological Association Institute. American Gastroenterological Association Institute guideline on the prevention and treatment of hepatitis B virus reactivation during immunosuppressive drug therapy. *Gastroenterology* **2015**, *148*, 215–219, quiz e16–e17. [CrossRef]
37. Sarmati, L.; Andreoni, M.; Antonelli, G.; Arcese, W.; Bruno, R.; Coppola, N.; Gaeta, G.B.; Galli, M.; Girmenia, C.; Mikulska, M.; et al. Recommendations for screening, monitoring, prevention, prophylaxis and therapy of hepatitis B virus reactivation in patients with haematologic malignancies and patients who underwent haematologic stem cell transplantation-a position paper. *Clin. Microbiol. Infect.* **2017**, *23*, 935–940. [CrossRef]
38. Huang, H.; Li, X.; Zhu, J.; Ye, S.; Zhang, H.; Wang, W.; Wu, X.; Peng, J.; Xu, B.; Lin, Y.; et al. Entecavir vs. lamivudine for prevention of hepatitis B virus reactivation among patients with untreated diffuse large B-cell lymphoma receiving R-CHOP chemotherapy: A randomized clinical trial. *JAMA* **2014**, *312*, 2521–2530. [CrossRef]

39. Zhang, M.Y.; Zhu, G.Q.; Shi, K.Q.; Zheng, J.N.; Cheng, Z.; Zou, Z.L.; Huang, H.H.; Chen, F.Y.; Zheng, M.H. Systematic review with network meta-analysis: Comparative efficacy of oral nucleos(t)ide analogues for the prevention of chemotherapy-induced hepatitis B virus reactivation. *Oncotarget* **2016**, *7*, 30642–30658. [CrossRef]
40. Nakaya, A.; Fujita, S.; Satake, A.; Nakanishi, T.; Azuma, Y.; Tsubokura, Y.; Hotta, M.; Yoshimura, H.; Ishii, K.; Ito, T.; et al. Delayed HBV reactivation in rituximab-containing chemotherapy: How long should we continue anti-virus prophylaxis or monitoring HBV-DNA? *Leuk. Res.* **2016**, *50*, 46–49. [CrossRef]
41. Wu, J.; Meng, Z.; Jiang, M.; Pei, R.; Trippler, M.; Broering, R.; Bucchi, A.; Sowa, J.P.; Dittmer, U.; Yang, D.; et al. Hepatitis B virus suppresses toll-like receptor-mediated innate immune responses in murine parenchymal and nonparenchymal liver cells. *Hepatology* **2009**, *49*, 1132–1140. [CrossRef] [PubMed]
42. Wu, J.; Lu, M.; Meng, Z.; Trippler, M.; Broering, R.; Szczeponek, A.; Krux, F.; Dittmer, U.; Roggendorf, M.; Gerken, G.; et al. Toll-like receptor-mediated control of HBV replication by nonparenchymal liver cells in mice. *Hepatology* **2007**, *46*, 1769–1778. [CrossRef] [PubMed]
43. Wu, J.; Huang, S.; Zhao, X.; Chen, M.; Lin, Y.; Xia, Y.; Sun, C.; Yang, X.; Wang, J.; Guo, Y.; et al. Poly(I:C) treatment leads to interferon-dependent clearance of hepatitis B virus in a hydrodynamic injection mouse model. *J. Virol.* **2014**, *88*, 10421–10431. [CrossRef] [PubMed]
44. Wu, J.; Han, M.; Li, J.; Yang, X.; Zhen, X.; Schlaak, J.F.; Yang, D.; Lu, M. Pattern Recognition Receptors and Liver Failure. *Crit. Rev. Immunol.* **2019**, *39*, 289–311. [CrossRef] [PubMed]
45. Neumann-Haefelin, C.; Thimme, R. Chronic hepatitis B virus infection: Current and future treatment strategies. *Bundesgesundheitsblatt Gesundh. Gesundh.* **2022**, *65*, 238–245. [CrossRef]
46. Zoulim, F.; Lenz, O.; Vandenbossche, J.J.; Talloen, W.; Verbinnen, T.; Moscalu, I.; Streinu-Cercel, A.; Bourgeois, S.; Buti, M.; Crespo, J.; et al. JNJ-56136379, an HBV Capsid Assembly Modulator, Is Well-Tolerated and Has Antiviral Activity in a Phase 1 Study of Patients with Chronic Infection. *Gastroenterology* **2020**, *159*, 521–533.e9. [CrossRef]
47. Bazinet, M.; Pantea, V.; Placinta, G.; Moscalu, I.; Cebotarescu, V.; Cojuhari, L.; Jimbei, P.; Iarovoi, L.; Smesnoi, V.; Musteata, T.; et al. Safety and Efficacy of 48 Weeks REP 2139 or REP 2165, Tenofovir Disoproxil, and Pegylated Interferon Alfa-2a in Patients with Chronic HBV Infection Naive to Nucleos(t)ide Therapy. *Gastroenterology* **2020**, *158*, 2180–2194. [CrossRef]
48. Yuen, M.F.; Locarnini, S.; Lim, T.H.; Strasser, S.I.; Sievert, W.; Cheng, W.; Thompson, A.J.; Given, B.D.; Schluep, T.; Hamilton, J.; et al. Combination treatments including the small-interfering RNA JNJ-3989 induce rapid and sometimes prolonged viral responses in patients with CHB. *J. Hepatol.* **2022**, *77*, 1287–1298. [CrossRef]
49. Gane, E.J.; Kim, H.J.; Visvanathan, K.; Kim, Y.J.; Nguyen, A.H.; Wallin, J.J.; Chen, D.Y.; McDonald, C.; Arora, P.; Tan, S.K.; et al. Safety, Pharmacokinetics, and Pharmacodynamics of the Oral TLR8 Agonist Selgantolimod in Chronic Hepatitis B. *Hepatology* **2021**, *74*, 1737–1749. [CrossRef]
50. Daffis, S.; Balsitis, S.; Chamberlain, J.; Zheng, J.; Santos, R.; Rowe, W.; Ramakrishnan, D.; Pattabiraman, D.; Spurlock, S.; Chu, R.; et al. Toll-Like Receptor 8 Agonist GS-9688 Induces Sustained Efficacy in the Woodchuck Model of Chronic Hepatitis B. *Hepatology* **2021**, *73*, 53–67. [CrossRef]
51. Boni, C.; Janssen, H.L.A.; Rossi, M.; Yoon, S.K.; Vecchi, A.; Barili, V.; Yoshida, E.M.; Trinh, H.; Rodell, T.C.; Laccabue, D.; et al. Combined GS-4774 and Tenofovir Therapy Can Improve HBV-Specific T-Cell Responses in Patients with Chronic Hepatitis. *Gastroenterology* **2019**, *157*, 227–241.e7. [CrossRef] [PubMed]
52. Zoulim, F.; Fournier, C.; Habersetzer, F.; Sprinzl, M.; Pol, S.; Coffin, C.S.; Leroy, V.; Ma, M.; Wedemeyer, H.; Lohse, A.W.; et al. Safety and immunogenicity of the therapeutic vaccine TG1050 in chronic hepatitis B patients: A phase 1b placebo-controlled trial. *Hum. Vaccin. Immunother.* **2020**, *16*, 388–399. [CrossRef] [PubMed]
53. Martinello, M.; Hajarizadeh, B.; Grebely, J.; Dore, G.J.; Matthews, G.V. HCV Cure and Reinfection Among People with HIV/HCV Coinfection and People Who Inject Drugs. *Curr. HIV/AIDS Rep.* **2017**, *14*, 110–121. [CrossRef] [PubMed]
54. Zeuzem, S. Treatment Options in Hepatitis C. *Dtsch. Arztebl. Int.* **2017**, *114*, 11–21. [CrossRef]
55. Simmons, B.; Saleem, J.; Heath, K.; Cooke, G.S.; Hill, A. Long-Term Treatment Outcomes of Patients Infected with Hepatitis C Virus: A Systematic Review and Meta-analysis of the Survival Benefit of Achieving a Sustained Virological Response. *Clin. Infect. Dis.* **2015**, *61*, 730–740. [CrossRef]
56. Calvaruso, V.; Craxi, A. Hepatic benefits of HCV cure. *J. Hepatol.* **2020**, *73*, 1548–1556. [CrossRef]
57. Calvaruso, V.; Cabibbo, G.; Cacciola, I.; Petta, S.; Madonia, S.; Bellia, A.; Tine, F.; Distefano, M.; Licata, A.; Giannitrapani, L.; et al. Incidence of Hepatocellular Carcinoma in Patients with HCV-Associated Cirrhosis Treated with Direct-Acting Antiviral Agents. *Gastroenterology* **2018**, *155*, 411–421.e4. [CrossRef]
58. Sarrazin, C.; Zimmermann, T.; Berg, T.; Hinrichsen, H.; Mauss, S.; Wedemeyer, H.; Zeuzem, S.; Deutsche Gesellschaft für Gastroenterologie; Verdauungs- und Stoffwechselkrankheiten (DGVS); Deutsche Gesellschaft für Pathologie e. V. (DGP) und Bundesverband Deutscher Pathologen (BDP); et al. Prophylaxe, Diagnostik und Therapie der Hepatitis-C-Virus(HCV)-Infektion. *Z. Gastroenterol.* **2020**, *58*, 1110–1131. [CrossRef]
59. Jiang, M.; Broering, R.; Trippler, M.; Poggenpohl, L.; Fiedler, M.; Gerken, G.; Lu, M.; Schlaak, J.F. Toll-like receptor-mediated immune responses are attenuated in the presence of high levels of hepatitis B virus surface antigen. *J. Viral Hepat.* **2014**, *21*, 860–872. [CrossRef]
60. Broering, R.; Lu, M.; Schlaak, J.F. Role of Toll-like receptors in liver health and disease. *Clin. Sci.* **2011**, *121*, 415–426. [CrossRef]

61. Rockstroh, J.K.; Lacombe, K.; Viani, R.M.; Orkin, C.; Wyles, D.; Luetkemeyer, A.F.; Soto-Malave, R.; Flisiak, R.; Bhagani, S.; Sherman, K.E.; et al. Efficacy and Safety of Glecaprevir/Pibrentasvir in Patients Coinfected with Hepatitis C Virus and Human Immunodeficiency Virus Type 1: The EXPEDITION-2 Study. *Clin. Infect. Dis.* **2018**, *67*, 1010–1017. [CrossRef]
62. Puoti, M.; Foster, G.R.; Wang, S.; Mutimer, D.; Gane, E.; Moreno, C.; Chang, T.T.; Lee, S.S.; Marinho, R.; Dufour, J.F.; et al. High SVR12 with 8-week and 12-week glecaprevir/pibrentasvir therapy: An integrated analysis of HCV genotype 1–6 patients without cirrhosis. *J. Hepatol.* **2018**, *69*, 293–300. [CrossRef] [PubMed]
63. Asselah, T.; Kowdley, K.V.; Zadeikis, N.; Wang, S.; Hassanein, T.; Horsmans, Y.; Colombo, M.; Calinas, F.; Aguilar, H.; de Ledinghen, V.; et al. Efficacy of Glecaprevir/Pibrentasvir for 8 or 12 Weeks in Patients with Hepatitis C Virus Genotype 2, 4, 5, or 6 Infection without Cirrhosis. *Clin. Gastroenterol. Hepatol.* **2018**, *16*, 417–426. [CrossRef] [PubMed]
64. Gane, E.; Lawitz, E.; Pugatch, D.; Papatheodoridis, G.; Brau, N.; Brown, A.; Pol, S.; Leroy, V.; Persico, M.; Moreno, C.; et al. Glecaprevir and Pibrentasvir in Patients with HCV and Severe Renal Impairment. *N. Engl. J. Med.* **2017**, *377*, 1448–1455. [CrossRef] [PubMed]
65. Zeuzem, S.; Foster, G.R.; Wang, S.; Asatryan, A.; Gane, E.; Feld, J.J.; Asselah, T.; Bourliere, M.; Ruane, P.J.; Wedemeyer, H.; et al. Glecaprevir-Pibrentasvir for 8 or 12 Weeks in HCV Genotype 1 or 3 Infection. *N. Engl. J. Med.* **2018**, *378*, 354–369. [CrossRef] [PubMed]
66. Brown, R.S., Jr.; Buti, M.; Rodrigues, L.; Chulanov, V.; Chuang, W.L.; Aguilar, H.; Horvath, G.; Zuckerman, E.; Carrion, B.R.; Rodriguez-Perez, F.; et al. Glecaprevir/pibrentasvir for 8 weeks in treatment-naive patients with chronic HCV genotypes 1–6 and compensated cirrhosis: The EXPEDITION-8 trial. *J. Hepatol.* **2020**, *72*, 441–449. [CrossRef]
67. Forns, X.; Lee, S.S.; Valdes, J.; Lens, S.; Ghalib, R.; Aguilar, H.; Felizarta, F.; Hassanein, T.; Hinrichsen, H.; Rincon, D.; et al. Glecaprevir plus pibrentasvir for chronic hepatitis C virus genotype 1, 2, 4, 5, or 6 infection in adults with compensated cirrhosis (EXPEDITION-1): A single-arm, open-label, multicentre phase 3 trial. *Lancet Infect. Dis.* **2017**, *17*, 1062–1068. [CrossRef]
68. Feld, J.J.; Jacobson, I.M.; Hezode, C.; Asselah, T.; Ruane, P.J.; Gruener, N.; Abergel, A.; Mangia, A.; Lai, C.L.; Chan, H.L.; et al. Sofosbuvir and Velpatasvir for HCV Genotype 1, 2, 4, 5, and 6 Infection. *N. Engl. J. Med.* **2015**, *373*, 2599–2607. [CrossRef]
69. Foster, G.R.; Afdhal, N.; Roberts, S.K.; Brau, N.; Gane, E.J.; Pianko, S.; Lawitz, E.; Thompson, A.; Shiffman, M.L.; Cooper, C.; et al. Sofosbuvir and Velpatasvir for HCV Genotype 2 and 3 Infection. *N. Engl. J. Med.* **2015**, *373*, 2608–2617. [CrossRef]
70. Everson, G.T.; Towner, W.J.; Davis, M.N.; Wyles, D.L.; Nahass, R.G.; Thuluvath, P.J.; Etzkorn, K.; Hinestrosa, F.; Tong, M.; Rabinovitz, M.; et al. Sofosbuvir with Velpatasvir in Treatment-Naive Noncirrhotic Patients with Genotype 1 to 6 Hepatitis C Virus Infection: A Randomized Trial. *Ann. Intern. Med.* **2015**, *163*, 818–826. [CrossRef]
71. Bourliere, M.; Gordon, S.C.; Flamm, S.L.; Cooper, C.L.; Ramji, A.; Tong, M.; Ravendhran, N.; Vierling, J.M.; Tran, T.T.; Pianko, S.; et al. Sofosbuvir, Velpatasvir, and Voxilaprevir for Previously Treated HCV Infection. *N. Engl. J. Med.* **2017**, *376*, 2134–2146. [CrossRef] [PubMed]
72. Roth, D.; Nelson, D.R.; Bruchfeld, A.; Liapakis, A.; Silva, M.; Monsour, H., Jr.; Martin, P.; Pol, S.; Londono, M.C.; Hassanein, T.; et al. Grazoprevir plus elbasvir in treatment-naive and treatment-experienced patients with hepatitis C virus genotype 1 infection and stage 4-5 chronic kidney disease (the C-SURFER study): A combination phase 3 study. *Lancet* **2015**, *386*, 1537–1545. [CrossRef] [PubMed]
73. Miao, Z.; Zhang, S.; Ou, X.; Li, S.; Ma, Z.; Wang, W.; Peppelenbosch, M.P.; Liu, J.; Pan, Q. Estimating the Global Prevalence, Disease Progression, and Clinical Outcome of Hepatitis Delta Virus Infection. *J. Infect. Dis.* **2020**, *221*, 1677–1687. [CrossRef] [PubMed]
74. Alfaiate, D.; Clement, S.; Gomes, D.; Goossens, N.; Negro, F. Chronic hepatitis D and hepatocellular carcinoma: A systematic review and meta-analysis of observational studies. *J. Hepatol.* **2020**, *73*, 533–539. [CrossRef] [PubMed]
75. Wranke, A.; Wedemeyer, H. Antiviral therapy of hepatitis delta virus infection—Progress and challenges towards cure. *Curr. Opin. Virol.* **2016**, *20*, 112–118. [CrossRef]
76. Wedemeyer, H.; Yurdaydin, C.; Dalekos, G.N.; Erhardt, A.; Cakaloglu, Y.; Degertekin, H.; Gurel, S.; Zeuzem, S.; Zachou, K.; Bozkaya, H.; et al. Peginterferon plus adefovir versus either drug alone for hepatitis delta. *N. Engl. J. Med.* **2011**, *364*, 322–331. [CrossRef]
77. Wedemeyer, H.; Yurdaydin, C.; Hardtke, S.; Caruntu, F.A.; Curescu, M.G.; Yalcin, K.; Akarca, U.S.; Gurel, S.; Zeuzem, S.; Erhardt, A.; et al. Peginterferon alfa-2a plus tenofovir disoproxil fumarate for hepatitis D (HIDIT-II): A randomised, placebo controlled, phase 2 trial. *Lancet Infect. Dis.* **2019**, *19*, 275–286. [CrossRef]
78. Heidrich, B.; Yurdaydin, C.; Kabacam, G.; Ratsch, B.A.; Zachou, K.; Bremer, B.; Dalekos, G.N.; Erhardt, A.; Tabak, F.; Yalcin, K.; et al. Late HDV RNA relapse after peginterferon alpha-based therapy of chronic hepatitis delta. *Hepatology* **2014**, *60*, 87–97. [CrossRef]
79. Wranke, A.; Serrano, B.C.; Heidrich, B.; Kirschner, J.; Bremer, B.; Lehmann, P.; Hardtke, S.; Deterding, K.; Port, K.; Westphal, M.; et al. Antiviral treatment and liver-related complications in hepatitis delta. *Hepatology* **2017**, *65*, 414–425. [CrossRef]
80. Roulot, D.; Brichler, S.; Layese, R.; BenAbdesselam, Z.; Zoulim, F.; Thibault, V.; Scholtes, C.; Roche, B.; Castelnau, C.; Poynard, T.; et al. Origin, HDV genotype and persistent viremia determine outcome and treatment response in patients with chronic hepatitis delta. *J. Hepatol.* **2020**, *73*, 1046–1062. [CrossRef]
81. Kamal, H.; Westman, G.; Falconer, K.; Duberg, A.S.; Weiland, O.; Haverinen, S.; Wejstal, R.; Carlsson, T.; Kampmann, C.; Larsson, S.B.; et al. Long-Term Study of Hepatitis Delta Virus Infection at Secondary Care Centers: The Impact of Viremia on Liver-Related Outcomes. *Hepatology* **2020**, *72*, 1177–1190. [CrossRef] [PubMed]

82. Niro, G.A.; Smedile, A.; Fontana, R.; Olivero, A.; Ciancio, A.; Valvano, M.R.; Pittaluga, F.; Coppola, N.; Wedemeyer, H.; Zachou, K.; et al. HBsAg kinetics in chronic hepatitis D during interferon therapy: On-treatment prediction of response. *Aliment. Pharm. Ther.* **2016**, *44*, 620–628. [CrossRef] [PubMed]
83. Yurdaydin, C.; Bozkaya, H.; Gurel, S.; Tillmann, H.L.; Aslan, N.; Okcu-Heper, A.; Erden, E.; Yalcin, K.; Iliman, N.; Uzunalimoglu, O.; et al. Famciclovir treatment of chronic delta hepatitis. *J. Hepatol.* **2002**, *37*, 266–271. [CrossRef]
84. Niro, G.A.; Ciancio, A.; Tillman, H.L.; Lagget, M.; Olivero, A.; Perri, F.; Fontana, R.; Little, N.; Campbell, F.; Smedile, A.; et al. Lamivudine therapy in chronic delta hepatitis: A multicentre randomized-controlled pilot study. *Aliment. Pharm. Ther.* **2005**, *22*, 227–232. [CrossRef] [PubMed]
85. Kabacam, G.; Onder, F.O.; Yakut, M.; Seven, G.; Karatayli, S.C.; Karatayli, E.; Savas, B.; Idilman, R.; Bozdayi, A.M.; Yurdaydin, C. Entecavir treatment of chronic hepatitis D. *Clin. Infect. Dis.* **2012**, *55*, 645–650. [CrossRef] [PubMed]
86. Heidrich, B.; Deterding, K.; Tillmann, H.L.; Raupach, R.; Manns, M.P.; Wedemeyer, H. Virological and clinical characteristics of delta hepatitis in Central Europe. *J. Viral Hepat.* **2009**, *16*, 883–894. [CrossRef]
87. Sureau, C.; Negro, F. The hepatitis delta virus: Replication and pathogenesis. *J. Hepatol.* **2016**, *64*, S102–S116. [CrossRef]
88. Urban, S.; Bartenschlager, R.; Kubitz, R.; Zoulim, F. Strategies to inhibit entry of HBV and HDV into hepatocytes. *Gastroenterology* **2014**, *147*, 48–64. [CrossRef]
89. Bogomolov, P.; Alexandrov, A.; Voronkova, N.; Macievich, M.; Kokina, K.; Petrachenkova, M.; Lehr, T.; Lempp, F.A.; Wedemeyer, H.; Haag, M.; et al. Treatment of chronic hepatitis D with the entry inhibitor myrcludex B: First results of a phase Ib/IIa study. *J. Hepatol.* **2016**, *65*, 490–498. [CrossRef]
90. Wedemeyer, H.; Schoneweis, K.; Bogomolov, P.; Blank, A.; Voronkova, N.; Stepanova, T.; Sagalova, O.; Chulanov, V.; Osipenko, M.; Morozov, V.; et al. Safety and efficacy of bulevirtide in combination with tenofovir disoproxil fumarate in patients with hepatitis B virus and hepatitis D virus coinfection (MYR202): A multicentre, randomised, parallel-group, open-label, phase 2 trial. *Lancet Infect. Dis.* **2023**, *23*, 117–129. [CrossRef]
91. Lampertico, P.; Roulot, D.; Wedemeyer, H. Bulevirtide with or without pegIFNalpha for patients with compensated chronic hepatitis delta: From clinical trials to real-world studies. *J. Hepatol.* **2022**, *77*, 1422–1430. [CrossRef] [PubMed]
92. Hamid, S.S.; Etzion, O.; Lurie, Y.; Bader, N.; Yardeni, D.; Channa, S.M.; Mawani, M.; Parkash, O.; Martins, E.B.; Gane, E.J. A phase 2 randomized clinical trial to evaluate the safety and efficacy of pegylated interferon lambda monotherapy in patients with chronic hepatitis delta virus infection. Interim results from the LIMT HDV study. *Hepatology* **2017**, *66*, 496A.
93. Etzion, O.; Hamid, S.S.; Lurie, Y.; Gane, E.; Bader, N.; Yardeni, D. End of study results from LIMT HDV study: 36% durable virologic response at 24 weeks post-treatment with pegylated interferon lambda monotherapy in patients with chronic hepatitis delta virus infection. *J. Hepatol.* **2019**, *70*, e32. [CrossRef]
94. Yurdaydin, C.; Keskin, O.; Kalkan, C.; Karakaya, F.; Caliskan, A.; Karatayli, E.; Karatayli, S.; Bozdayi, A.M.; Koh, C.; Heller, T.; et al. Optimizing lonafarnib treatment for the management of chronic delta hepatitis: The LOWR HDV-1 study. *Hepatology* **2018**, *67*, 1224–1236. [CrossRef]
95. Yurdaydin, C.; Kalkan, C.; Karakaya, F.; Caliskan, A.; Karatayli, S.; Keskin, O.; Idilman, R.; Bozdayi, A.M.; Koh, C.; Heller, T.; et al. Subanalysis of the LOWR HDV-2 study reveals high response rates to lonafarnib in patients with low viral loads. *J. Hepatol.* **2018**, *68*, S89. [CrossRef]
96. Bazinet, M.; Pantea, V.; Cebotarescu, V.; Cojuhari, L.; Jimbei, P.; Albrecht, J.; Schmid, P.; Le Gal, F.; Gordien, E.; Krawczyk, A.; et al. Safety and efficacy of REP 2139 and pegylated interferon alfa-2a for treatment-naive patients with chronic hepatitis B virus and hepatitis D virus co-infection (REP 301 and REP 301-LTF): A non-randomised, open-label, phase 2 trial. *Lancet Gastroenterol. Hepatol.* **2017**, *2*, 877–889. [CrossRef]
97. Bazinet, M.; Pantea, V.; Cebotarescu, V.; Cojuhari, L.; Jimbei, P.; Anderson, M.; Gersch, J.; Holzmayer, V.; Elsner, C.; Krawczyk, A.; et al. Persistent Control of Hepatitis B Virus and Hepatitis Delta Virus Infection Following REP 2139-Ca and Pegylated Interferon Therapy in Chronic Hepatitis B Virus/Hepatitis Delta Virus Coinfection. *Hepatol. Commun.* **2021**, *5*, 189–202. [CrossRef]

Journal of Personalized Medicine

Article

Factors Positively Correlated with Hepatitis B Surface Antigen Seroconversion in Chronic Hepatitis B

Matthias Buechter [1,2,*], Arne Maria Günther [1], Paul Manka [3], Guido Gerken [1,4] and Alisan Kahraman [1,5]

1 Department of Gastroenterology and Hepatology, University Clinic of Essen, University of Duisburg-Essen, 45147 Essen, Germany; arne1604@web.de (A.M.G.); guido@clausgerken.de (G.G.); alisan.kahraman@max-grundig-klinik.de (A.K.)
2 Department of Gastroenterology and Hepatology, Elisabeth Hospital, 58638 Iserlohn, Germany
3 Department of Internal Medicine, University Hospital Knappschaftskrankenhaus, Ruhr-University Bochum, 44801 Bochum, Germany; paul.manka@kk-bochum.de
4 Department of Gastroenterology and Hepatology, Helios Clinic, 42549 Velbert, Germany
5 Department of Gastroenterology and Hepatology, Max Grundig Clinic, 77815 Bühl, Germany
* Correspondence: m.buechter@kkimk.de

Abstract: Background and Aims: Chronic hepatitis B virus (HBV) infection is a global public health challenge since more than 250 million individuals are affected worldwide. Since different treatment modalities are available and not all patients are candidates for antiviral treatment, biomarkers that potentially predict the possibility of HBsAg clearance and seroconversion may be useful in clinical practice. Patients and methods: In this retrospective study, we aimed to identify factors positively correlated with HBsAg seroconversion in a large cohort of 371 chronic hepatitis B patients treated at a German tertial center between 2005 and 2020. Results: Seroconversion occurred in 25/371 (6.7%) and HBsAg loss in 29/371 patients (7.8%) with chronic HBV infection. Antiviral therapy was associated with a lower chance of seroconversion (seroconversion antiviral therapy 14/260 (5.4%) vs. therapy-naïve patients 11/111 (9.9%), $p = 0.027$). Seroconversion rates were higher in patients with (very) low titers of HBV DNA (best cut-off value 357 IU/mL) and quantitative HBsAg. The best cut-off value with regard to seroconversion was 357 IU/mL for HBV DNA (AUC 0.693 (95%-CI 0.063–0.422), sensitivity 0.714, specificity 0.729; $p < 0.0005$) and 33,55 IU/mL for HBsAg (AUC 0.794 (95%-CI 0.651–0.937), sensitivity 0.714, specificity 0.949; $p < 0.0005$). However, male gender was positively associated with seroconversion (seroconversion: males 7.6% vs. females 2.7%, $p = 0.036$). Conclusions: Treatment-naïve male chronic HBV patients with low viral load and inflammatory activity have the best chance to achieve seroconversion. In the absence of cirrhosis, antiviral therapy should therefore not be performed in this patient collective.

Keywords: chronic hepatitis b; HBsAg seroconversion; seroclearance; hepatocellular carcinoma

check for updates

Citation: Buechter, M.; Günther, A.M.; Manka, P.; Gerken, G.; Kahraman, A. Factors Positively Correlated with Hepatitis B Surface Antigen Seroconversion in Chronic Hepatitis B. *J. Pers. Med.* **2024**, *14*, 390. https://doi.org/10.3390/jpm14040390

Academic Editors: Giovanni Squadrito and Konstantinos Tziomalos

Received: 25 January 2024
Revised: 25 March 2024
Accepted: 27 March 2024
Published: 5 April 2024

1. Introduction

Chronic hepatitis B virus (HBV) infection is a global public health challenge since more than 250 million individuals are affected worldwide. Approximately 45% of the world population lives in areas of high endemicity, including many African and Asian countries. Patients who are infected with HBV have an enhanced risk of complications such as the development of liver cirrhosis, liver decompensation and hepatocellular carcinoma (HCC), increasing morbidity and mortality dramatically. Although the majority of patients do not develop complications, it is estimated that during their lifetimes, 15–40% may develop serious sequelae of infection [1–4].

After exposure to HBV infection, proper management is necessary. With clinical outcomes from inactive carrier state to end-stage liver disease, close monitoring with periodic evaluation is required to prevent future complications.

Hepatitis B surface antigen (HBsAg) loss (seroclearance) or seroconversion is the ideal endpoint of antiviral therapy since this state implies functional cure. A negativation of HBsAg during the course of HBV infection usually confers an excellent long-term prognosis in the absence of cirrhosis or concurrent infections with hepatitis C or D. Thus, different treatment strategies have been explored to improve patients' outcomes [5,6]. Nucleos(t)ide analogs (NAs), which suppress HBV DNA replication via inhibiting the reverse transcription of pregenomic RNA into HBV DNA, are considered the first-line treatment option. However, NAs fail to clear the HBV covalently closed circular DNA (cccDNA) and RNA replication intermediates. Thus, long-term treatment is needed for sustained viral suppression in chronic HBV patients. Although the new NAs are supposed to be drugs with a high genetic barrier to resistance and therefore reliably facilitate sustained viral suppression, functional cure (e.g., HBsAg loss or seroconversion) is rare due to the long-term persistence of viral DNA in the hepatocytes of treated patients [7–9]. The other cornerstone in hepatitis B treatment is interferon and its pegylated form, pegylated interferon α (PEG IFN α). PEG IFN α consists of a group of cytokines with antiviral activity; it mainly acts as an immunomodulator and enhances the host cell-mediated immune response. Although PEG IFN α has restricted efficacy in only a proportion of patients, responders to therapy may maintain a virologic response after drug withdrawal, and permanent HBeAg or even HBsAg clearance can be achieved. In contrast to NAs, the finite treatment duration (e.g., 6–12 months) offers an advantage to patients. However, its efficacy with regard to functional cure is still not satisfactory (seroclearance rates < 10%). In addition, multiple side-effects limit the use of PEG IFN α and lead to the frequent discontinuation of therapy in a substantial proportion of patients [10–12].

Both radiologic and serum biomarkers may yield valuable information on how to assess clinical disease evolution and associate available therapies. Viral markers such as HBV DNA, HBV RNA, quantitative HBsAg, and HBeAg are currently used to predict therapeutic response for chronic HBV patients undergoing NA therapy [5,7,9]. However, since different treatment modalities are available and not all patients are candidates for antiviral treatment, biomarkers that potentially predict the possibility of HBsAg clearance and seroconversion may be useful in clinical practice. In this retrospective study, we aimed to identify factors positively correlated with HBsAg seroconversion in a large cohort of chronic hepatitis B patients treated at a German tertial center between 2005 and 2020.

2. Patients and Methods

2.1. Patient Information and Ethical Considerations

In this retrospective study between 06/2005 and 02/2020, a total of 371 consecutive patients with chronic HBV infection were included. Chronic hepatitis B was defined as the persistence of hepatitis b surface antigen (HBsAg) for 6 months or more after acute infection with hepatitis B virus [13,14]. Study inclusion therefore necessitated two positive HBsAg results at an interval of at least 6 months. The second positive HBsAg verification was defined as the baseline date for study inclusion, data collection and follow-up. Consequently, patients without evidence for HBsAg persistence > 6 months were excluded from the study. The University Clinic of Essen ethics committee approved the retrospective, anonymous analysis of the data (17-7655-BO), and the study was conducted according to the principles expressed in the Declaration of Helsinki. All patients gave their written informed consent prior to study inclusion.

2.2. Data Collection, Laboratory Parameters, Antiviral Therapy and Definition of Seroconversion

The following epidemiological and laboratory parameters were recorded: age, body mass index (BMI, kg/m^2), existence or occurrence of hepatocellular carcinoma (HCC), HBV-DNA (IU/mL), HBV genotype, HBsAg (IU/mL), anti-HBs (IU/mL), HBeAg (IU/mL), anti-HBe, gamma-GT (U/L), aspartate aminotransferase (AST; U/L), alanine aminotransferase (ALT; U/L), total bilirubin (mg/dL), cholesterol (mg/dL), high-density lipoprotein (HDL;

mg/dL), low-density lipoprotein (LDL; mg/dL), triglyceride (mg/dL), plasma glucose (mg/dL) and glycated hemoglobin (HbA1c; %).

Structural changes in the liver parenchyma (fibrosis and steatosis) were assessed by transient elastography (TE). The measurement of liver stiffness (LSM, kPa) and controlled attenuation parameter (CAP, dB/m) was performed by Fibroscan (Echosens, Paris, France) according to the manufacturer's recommendations. TE was performed with the patient in the supine position with the right arm in maximal abduction. The TE probe was used via the intercostal spaces with the probe on the right lobe of the liver. Each set of examinations was assisted by B-mode ultrasound (APLIO 500, Toshiba, Tokio, Japan) for the identification of a feasible probe position and the exclusion of perihepatic ascites. The median of at least 10 LSM values expressed in kPa was used as the representative measurement. The success rate was calculated as the number of valid measurements divided by the number of total measurements. According to the manufacturer's recommendations, only patients with an interquartile range (IQR) < 30% of the median value and a success rate >60% were included in the analysis. According to the actual literature, significant fibrosis and liver cirrhosis were defined as LSM > 8 kPa and > 15 kPa, respectively. In addition, a CAP value > 222 dB/m was defined as significant steatosis [15–24].

Histology was obtained either percutaneously or via mini-laparoscopy-guided liver biopsy in some patients with, for example, overlap with metabolic dysfunction-associated steatotic liver disease (MASLD; diabetics with obesity) or possible overlap with autoimmune hepatitis, since hepatitis B sometimes triggers autoimmune phenomena (LKM positivity). However, liver biopsy was not routinely performed in our department.

Antiviral treatment was performed according to the German guideline recommendations ("Deutsche Gesellschaft für Gastroenterologie, Verdauungs- und Stoffwechselkrankheiten"; DGVS) and required HBV DNA > 2000 IU/mL, elevated ALT and/or at least moderate histological lesions, while all cirrhotic patients with detectable HBV DNA were treated. Antiviral therapy was performed by the use of either nucleos(t)ide analog (NA) to suppress HBV replication to induce the stabilization of HBV-induced liver disease and to prevent disease progression and HCC development or pegylated interferon-α to induce long-term immunological control with a finite duration treatment. However, drug treatment selection was based on individual parameters such as the severity of liver disease, pretreatment, co-morbidities, response probability, patients' will, contraindications and local expertise according to published DGVS guidelines. Besides interferon-α, the following NAs were used: lamivudine, telbivudine, adefovir, entecavir, and tenofovir. First-line treatment was usually performed with drugs with good potency and high genetic barriers to drug treatment (e.g., tenofovir, entecavir). Patient monitoring followed DGVS recommendations for chronic HBV infection (determination of ALT, HBV DNA, HBeAg (if initially positive), HBsAg; sonography for HCC screening every 6 months). The general goal of antiviral HBV drug treatment was to suppress HBV DNA (below the detectable limit) and normalize ALT. In the case of an insufficient response, resistant variants were tested. Seroconversion was defined as the loss of HBsAg and the development of anti-HBsAg antibodies during follow-up [13,14,25,26].

2.3. Statistical Analysis

Statistical analysis was performed using SPSS (IBM, Chicago, IL, USA). For descriptive statistics, medians and IQRs were determined. All variables were tested for normal distribution with the Kolmogorov–Smirnov test, the Shapiro–Wilk test, and calculation of skew and kurtosis. The Mann–Whitney U test was used to compare differences between independent groups. Categorical data were tested with the chi-square test, and the Kruskal–Wallis test was used for multiple comparisons. To adjust for several risk factors, the multivariate logistic analysis was performed with all the variables found significant in univariate analysis in a single step. A p value < 0.05 was considered statistically significant.

3. Results

3.1. Demographic and Laboratory Data

Between 2005 and 2020, a total of 371 patients with chronic HBV infection and a well-documented long-term course were included in this retrospective study. The majority of the study population was male (n = 224, 60.4%), the mean age at initial diagnosis 48.08 ± 13.41 [23–84] years and the mean observation period 175.94 months. In addition, 10/371 patients (2.7%) were co-infected with hepatitis delta virus. Laboratory data on initial diagnosis are presented in Table 1, and the frequency of HBV genotypes is presented in Table 2.

Table 1. Laboratory data on initial diagnosis.

Parameter	Value (Median; IQR)	Reference Value
HBV DNA [IU/mL]	3329 (356; 759626)	<10 IU/mL
HBsAg [IU/mL]	3823 (729; 12502)	<5 IU/mL
γGT [U/L]	23 (14; 43)	<60 U/L
AST [U/L]	29 (22; 42)	<35 U/L
ALT [U/L]	37 (24; 65)	<45 U/L
bilirubin [mg/dL]	0.6 (0.5; 0.8)	0.1–1.2 mg/dL
cholesterol [mg/dL]	186 (161; 215)	<250 mg/dL
HDL [mg/dL]	50 (40; 63)	40–60 mg/dL
LDL [mg/dL]	114 (91; 139)	<150 mg/dL
triglycerides [mg/dL]	96 (71; 141)	<150 mg/dL
glucose [mg/dL]	88 (81; 95)	60–99 mg/dL
HbA1c [%]	5.5 (5.2; 5.8)	<5.7%

Table 2. Distribution of HBV genotypes among the study population.

Genotype	Frequency (n)	Proportion (%)
A	43	11.6
B	9	2.4
C	4	1.1
D	110	29.6
E	9	2.4
F	1	0.3
G	1	0.3
H	1	0.3
Unknown	193	52.0
Total	371	100.0

3.2. HCC Incidence, Liver Elastography and CAP

Liver stiffness measurement (LSM) was available in 337/371 patients (90.8%). In total, 37/337 patients (11.0%) had liver cirrhosis (LSM > 15 kPa) and 85/337 patients (25.2%) had significant fibrosis (LSM > 8 kPa). CAP measurement was available in 256/371 patients (69.0%). Significant steatosis (CAP > 222 dB/m) was present in 103/256 patients (40.2%). Hepatocellular carcinoma (HCC) occurred in 8/371 cases (2.2%). LSM was available in six out of eight patients (cirrhosis: n = 1 (LSM 32.0 kPa), significant fibrosis n = 2 (LSM 11.6 kPa and 13.4 kPa), below cut-off n = 3 (LSM 4.4 kPa, 6.6 kPa, and 7.7 kPa)).

3.3. *Antiviral Therapy*

Antiviral therapy was performed in 260/371 patients (70.1%). In 117 out of 260 patients (45.0%), more than one antiviral substance was administered. A total of 58 patients (15.6%) received pegylated interferon for a mean timespan of 12.6 ± 8.6 months. Seroconversion was achieved in 4/58 of these patients (6.9%). Oral antiviral therapy was conducted by the use of tenofovir (n = 165), entecavir (n = 93), lamivudine (n = 63), adefovir (n = 55) and telbivudine (n = 29). Seroconversion was observed in 3.8% of patients (10/260) receiving oral antiviral therapies.

3.4. *Factors Associated with Seroconversion*

Seroconversion (the development of anti-HB antibodies) occurred in 25/371 (6.7%) and HBsAg loss in 29/371 patients (7.8%) with chronic HBV infection. The mean timespan from first diagnosis to seroconversion was 146 [7–478] months. The age at first diagnosis was not associated with seroconversion. However, male gender was positively associated with seroconversion (seroconversion: males 7.6% vs. females 2.7%, $p = 0.036$).

3.4.1. Antiviral Therapy

Antiviral therapy was associated with a lower chance of seroconversion (seroconversion antiviral therapy 14/260 (5.4%) vs. therapy-naïve patients 11/111 (9.9%), $p = 0.027$). Although the rate of seroconversion was higher in the PEG-IFN (n = 4/58, 6.9%) group than the oral therapy group (n = 10/206, 4.9%), this difference did not reach statistical significance ($p > 0.05$). Data regarding seroconversion rates according to the presence or absence of antiviral therapy are presented in Table 3.

Table 3. Seroconversion rates according to presence or absence of antiviral therapy.

Therapy	Number of Patients Achieving Seroconversion	Number of Patients under That Specific Therapy	Fraction of Patients Achieving Seroconversion	Mean Timespan from Initial Diagnosis to Seroconversion in Months	Standard Deviation in Months
No therapy	11	111 (29.9%)	9.9%	167.5	148.2
Interferone	4	58 (15.6%)	6.9%	198	51.2
Telbivudine	2	29 (7.8%)	6.9%	93	48.1
Lamivudine	3	63 (17%)	4.8%	174.3	60.1
Any therapy (**NRTI+ *NNRTI+ Interferone)	10	260 (70.1%)	3.8%	152.7	78
Adefovir	2	55 (14.8%)	3.6%	140.5	19.1
**NRTI + *NNRTI	8	237 (63.9%)	3.3%	143.3	81.7
Entecavir	3	93 (25.1%)	3.2%	106.3	86.6
Tenofovir	4	165 (44.5%)	2.4%	146.3	77

*NNRTI, non-nucleoside reverse transcriptase inhibitor. **NRTI, nucleoside reverse transcriptase inhibitor.

3.4.2. HBV-Specific and Liver-Specific Laboratory Parameters and Liver Fibrosis Measured by Transient Elastography (TE)

Seroconversion rates were higher in patients with (very) low titers of HBV DNA and quantitative HBsAg. The best cut-off value with regard to seroconversion was 357 IU/mL for HBV DNA (AUC 0.693 (95%-CI 0.063–0.422), sensitivity 0.714, specificity 0.729; $p < 0.0005$) and for HBsAg, 33.55 IU/mL (AUC 0.794 (95%-CI 0.651–0.937), sensitivity 0.714, specificity 0.949; $p < 0.0005$). Corresponding ROC curves are demonstrated in Figures 1 and 2. However, the presence or absence of HBe antigen and anti-HBe antibodies was not a predictive factor regarding seroconversion.

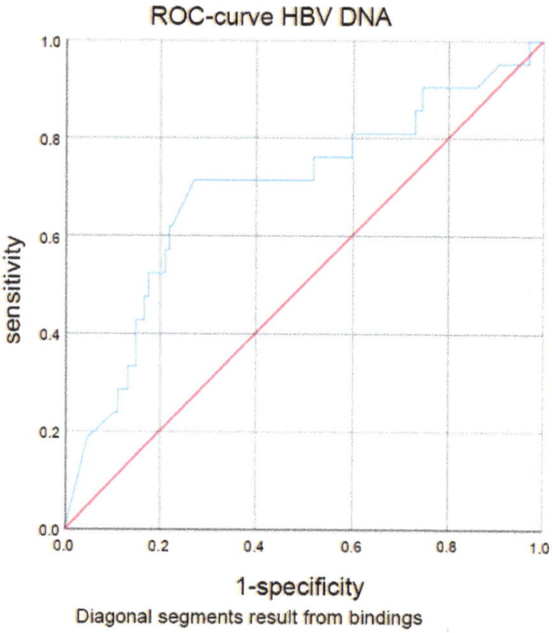

Figure 1. HBV DNA ROC curve according to seroconversion (cut-off 357 IU/mL, AUC 0.693 (95%-CI 0.063–0.422), sensitivity 0.714, specificity 0.729; $p < 0.0005$).

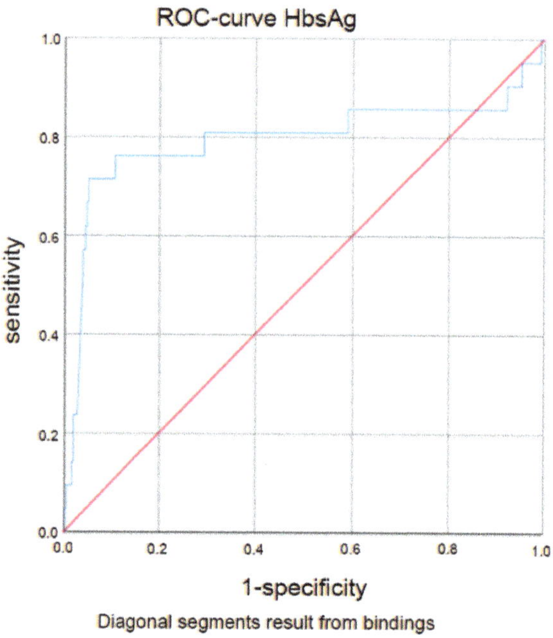

Figure 2. Quantitative HBsAg ROC curve according to seroconversion (cut-off 33.55 IU/mL, AUC 0.794 (95%-CI 0.651–0.937), sensitivity 0.714, specificity 0.949; $p < 0.0005$).

Furthermore, γ-glutamyltransferase (GGT) was higher in patients attaining seroconversion. Statistical analysis, however, failed to reach significance, but showed a positive trend (mean GGT seroconversion 173.20 IU/mL vs. no seroconversion 44.89 IU/mL, $p = 0.052$). In addition, level of serum aspartat aminotransferase (AST), alanin aminotransferase (ALT) and bilirubin was not associated with seroconversion in chronic HBV patients.

Interestingly, statistical analysis revealed a positive correlation between elevated liver stiffness and subsequent seroconversion (mean LSM seroconversion 8.2 ± 8.4 (3.3–32.4) kPa vs. mean LSM no seroconversion 5.4 ± 5.5 (3.2–43.5) kPa, $p = 0.046$) (Figure 3). When applying the previously cited cut-offs for significant fibrosis (>8 kPa) and cirrhosis (>15 kPa), subgroup analysis also showed a positive correlation (seroconversion significant fibrosis 11.0% vs. no significant fibrosis 3.4%, $p = 0.008$; seroconversion cirrhosis 16.2% vs. no cirrhosis 4.3%, $p = 0.011$). However, ROC analysis revealed an optimal LSM cut-off value with regard to seroconversion of 7.65 kPa (AUC 0.664, 95%-CI 0.528–0.800, sensitivity 0.579, specificity 0.7676, $p = 0.016$).

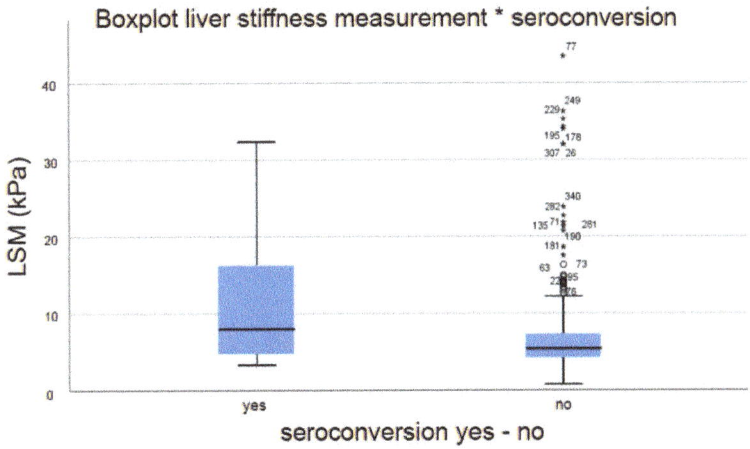

Figure 3. Correlation between liver stiffness measurement (LSM) and subsequent seroconversion (n = 337).

3.4.3. Steatosis, Lipid Metabolism, Diabetes Mellitus and Obesity Were Not Associated with Seroconversion

Parameters associated with lipid metabolism disorders, metabolic dysfunction-associated steatotic liver disease (MASLD) and metabolic dysfunction-associated steatohepatitis (MASH) did not influence the probability of attaining seroconversion. The following parameters were analyzed: body mass index (BMI; seroconversion BMI > 25 kg/m^2 14/197 patients (7.11%) vs. seroconversion BMI < 25 kg/m^2 8/174 patients (4.60%), $p = 0.20$), concurrent diabetes mellitus/HbA1c > 5.7% (seroconversion 3/21 patients (14.29%) vs. no seroconversion 18/350 patients (5.14%), $p = 0.095$) and controlled attention parameter (CAP; CAP seroconversion 261,33 dB/m vs. CAP no seroconversion 240,51 dB/m, $p = 0.218$). Likewise, lipid metabolism parameters (serum cholesterol, triglycerides, high-density lipoprotein (HDL) and low-density lipoproteins (LDL)) had no influence on seroconversion ($p > 0.05$).

3.4.4. Gender, GGT and Antiviral Therapy Were Independently Associated with Seroconversion

Multivariate logistic regression analysis was used to identify independent predictors of seroconversion. Significant parameters from the univariate analysis were included (antiviral therapy, gender, GGT, LSM, HBV-DNA, HBsAg). However, gender (OR: 0.233; CI: 0.060–0.898; $p = 0.034$), GGT (OR 0.993; CI: 0.990–0.997; $p < 0.001$) and antiviral therapy (OR: 3.97; CI: 1.302–11.722; $p = 0.015$) were independently associated with seroconversion.

4. Discussion

Functional cure in terms of HBsAg loss and the development of anti-HB antibodies (seroconversion) should be the ultimate goal in the treatment of patients with chronic hepatitis B with regard to drug withdrawal safety and improvements in prognosis. Although first-line drug therapy with NAs can suppress virus replication reliably (and thereby reduce the risk of complications such as the development of HCC and decompensation), lifelong therapy is usually required since the viral cccDNA persists in the nucleus of hepatocytes [27,28]. In selected patients, subcutaneous applied PEG IFN α represents a therapeutic alternative with the option of temporary limited treatment and the elimination of HBV. However, its use is often restricted by relevant side effects which leads to drug discontinuation in a significant proportion of patients [11,29–32].

Various algorithms regarding therapy indication in chronic HBV have been proposed, such as that by the American Association for the Study of the Liver Diseases (AASLD), the European Association for the Study of the Liver (EASL) or the Asian Pacific Association for the Study of the Liver (APASL). Although the recommendations slightly differ, antiviral treatment is usually indicated when chronic active hepatitis B disease is evident and at least one of the following criteria is fulfilled: (i) presence of advanced fibrosis or cirrhosis, (ii) HBV DNA level above at least 2000 IU/mL and (iii) persistently or repeatedly abnormal ALT/AST levels [13,14,26].

However, since different treatment modalities are available and not all patients are candidates for antiviral treatment, biomarkers that potentially predict the possibility of HBsAg clearance and seroconversion may be useful in clinical practice. In this retrospective study, we aimed to identify factors positively correlated with HBsAg seroconversion in a large cohort of 371 patients with chronic HBV infection treated at the University Hospital Essen between 2005 and 2020. In contrast to other studies, which were conducted in regions where HBV infection is endemic and usually acquired during the perinatal period or early infancy, Germany has low or intermediate patterns of endemicity. In population-based studies, approximately 0.3% of adults and 0.2% of infants were infected with HBV (HBsAg positive). In these studies, anti-HBc prevalence was 5.1% for adults and 0.5% for infants [1].

The overall HBsAg seroconversion rate in the mean observation time of approximately 176 months was relatively low, being 6.7% (25/371) of patients. However, these data match with data from the actual literature, in which short- and long-term HBsAg seroclearance and seroconversion rates are stated to be between 4.2% and 20.6% [33–37]. Interestingly, HBsAg seroconversion was significantly higher in therapy-naïve patients than in patients treated with antiviral drugs (11/111 patients (9.9%) vs. 14/260 patients (5.4%), $p = 0.027$). Subgroup analysis revealed that the seroconversion rate was higher in the PEG-IFN group (n = 4/58, 6.9%) than in the oral therapy group (n = 10/206, 4.9%), although this difference did not reach statistical significance ($p > 0.05$). However, antiviral therapy with NAs is very potent and usually leads to a rapid reduction in viral load and thereby in (systemic) inflammation. The consequently reduced confrontation/interaction between the virus and the host's immune system might be an explanation for these findings with regard to pathogenesis.

Several studies have shown that serum HBV-DNA and quantitative HBsAg levels were the most significant indicators of seroclearance or seroconversion [34,38–42]. Similarly, in the present study, we observed that low HBV-DNA and quantitative HBsAg levels were associated with seroconversion, while the best cut-off values were 357 IU/mL and 34 IU/mL, respectively. Consequently, in selected patients with low HBsAg and HBV-DNA levels without cirrhosis, it might be useful to discontinue antiviral therapy with NAs to achieve spontaneous seroconversion. Fang et al. found that HBeAg-negative patients without cirrhosis and low HBsAg levels at the end of treatment who previously received entecavir or tenofovir had a high HBsAg loss rate after the discontinuation of treatment [43]. In addition, Choi and colleagues demonstrated that HBsAg loss occurred in 6.8% of HBeAg-negative HBV patients following NAs cessation [44]. Similarly, in the so-called "STOP-NUC trial", a multicenter randomized-controlled trial, the research group of

Thomas Berg and colleagues could demonstrate that stopping long-term NA treatment can induce a functional cure in HBeAg-negative patients with low HBsAg levels < 1000 IU/mL at the time point of NA treatment cessation [45]. The authors therefore conclude that finite therapy can be considered for chronic HBV patients on NA therapy. However, our data support the idea that a low viral load is an important factor for HBV cure and therapy decisions.

Some reports have demonstrated that males with chronic HBV infection seem to be more likely than females to experience seroclearance or seroconversion [34,39,46]. Likewise, male gender was associated with seroconversion in our cohort (seroconversion: males 7.6% vs. females 2.7%, $p = 0.036$).

Multivariate analysis showed statistically significant results with regard to gender (OR: 0.233; CI: 0.060–0.898; $p = 0.034$), GGT (OR 0.993; CI: 0.990–0.997; $p < 0.001$) and antiviral therapy (OR: 3.97; CI: 1.302–11.722; $p = 0.015$), which were independently associated with seroconversion.

Our data clearly indicate that treatment-naïve male chronic HBV patients with low viral load and inflammatory activity (elevated liver enzymes) should not receive antiviral therapy in the absence of liver cirrhosis, since they have a high chance of achieving spontaneous HBV seroclearance and seroconversion. As a future perspective, an increased understanding of trained immunity and how HBV establishes a permissive state in the host that bears no characteristics of immunologic tolerance may allow for the development of therapies targeted towards patients not currently indicated for treatment affecting both therapy initiation and termination [47].

Interestingly, a great proportion of individuals with chronic HBV infection showed signs of significant hepatic steatosis (CAP measurement > 222 dB/m; n = 103/256 (40.2%)) in terms of overlap with MASLD. Huang and colleagues investigated a great cohort of 4084 chronic HBV patients and found that in untreated HBeAg-negative patients, concurrent MASLD is associated with higher rates of seroclearance and seroconversion. Likewise, Li et al. retrospectively studied 6786 patients and identified that fatty liver was significantly and independently associated with a greater chance of achieving HBsAg seroclearance. The authors therefore conclude that metabolic dysfunctions have additive effects on the functional cure of chronic hepatitis B [48,49]. However, steatosis or related metabolic disorders were not associated with seroconversion in our cohort.

Chronic hepatitis B patients are at increased risk for hepatocellular carcinoma (HCC), especially when an advanced stage of liver fibrosis or cirrhosis is present. The risk for developing HCC increases from 6 to 37 times compared to control subjects. According to the actual literature, HCC incidence rates in chronic HBV subjects vary between 2 and 37% for overall incidence and 0.4 and 3% for annual incidence depending on the presence or absence of liver cirrhosis. There is solid evidence supporting the fact that, in addition to cirrhosis, the viral load reflected by HBV DNA and quantitative HBsAg levels is a strong risk factor for HCC development. Therefore, NA treatment can significantly reduce the incidence of HCC, though it does not completely eliminate the risk of HCC [50–55]. EASL and DGVS guidelines for HBV and HCC recommend HCC screening by abdominal ultrasound every six months with the optional determination of alpha-fetoprotein in chronic HBV patients with enhanced HCC risk (patients with advanced fibrosis or cirrhosis) [13,25]. Among our cohort, HCC development during a mean observation time of >175 months was rare (overall HCC incidence 2.2%) in chronic HBV patients, although a significant proportion had advanced fibrosis (25.2%) or cirrhosis (11.0%). The pathogenesis of HBV-induced HCC is thought to be multifactorial with both direct and indirect mechanisms. HBV-related HCC can also arise in non-cirrhotic livers, supporting the notion that HBV plays a direct role in liver transformation by triggering both common and etiology-specific oncogenic pathways in addition to stimulating the host immune response and driving liver chronic necroinflammation [2]. In any case, among our patient collective, 3/8 patients (37.5%) developed HCC without having cirrhosis. In our opinion, it is therefore reasonable to perform HCC screening in chronic HBV patients irrespective of whether fibrosis or cirrhosis is present.

Another interesting aspect of our study was the positive correlation between elevated liver stiffness and subsequent seroconversion (LSM seroconversion 8.2 ± 8.4 (3.3–32.4) kPa vs. LSM no seroconversion 5.4 ± 5.5 (3.2–43.5) kPa, $p = 0.046$ (Figure 3)), which applied to both significant fibrosis (seroconversion significant fibrosis 11.0% vs. no significant fibrosis 3.4%, $p = 0.008$) and cirrhosis (seroconversion cirrhosis 16.2% vs. no cirrhosis 4.3%, $p = 0.011$). The actual literature hardly offers any scientific data with regard to this consideration. Ming-Lun and colleagues, who investigated 81 patients dually infected with hepatitis B and C, found that baseline cirrhosis was a significant factor associated with HBsAg seroclearance [56]. However, the role of liver fibrosis in chronic HBV with regard to seroconversion still remains unknown.

We are aware of the limitations of our study, the most important of them being that it was a retrospectively performed single-center study.

In summary, we could demonstrate that treatment-naïve male chronic HBV patients with low viral loads and inflammatory activity (elevated liver enzymes) have the best chance of achieving seroconversion. The indication for antiviral therapy should therefore be provided carefully under these circumstances. When treated and surveilled at an expert center for liver diseases and according to the actual guidelines, HCC development seems to be relatively low though advanced fibrosis or cirrhosis is present.

Author Contributions: Conceptualization, G.G. and A.K.; methodology, A.M.G. and P.M.; software, A.M.G. and P.M.; validation, M.B., G.G. and A.K.; formal analysis, A.M.G. and P.M.; investigation, M.B., G.G. and A.K.; resources, A.M.G. and P.M.; data curation, A.M.G.; writing—original draft preparation, M.B.; writing—review and editing, M.B., A.M.G., P.M., G.G. and A.K.; visualization, M.B., G.G. and A.K.; supervision, M.B., G.G. and A.K.; project administration, M.B.; funding acquisition, not applicable. All authors have read and agreed to the published version of the manuscript.

Funding: This research received no external funding.

Institutional Review Board Statement: The study was conducted in accordance with the Declaration of Helsinki and approved by the University Clinic of Essen ethics committee (17-7655-BO).

Informed Consent Statement: Informed consent was obtained from all subjects involved in the study.

Data Availability Statement: Data will be made available by the corresponding author upon reasonable request.

Conflicts of Interest: The authors declare no conflicts of interest.

Abbreviations

ALT: alanin aminotransferase; AST, aspartate aminotransferase; AUC, area under the curve; CAP, controlled attenuation parameter; ccc DNA, closed circular DNA; GGT, γ-glutamyltransferase; HbA1c, glycated hemoglobin; HBeAg, hepatitis B e-antigen; HBsAg, hepatitis B surface antigen; HBV, hepatitis B virus; HBV DNA, hepatitis B virus deoxyribonucleic acid; HBV RNA, hepatitis B virus ribonucleic acid; HCC, hepatocellular carcinoma; HDL, high-density lipoprotein; LDL, low-density lipoprotein; LSM, liver stiffness measurement; MASH, metabolic dysfunction-associated steatohepatitis; MASLD, metabolic dysfunction-associated steatotic liver disease; NA, nucleos(t)ide analog; PEG IFN α, pegylated interferon α; TE, transient elastography.

References

1. Dudareva, S.; Faber, M.; Zimmermann, R.; Bock, C.T.; Offergeld, R.; Steffen, G.; Enkelmann, J. Epidemiology of viral hepatitis A to E in Germany. *Bundesgesundheitsblatt Gesundheitsforschung Gesundheitsschutz* **2022**, *65*, 149–158. [CrossRef] [PubMed]
2. Ligat, G.; Schuster, C.; Baumert, T.F. Hepatitis B Virus Core Variants, Liver Fibrosis, and Hepatocellular Carcinoma. *Hepatol. Baltim. Md.* **2019**, *69*, 5–8. [CrossRef] [PubMed]

3. Ott, J.; Stevens, G.; Groeger, J.; Wiersma, S. Global epidemiology of hepatitis B virus infection: New estimates of age-specific HBsAg seroprevalence and endemicity. *Vaccine* **2012**, *30*, 2212–2219. [CrossRef] [PubMed]
4. Schweitzer, A.; Horn, J.; Mikolajczyk, R.T.; Krause, G.; Ott, J.J. Estimations of worldwide prevalence of chronic hepatitis B virus infection: A systematic review of data published between 1965 and 2013. *Lancet Lond Engl.* **2015**, *386*, 1546–1555. [CrossRef] [PubMed]
5. Zhong, Y.-W.; Shi, Y.-M.; Chu, F.; Liu, J.; Shi, C.; Xu, J.-J.; Liu, P.; Bai, Y.-J.; Xiao, X.-H.; Zhang, X.-C.; et al. Prediction for HBsAg seroconversion in children with chronic hepatitis B. *BMC Infect. Dis.* **2021**, *21*, 1211. [CrossRef] [PubMed]
6. Ko, C.; Chakraborty, A.; Chou, W.-M.; Hasreiter, J.; Wettengel, J.M.; Stadler, D.; Bester, R.; Asen, T.; Zhang, K.; Wisskirchen, K.; et al. Hepatitis B virus genome recycling and de novo secondary infection events maintain stable cccDNA levels. *J. Hepatol.* **2018**, *69*, 1231–1241. [CrossRef] [PubMed]
7. Wang, Y.; Liao, H.; Deng, Z.; Liu, Y.; Bian, D.; Ren, Y.; Yu, G.; Jiang, Y.; Bai, L.; Liu, S.; et al. Serum HBV RNA predicts HBeAg clearance and seroconversion in patients with chronic hepatitis B treated with nucleos(t)ide analogues. *J. Viral Hepat.* **2022**, *29*, 420–431. [CrossRef] [PubMed]
8. Lu, F.; Wang, J.; Chen, X.; Xu, D.; Xia, N. Potential use of serum HBV RNA in antiviral therapy for chronic hepatitis B in the era of nucleos(t)ide analogs. *Front. Med.* **2017**, *11*, 502–508. [CrossRef] [PubMed]
9. Luo, H.; Zhang, X.-X.; Cao, L.-H.; Tan, N.; Kang, Q.; Xi, H.-L.; Yu, M.; Xu, X.-Y. Serum hepatitis B virus RNA is a predictor of HBeAg seroconversion and virological response with entecavir treatment in chronic hepatitis B patients. *World J. Gastroenterol.* **2019**, *25*, 719–728. [CrossRef] [PubMed]
10. Dusheiko, G. Treatment of HBeAg positive chronic hepatitis B: Interferon or nucleoside analogues. *Liver Int. Off. J. Int. Assoc. Study Liver* **2013**, *33* (Suppl. S1), 137–150. [CrossRef] [PubMed]
11. Ye, J.; Chen, J. Interferon and Hepatitis B: Current and Future Perspectives. *Front. Immunol.* **2021**, *12*, 733364. [CrossRef] [PubMed]
12. Hui, A.Y.; Chan, H.L.; Cheung, A.Y.; Cooksley, G.; Sung, J.J. Systematic review: Treatment of chronic hepatitis B virus infection by pegylated interferon. *Aliment. Pharmacol. Ther.* **2005**, *22*, 519–528. [CrossRef] [PubMed]
13. European Association for the Study of the Liver. EASL 2017 Clinical Practice Guidelines on the management of hepatitis B virus infection. *J. Hepatol.* **2017**, *67*, 370–398. [CrossRef] [PubMed]
14. Terrault, N.A.; Lok, A.S.F.; McMahon, B.J.; Chang, K.-M.; Hwang, J.P.; Jonas, M.M.; Brown, R.S., Jr.; Bzowej, N.H.; Wong, J.B. Update on prevention, diagnosis, and treatment of chronic hepatitis B: AASLD 2018 hepatitis B guidance. *Hepatol Baltim Md.* **2018**, *67*, 1560–1599. [CrossRef] [PubMed]
15. Barr, R.G.; Ferraioli, G.; Palmeri, M.L.; Goodman, Z.D.; Garcia-Tsao, G.; Rubin, J.; Garra, B.; Myers, R.P.; Wilson, S.R.; Rubens, D.; et al. Elastography Assessment of Liver Fibrosis: Society of Radiologists in Ultrasound Consensus Conference Statement. *Radiology* **2015**, *276*, 845–861. [CrossRef] [PubMed]
16. Buechter, M.; Gerken, G. Liver Function-How to Screen and to Diagnose: Insights from Personal Experiences, Controlled Clinical Studies and Future Perspectives. *J. Pers. Med.* **2022**, *12*, 1657. [CrossRef] [PubMed]
17. Castera, L. Noninvasive methods to assess liver disease in patients with hepatitis B or C. *Gastroenterology* **2012**, *142*, 1293–1302.e4. [CrossRef] [PubMed]
18. Chen, J.; Wu, D.; Wang, M.; Chen, E.; Bai, L.; Liu, C.; Liao, J.; Tang, H. Controlled attenuation parameter for the detection of hepatic steatosis in patients with chronic hepatitis B. *Infect. Dis.* **2016**, *48*, 670–675. [CrossRef] [PubMed]
19. De Franchis, R. Expanding consensus in portal hypertension: Report of the Baveno VI Consensus Workshop: Stratifying risk and individualizing care for portal hypertension. *J. Hepatol.* **2015**, *63*, 743–752. [CrossRef] [PubMed]
20. Dietrich, C.F.; Bamber, J.; Berzigotti, A.; Bota, S.; Cantisani, V.; Castera, L. EFSUMB Guidelines and Recommendations on the Clinical Use of Liver Ultrasound Elastography, Update 2017 (Long Version). *Ultraschall Med.-Eur. J. Ultrasound* **2017**, *38*, e16–e47.
21. Fang, C.; Lim, A.; Sidhu, P.S. Ultrasound-based liver elastography in the assessment of fibrosis. *Clin. Radiol.* **2020**, *75*, 822–831. [CrossRef]
22. Ferraioli, G. Review of Liver Elastography Guidelines. *J. Ultrasound Med. Off. J. Am. Inst. Ultrasound Med.* **2019**, *38*, 9–14. [CrossRef] [PubMed]
23. Ferraioli, G. Ultrasound Techniques for the Assessment of Liver Stiffness: A Correct Terminology. *Hepatol. Baltim. Md.* **2019**, *69*, 461. [CrossRef] [PubMed]
24. Loomba, R.; Adams, L.A. Advances in non-invasive assessment of hepatic fibrosis. *Gut* **2020**, *69*, 1343–1352. [CrossRef] [PubMed]
25. Cornberg, M.; Sandmann, L.; Protzer, U. Not Available. *Z. Gastroenterol.* **2021**, *59*, 691–776. [PubMed]
26. Sarin, S.K.; Kumar, M.; Lau, G.K.; Abbas, Z.; Chan, H.L.Y.; Chen, C.J.; Chen, D.S.; Chen, H.L.; Chen, P.J.; Chien, R.N.; et al. Asian-Pacific clinical practice guidelines on the management of hepatitis B: A 2015 update. *Hepatol. Int.* **2016**, *10*, 1–98. [CrossRef]
27. Lee, H.W.; Lee, J.S.; Ahn, S.H. Hepatitis B Virus Cure: Targets and Future Therapies. *Int. J. Mol. Sci.* **2020**, *22*, 213. [CrossRef]
28. Nassal, M. HBV cccDNA: Viral persistence reservoir and key obstacle for a cure of chronic hepatitis B. *Gut* **2015**, *64*, 1972–1984. [CrossRef]
29. Trépo, C.; Chan, H.L.Y.; Lok, A. Hepatitis B virus infection. *Lancet Lond. Engl.* **2014**, *384*, 2053–2063. [CrossRef] [PubMed]
30. Wan, M.B.; Weng, X.H. Expert recommendations on the application of interferon for chronic hepatitis B. *J. Dig. Dis.* **2013**, *14*, 626–637. [CrossRef]

31. Viganò, M.; Grossi, G.; Loglio, A.; Lampertico, P. Treatment of hepatitis B: Is there still a role for interferon? *Liver Int. Off. J. Int. Assoc. Study Liver* **2018**, *38* (Suppl. S1), 79–83. [CrossRef] [PubMed]
32. Konerman, M.A.; Lok, A.S. Interferon Treatment for Hepatitis B. *Clin. Liver Dis.* **2016**, *20*, 645–665. [CrossRef] [PubMed]
33. Zhang, Y.; Chen, B.; Wang, L.; Chi, J.; Song, S.; Liu, M.; Zhao, Z. HBsAg seroclearance or seroconversion induced by peg-interferon alpha and lamivudine or adefovir combination therapy in chronic hepatitis B treatment: A meta-analysis and systematic review. *Rev. Esp. Enferm. Dig.* **2016**, *108*, 263–270. [CrossRef]
34. Ferreira, S.C.; Chachá, S.G.F.; Souza, F.F.; Teixeira, A.C.; Santana, R.C.; Villanova, M.G.; Zucoloto, S.; Ramalho, L.N.; Perdoná, G.S.; Passos, A.D.; et al. Factors associated with spontaneous HBsAg clearance in chronic hepatitis B patients followed at a university hospital. *Ann. Hepatol.* **2014**, *13*, 762–770. [CrossRef] [PubMed]
35. Song, A.; Lin, X.; Chen, X. Functional cure for chronic hepatitis B: Accessibility, durability, and prognosis. *Virol. J.* **2021**, *18*, 114. [CrossRef] [PubMed]
36. Zhou, K.; Contag, C.; Whitaker, E.; Terrault, N. Spontaneous loss of surface antigen among adults living with chronic hepatitis B virus infection: A systematic review and pooled meta-analyses. *Lancet Gastroenterol. Hepatol.* **2019**, *4*, 227–238. [CrossRef]
37. Yeo, Y.H.; Ho, H.J.; Yang, H.-I.; Tseng, T.C.; Hosaka, T.; Trinh, H.N. Factors Associated With Rates of HBsAg Seroclearance in Adults With Chronic HBV Infection: A Systematic Review and Meta-analysis. *Gastroenterology* **2019**, *156*, 635–646.e9. [CrossRef]
38. Tseng, T.-C.; Liu, C.-J.; Yang, H.-C.; Su, T.-H.; Wang, C.-C.; Chen, C.-L.; Kuo, S.F.-T.; Liu, C.-H.; Chen, P.-J.; Chen, D.-S.; et al. Determinants of spontaneous surface antigen loss in hepatitis B e antigen-negative patients with a low viral load. *Hepatol. Baltim. Md.* **2012**, *55*, 68–76. [CrossRef]
39. Liu, J.; Yang, H.-I.; Lee, M.-H.; Lu, S.; Jen, C.; Wang, L.; You, S.; Iloeje, U.H.; Chen, C. Incidence and determinants of spontaneous hepatitis B surface antigen seroclearance: A community-based follow-up study. *Gastroenterology* **2010**, *139*, 474–482. [CrossRef]
40. Yuen, M.-F.; Wong, D.K.-H.; Sablon, E.; Tse, E.; Ng, I.O.-L.; Yuan, H.-J.; Siu, C.-W.; Sander, T.J.; Bourne, E.J.; Hall, J.G.; et al. HBsAg seroclearance in chronic hepatitis B in the Chinese: Virological, histological, and clinical aspects. *Hepatol. Baltim. Md.* **2004**, *39*, 1694–1701. [CrossRef] [PubMed]
41. Arai, M.; Togo, S.; Kanda, T.; Fujiwara, K.; Imazeki, F.; Yokosuka, O. Quantification of hepatitis B surface antigen can help predict spontaneous hepatitis B surface antigen seroclearance. *Eur. J. Gastroenterol. Hepatol.* **2012**, *24*, 414–418. [CrossRef] [PubMed]
42. Chen, Y.-C.; Jeng, W.-J.; Chu, C.-M.; Liaw, Y.F. Decreasing levels of HBsAg predict HBsAg seroclearance in patients with inactive chronic hepatitis B virus infection. *Clin. Gastroenterol. Hepatol. Off. Clin. Pract. J. Am. Gastroenterol. Assoc.* **2012**, *10*, 297–302. [CrossRef] [PubMed]
43. Fang, H.-W.; Hu, T.-H.; Wang, J.-H.; Hung, C.H.; Lu, S.N.; Chen, C.H. Post-treatment HBsAg decline predicts high rate of HBsAg loss after stopping entecavir or tenofovir in HBeAg-negative patients without retreatment. *Dig. Liver Dis. Off. J. Ital. Soc. Gastroenterol. Ital. Assoc. Study Liver* **2023**, *55*, 1223–1229. [CrossRef] [PubMed]
44. Choi, H.S.J.; Hirode, G.; Chen, C.-H.; Su, T.-H.; Seto, W.-K.; Van Hees, S.; Papatheodoridi, M.; Lens, S.; Wong, G.L.; Brakenhoff, S.M.; et al. Differential Relapse Patterns After Discontinuation of Entecavir vs Tenofovir Disoproxil Fumarate in Chronic Hepatitis B. *Clin. Gastroenterol. Hepatol. Off. Clin. Pract. J. Am. Gastroenterol. Assoc.* **2023**, *21*, 1513–1522.e4. [CrossRef] [PubMed]
45. van Bömmel, F.; Stein, K.; Heyne, R.; Petersen, J.; Buggisch, P.; Berg, C. A multicenter randomized-controlled trial of nucleos(t)ide analogue cessation in HBeAg-negative chronic hepatitis B. *J. Hepatol.* **2023**, *78*, 926–936. [CrossRef] [PubMed]
46. Liaw, Y.F.; Sheen, I.S.; Chen, T.J.; Chu, C.M.; Pao, C.C. Incidence, determinants and significance of delayed clearance of serum HBsAg in chronic hepatitis B virus infection: A prospective study. *Hepatol. Baltim. Md.* **1991**, *13*, 627–631. [CrossRef]
47. Gish, R.G.; Given, B.D.; Lai, C.-L.; Locarnini, S.A.; Lau, J.Y.; Lewis, D.L.; Schluep, T. Chronic hepatitis B: Virology, natural history, current management and a glimpse at future opportunities. *Antiviral Res.* **2015**, *121*, 47–58. [CrossRef] [PubMed]
48. Huang, S.-C.; Su, T.-H.; Tseng, T.-C.; Chen, C.-L.; Hsu, S.-J.; Liu, C.-H.; Liao, S.-H.; Hong, C.-M.; Lan, T.-Y.; Yang, H.-C.; et al. Metabolic Dysfunction-Associated Steatotic Liver Disease Facilitates Hepatitis B Surface Antigen Seroclearance and Seroconversion. *Clin. Gastroenterol. Hepatol.* **2024**, *22*, 581–590.e6. [CrossRef] [PubMed]
49. Li, J.; Yang, H.I.; Yeh, M.L.; Le, M.H.; Le, A.K.; Yeo, Y.H.; Dai, C.-Y.; Barnett, S.; Zhang, J.Q.; Huang, J.-F.; et al. Association Between Fatty Liver and Cirrhosis, Hepatocellular Carcinoma, and Hepatitis B Surface Antigen Seroclearance in Chronic Hepatitis B. *J. Infect. Dis.* **2021**, *224*, 294–302. [CrossRef] [PubMed]
50. Gao, X.; Yang, H.-I.; Trinh, H.; Jeong, D.; Li, J.; Zhang, J.; Le, A.; Hoang, J.; Nguyen, P.; Henry, L.; et al. Antiviral therapy and hepatocellular carcinoma risk in hepatitis B patients with cirrhosis. *Eur. J. Gastroenterol. Hepatol.* **2020**, *32*, 1207–1211. [CrossRef]
51. Gnyawali, B.; Pusateri, A.; Nickerson, A.; Jalil, S.; Mumtaz, K. Epidemiologic and socioeconomic factors impacting hepatitis B virus and related hepatocellular carcinoma. *World J. Gastroenterol.* **2022**, *28*, 3793–3802. [CrossRef] [PubMed]
52. Liu, M.; Zhao, T.; Zhang, Y.; Zhang, A.-M.; Geng, J.; Xia, X. The incidence of hepatocellular carcinoma and clearance of hepatitis B surface for CHB patients in the indeterminate phase: A systematic review and meta-analysis. *Front. Cell Infect. Microbiol.* **2023**, *13*, 1226755. [CrossRef] [PubMed]
53. Papatheodoridis, G.V.; Lampertico, P.; Manolakopoulos, S.; Lok, A. Incidence of hepatocellular carcinoma in chronic hepatitis B patients receiving nucleos(t)ide therapy: A systematic review. *J. Hepatol.* **2010**, *53*, 348–356. [CrossRef] [PubMed]
54. Roberts, S.K.; Majeed, A.; Kemp, W. Controversies in the Management of Hepatitis B: Hepatocellular Carcinoma. *Clin. Liver Dis.* **2021**, *25*, 785–803. [CrossRef] [PubMed]

55. Li, Y.-W.; Yang, F.-C.; Lu, H.-Q.; Zhang, J.-S. Hepatocellular carcinoma and hepatitis B surface protein. *World J. Gastroenterol.* **2016**, *22*, 1943–1952. [CrossRef] [PubMed]
56. Yeh, M.L.; Hung, C.H.; Huang, J.F.; Liu, C.-J.; Lee, C.-M.; Dai, C.-Y.; Wang, J.-H.; Lin, Z.-Y.; Lu, S.-N.; Hu, T.-H.; et al. Long-term effect of interferon plus ribavirin on hepatitis B surface antigen seroclearance in patients dually infected with hepatitis B and C viruses. *PLoS ONE* **2011**, *6*, e20752. [CrossRef] [PubMed]

Journal of *Personalized Medicine*

MDPI

Review

Prevention in Hepatology

Ana-Maria Muñoz-Restrepo [1,2], Maria-Cristina Navas [2], Jimmy Daza [1], Marcos Girala [3], Ezequiel Ridruejo [4], Guido Gerken [5] and Andreas Teufel [1,6,*]

1 Department of Medicine II, Division of Hepatology, Medical Faculty Mannheim, Heidelberg University, 68167 Mannheim, Germany; ana.munoz10@udea.edu.co (A.-M.M.-R.); jimmy.daza@medma.uni-heidelberg.de (J.D.)
2 Grupo Gastrohepatología, Facultad de Medicina, Universidad de Antioquia, UdeA, Calle 70 # 52-21, Medellin 050010, Colombia; maria.navas@udea.edu.co
3 Department of Hepatology, Universidad Nacional de Asunción, San Lorenzo 111421, Paraguay; marcosgirala@yahoo.com
4 Hepatology Section, Department of Medicine, Center for Medical Education and Clinical Research, Norberto Quirno CEMIC, Buenos Aires 1431, Argentina; eridruejo@cemic.edu.ar
5 Department of Gastroenterology, University Hospital Essen, 45147 Essen, Germany; gugerken@icloud.com
6 Clinical Cooperation Unit Healthy Metabolism, Center for Preventive Medicine, and Digital Health (CPD), Medical Faculty Mannheim, Heidelberg University, 69117 Mannheim, Germany
* Correspondence: andreas.teufel@medma.uni-heidelberg.de

Abstract: The prevention of liver disease has improved significantly in the last few decades, to the point that it can now be considered a true success story. The wide variety of interventions, including comprehensive vaccination strategies, novel medications, lifestyle changes, and even preventive surgeries, have reduced the morbidity and mortality of chronic liver diseases. However, the prevalence of chronic liver diseases is increasing worldwide. Currently, fatty liver disease alone is estimated to be present in as much as 30% of the adult population. Furthermore, there is a trend towards increasing incidences of chronic hepatitis B, and a global lack of success in efforts to eliminate chronic hepatitis C. Thus, improving and efficiently rolling out existing and successful prevention strategies for chronic liver diseases will play an essential role in healthcare throughout the upcoming decades. In this review, we summarize the current options and concepts for preventing chronic liver diseases, highlight their limitations, and provide an outlook on probable future developments to improve awareness, integrated care, and the analysis of big data.

Keywords: liver disease; hepatology; prevention; management

Citation: Muñoz-Restrepo, A.-M.; Navas, M.-C.; Daza, J.; Girala, M.; Ridruejo, E.; Gerken, G.; Teufel, A. Prevention in Hepatology. *J. Pers. Med.* **2024**, *14*, 132. https://doi.org/10.3390/jpm14020132

Academic Editor: Johann Von Felden

Received: 26 November 2023
Revised: 1 January 2024
Accepted: 9 January 2024
Published: 23 January 2024

1. Background

Globally, chronic liver disease (CLD) imposes a substantial health burden, resulting in approximately two million deaths yearly (accounting for 3.5% of global mortality). Half of these deaths are attributed to Hepatocellular Carcinoma (HCC), and the other half to complications of cirrhosis [1]. A major issue is that cirrhosis often goes undetected in many patients until complications and liver cancer develop, making it the eleventh leading cause of death. Furthermore, the global health burden of CLD is expected to continue to increase, with a consistent 5% rise observed since 2000 [2].

Cirrhosis also ranks among the leading causes of disability-associated life years (DALYs) among individuals aged 50 to 74 years old [3]. The economic burden of chronic liver diseases (CLDs) is often underestimated, though some studies have tried to estimate these costs. For instance, in the case of HCC, the annual cost per patient in the United States increased in 2010, ranging from USD 133,000 for stage zero to USD 467,000 for stage D based on the Barcelona Clinic Liver Cancer Criteria (BCLC) [4].

As for the underlying causes, the estimated total cost for chronic hepatitis C (CHC) before the introduction of direct-acting antivirals (DAAs) was USD 10.6 billion, while non-alcoholic steatohepatitis (NASH) incurred a cost of USD 103 billion annually. Furthermore, the estimated 3-year healthcare cost per patient who underwent liver transplantation (LT) was USD 539,955 [5] and costs were constantly rising over the years [6].

Hence, the prevention of liver disease and its complications holds paramount importance both from a medical and economic perspective, benefiting not only patients, but also society at large.

2. Diverse Etiologies, Common End Stage

All liver diseases share a common end-stage outcome: fibrosis and cirrhosis, which can lead to various complications, including liver failure, hepatocellular carcinoma (HCC), and esophageal variceal bleeding, among others. Inflammation is the primary mechanism driving this progression, followed by parenchymal necrosis, activated fibrogenesis, angiogenesis, and profound vascular changes [7].

An imbalance between pro-fibrogenic and anti-fibrogenic mechanisms in the liver can result in excessive extracellular matrix production and alterations in the hepatic angioarchitecture [8]. The risk of developing these complications is influenced by several factors, including lifestyle modifications such as alcohol cessation, weight loss, and the management of infections (e.g., through hepatitis B vaccination and inflammation control) [2].

While the prevalence and incidence of common liver diseases vary significantly across the globe, the most widespread causes of cirrhosis worldwide include chronic alcohol consumption, metabolic-associated fatty liver disease (the revised term for non-alcoholic fatty liver disease), chronic viral hepatitis (with a particular emphasis on B and C), and autoimmune factors. It is important to note that a single patient may have more than one contributing cause, potentially accelerating the progression of the disease, even beyond what the presence of individual diseases and comorbidities would suggest [2].

Over the past two decades, efforts to assess chronic liver diseases (CLDs) have often concentrated on their consequences, particularly fibrosis, rather than their root causes. Thus, the need to raise public awareness of the different causes of fibrosis and cirrhosis—including fatty liver disease and alcohol consumption—and a shift of the management strategies towards preventive medicine as opposed to merely treating the complications are long overdue, especially considering that plenty of evidence suggests that early detection and preventive care could alter the future of the over two million patients who succumb annually to chronic liver diseases [9–11].

3. Hepatitis B Virus (HBV)

Among the causes of cirrhosis, hepatitis B is a life-threatening liver infection and a major global health problem caused by the hepatitis B virus (HBV). It can lead to chronic infection, primarily associated with vertical transmission in African countries and sexual transmission in Western countries [12,13]. In 2015, 3.5% of the world's population was living with this chronic infection, with 68% in the Western Pacific and African regions. In Europe, chronic hepatitis B (CHB) affects 15 million people and leads to 56,000 deaths annually [14].

The primary goal of HBV infection treatment is to achieve a functional cure, measured by the long-term loss of hepatitis B surface antigen (HBsAg) with or without serocoversion and undetectable HBV DNA after therapy interruption. However, despite the availability of various treatments and prolonged patient care, less than 5% of individuals were HBsAg negative after 12 months of treatment [14]. The risk of up to 40% cirrhosis development in untreated patients underscores the need to promote and extend prevention programs, particularly through vaccination [2].

Several risk factors contribute to the progression of HBV infection into HCC. Viral factors such as persistent positive serum HBsAg are highly significant, as shown by a prospective study conducted in Taiwan involving 22,707 individuals, which demonstrated a relative risk of 0.66 among HBsAg+ [15–17]. Another study reported that, in HBsAg—-positive individuals, the joint presence of HBeAg—a marker for high HBV replication levels—increased the risk of HCC three- to sixfold compared to the HBeAg-negative population [18]. Additionally, in individuals with CHB, serum HBV DNA levels seem to predict higher rates of HCC, as patients with high counts (>10^4 copies/mL) face an increased risk of HCC in the long-term follow-up [19]. Moreover, factors such as lifestyle are also important to consider, as chronic alcohol consumption, exposure to aflatoxin and smoking have also shown an increase in HCC risk [15–17].

The risk of developing chronic HBV infection in susceptible individuals depends mainly on the age of acquisition, with a decreasing tendency as the age of infection increases [20]. Among the crucial risk factors to evaluate in primary prevention, the maternal effect stands out. Neonates born to HBsAg+ mothers have a 30-fold higher risk of developing HCC compared to those born to HBsAg− mothers [21]. Hence, practical strategies for hepatitis B infection control and prevention are crucial from an early age. Infant vaccination stands out as the most effective strategy to achieve the goal of the 'Elimination of viral hepatitis by 2030,' which is one of the international Sustainable Development Goals.

Since 1991, the WHO has recommended the inclusion of HBV vaccination in national immunization programs [22]. By the end of 2019, the HBV vaccine had been introduced nationwide in 189 (97%) countries. However, vaccination coverage varies significantly across WHO regions. The regions of the Western Pacific, the Americas, and Southeast Asia exceed the global average, while the European, Eastern Mediterranean, and African regions fall below it [23].

In Western countries endemic for hepatitis B, the typical schedule for the HBV vaccine includes a monovalent birth dose administered within 24 h of birth to all newborns, effectively preventing perinatal transmission. This is followed by three doses of an HBV vaccine at 2, 4, and 6 months [2]. While immunization schedules may vary globally, all have demonstrated the ability to induce seroprotection levels exceeding 95% in healthy infants, children, and young adults [23].

Universal vaccination programs have effectively reduced the global rate of chronic HBV infection to one-tenth of the pre-vaccination era. In Taiwan, for example, several epidemiologic surveys of serum HBV markers showed a significant decrease in HBsAg positivity, dropping from 10% to 0.6–0.7% after the introduction of vaccination [15–17] and even 0% in 3203 children aged 5–10 years in a recently study carried out in Colombia [24].

Similar patterns have been observed in Gambia and Korea. In Western countries, the incidence of HCC related to CHB has also declined since the 2000s, thanks to national vaccination programs and recommendations [25]. As a result, the incidence of CHB infection among children under five years old is now low, primarily attributed to the implementation of universal neonatal HBV vaccination programs.

One of the primary etiologic factors leading to HCC is HBV, highlighting the significant role of primary prevention from childhood to early adulthood. A 20-year follow-up study in Taiwan demonstrated that the HCC incidence was markedly lower among vaccinated children compared to those in unvaccinated birth cohorts (35.9%). Furthermore, among those who were vaccinated, the development of HCC was statistically associated with an incomplete HBV vaccination schedule [21].

However, while the aforementioned strategy for assessing HCC related to HBV is crucial, a population-level HBV vaccination program is expected to have a limited impact over the next two or three decades. Thus, a significant risk persists among those born before the vaccine became available. In these cases, secondary prevention through antiviral agents has emerged as a vital approach to reducing the short-term incidence of HCC. Currently, approved treatments include nucleos(t)ide analogs and interferons [26].

Some current practice guidelines fall short of addressing HCC prevention in cases without inflammation or liver fibrosis. This underscores the importance of increasing awareness among healthcare providers regarding the risk of HBV-associated hepatocarcinogenesis, which can occur independently of fibrosis [26].

In the case of nucleos(t)ide analogs, both Entecavir and Tenofovir Disoproxil Fumarate (TDF) have demonstrated similar efficacy in reducing HCC rates related to HBV, as shown in historical cohort studies. However, a recent meta-analysis involving 42,939 patients from

Korea, Taiwan, and Hong Kong, alongside several smaller meeting reports, has sparked a debate on differing HCC rates under Entecavir (ETV) and Tenofovir Disoproxil Fumarate (TDF) treatment by favoring the latter, as TDF-treated patients had a significantly lower risk of developing HCC compared to ETV-treated patients [27]. Tenofovir Alafenamide (TAF), a more recent introduction, has also shown promising results. However, it requires further investigation due to the absence of long follow-up data [1].

Another preventive measure includes nucleos(t)ide treatment for highly viremic mothers during pregnancy. Tenofovir is preferred for its efficacy, safety in pregnant women, and low resistance rates. In a well-controlled prospective trial involving pregnant women in Taiwan, Tenofovir treatment reduced HBsAg positivity in infants, decreasing the prevalence from 10.71% to 1.54% [28].

In the case of HBV, liver cancer prevention can be categorized into three levels: primary, secondary, and tertiary [15]. Primary prevention is performed through universal HBV vaccine programs, aiming to prevent both the mother-to-child and horizontal transmission of HBV infection and representing the safest and most effective approach to preventing liver cancer [15–17]. Secondary prevention centers on patients with chronic hepatitis B by using antiviral agents to reduce viral load and then liver injury and fibrosis, as indicated by markers such as the normalization of transaminases [15–17]. Finally, tertiary prevention includes patients who have successfully undergone HCC treatment for whom antiviral agents are used to prevent HCC recurrence [15–17]. However, there remains a need to further reduce the risk of HCC by managing modifiable factors, including metabolic syndrome, aflatoxin exposure, heavy drinking, smoking, and other comorbidities that can contribute to liver inflammation. Equally crucial is the emphasis on minimizing high-risk behaviors related to blood and injection safety, not only because of their added risk to CHB, but also because they can independently lead to end-stage liver disease.

4. Hepatitis C Virus (HCV)

Another significant contributor to liver disease is hepatitis C. According to the Polaris Observatory Collaborators, over 56.8 million people worldwide are living with chronic hepatitis C, resulting in more than 400,000 deaths each year [29,30].

The prevalence of hepatitis C varies significantly by geographic location. The highest rates of infection are observed in low- and middle-income countries in Africa and Asia. Mongolia, for example, has a prevalence of over 4%, while in the Eastern Mediterranean region, it is close to 2%; in Europe, nearly 1.5%; and in the Western Pacific region and the Americas, the estimated prevalence is less than 1% [29].

The modes of HCV transmission vary based on regional factors and risk profiles. In high-income countries, the primary route is through injecting drug use, whereas in low-income countries, transmission often occurs through contaminated medical procedures and blood transfusions. Other potential avenues of transmission include unprotected sexual contact and mother-to-child transmission during childbirth [31].

As with HBV, it is imperative to implement effective preventive measures for HCV, not only due to the possibility of chronic infection, but also of the risk of developing liver tumors. In HCV patients, the incidence of HCC can increase by 10–20-fold [32], being responsible for approximately 30–50% of HCC cases worldwide [33]. Moreover, despite the changes in the epidemiology of HCC over the years, with metabolic-associated fatty liver disease gaining prominence, chronic hepatitis C remained the second leading cause of liver transplantation in the United States for men in 2019 and the third leading cause for women [34].

According to the natural history of HCV infection, up to 80% of individuals do not achieve spontaneous viral clearance, and in 20% of those patients, it progresses to cirrhosis [35]. Chronic hepatitis C often remains asymptomatic for several years, leading to a delay in diagnosis and treatment. Patients typically seek clinical care when symptoms related to complications of cirrhosis or HCC itself become evident, accounting for approximately 15% of cases [36].

Direct acting antiviral (DAA) therapy has revolutionized the treatment of HCV infection, offering a cure for most patients. Combinations of two (Sofosbuvir/Velpatasvir, Glecaprevir/Pibrentasvir) or three (Sofosbuvir/Velpatasvir/Voxilaprevir, used in treatment failure) DAAs have consistently achieved an overall SVR rate exceeding 95%. These treatments offer excellent tolerability and safety, which previous therapies did not, irrespective of factors such as genotype, fibrosis stage, intravenous drug abuse, or psychiatric comorbidities [32].

Attaining SVR through DAA therapy is strongly associated with a reduced risk of developing HCC, making SVR a pivotal factor in decreasing HCC incidence. Some meta-analyses of DAAs, especially pan-genotypic ones, have shown that achieving SVR can reduce the HCC risk by 50–80% [37]. It is essential to recognize that the risk of HCC does not return to baseline levels, particularly for patients with characteristics that are less favorable for achieving SVR [38].

As a result, international guidelines recommend post-SVR surveillance, including liver imaging and alpha-fetoprotein (AFP) tests every six months for cirrhotic patients. EASL guidelines extend this recommendation to individuals with advanced fibrosis (F3). The risk factors for post-SVR HCC development include age, male gender, lower baseline albumin, higher bilirubin levels, an FIB-4 score > 3.25, hepatitis B coinfection, and liver stiffness post-SVR ≥ 20 kPa [39].

Some studies have examined the impact of SVR achieved with DAAs on hepatic fibrosis in patients with chronic hepatitis C. These studies found that fibrosis improved by at least one stage in 56% of patients after a 15-month follow-up, and, notably, cirrhosis reversed in 29% of patients [40].

Another crucial consideration is the identification of surrogate markers predictive of fibrosis. For instance, splenomegaly was found to be a negative predictor of fibrotic improvement in cirrhotic patients who achieved SVR, in contrast to a platelet count greater than $152 \times 10^9/L$, which served as a sensitive and specific marker for fibrosis regression [39]. It is essential to recognize the potential for fibrosis reversal, particularly in high-risk patients (those with advanced liver fibrosis—F3 or cirrhosis—at the time of DAA treatment). Doing so can lead to improved clinical outcomes, including a reduction in hepatic decompensation and complications, and a decreased risk of HCC development [40].

Preventive strategies for HCC in patients with HCV also include lifestyle modifications, such as reducing alcohol consumption and maintaining a healthy weight, which can help reduce the risk of HCC in these patients. Additionally, the management of comorbidities such as diabetes can also help reduce the risk of HCC [32].

Screening for HCC in patients with HCV is of utmost importance for early detection and treatment. Currently, ultrasound (US) serves as the established surveillance modality and is acknowledged as the most suitable imaging technique for HCC surveillance according to all international guidelines [41]. Cost-effectiveness studies have demonstrated that ultrasound-based surveillance every 6 months enhances quality-adjusted life expectancy at reasonable costs.

The early detection of HCC allows for curative treatments, such as resection, liver transplantation, or regional therapies, all of which significantly improve outcomes in patients with HCV [42].

In conclusion, preventing HCC in patients with HCV infection is crucial for reducing the global burden of liver cancer and its impact on communities. The effective, cost-efficient treatment of HCV, increased investment in screening and diagnosis, and lifestyle modifications all represent powerful strategies for preventing HCC in these patients. Further research is essential to enhance outcomes and alleviate the HCC burden in HCV-infected individuals.

5. Prevention of Fatty Liver Disease

The prevalence of fatty liver disease has also increased significantly in recent years. Most recent meta-analyses estimate the global prevalence for MASLD as being between

30% and 38% among the adult population [43]. As with any other chronic liver disease, MASLD, and particularly MASH, may also eventually lead to liver fibrosis in 35% of patients. Per year, approximately 2–5% of these patients will be diagnosed with liver cirrhosis [44]. Not surprisingly, MASLD is the fastest rising etiology of cirrhosis associated with acute-on-chronic liver disease (ACLF) among patients listed for liver transplantation in the US [45]. Also, of all patients with MASLD-derived cirrhosis every year, 2–3% will develop liver cancer [44].

However, fatty liver disease not only leads to cirrhosis, ACLF, liver cancer, and other liver-related complications. The disease is also closely linked to metabolic syndrome, and patients with fatty liver disease have a high rate of cardiovascular co-morbidities [46]. Most importantly, cohort studies clearly demonstrated that cardiovascular disease is the most common cause of death in patients with MASLD [47,48]. Also, in MASLD patients undergoing coronary angiography, the disease was significantly and independently correlated with the severity of coronary artery disease [49,50]. Furthermore, MASLD was repeatedly associated with an increased risk of stroke [51].

HCC development is a significant major threat to patients with MASLD. It is essential to acknowledge that MASLD-associated HCC may develop in patients with or without cirrhosis and that the rate of HCC in non-cirrhotic patients may be higher compared to patients with other chronic liver diseases, e.g., chronic viral hepatitis C. As these patients may not receive a standard surveillance ultrasound very six months, the diagnosis of these cancers may be significantly delayed, further aggravating the risk of a lethal course of the disease and considerably impacting the 5-year survival [52].

Generally, a cure for the disease could be fairly easy, as a body weight reduction of 7–10% improved liver fat content, MASH, and fibrosis [53]. Furthermore, intervention studies repetitively demonstrated the efficacy of weight loss in managing metabolic parameters and fatty liver disease [54]: blood pressure, insulin resistance, and muscle strength, among others, are significantly improved with weight loss, factors that are closely linked to metabolic syndrome [55].

However, weight loss is challenging for most patients, and long-term weight loss maintenance remains highly difficult. Some reports have reported most of the weight being regained after five years by a considerable number of patients [56,57].

At the same time, the medical treatment of MASLD also remains surprisingly challenging and, currently, no drugs are available or approved for the medical treatment of the disease. Just recently, a randomized phase III clinical trial of Resmetirom, targeting the thyroid hormone receptor (THR)-b, demonstrated positive effects on the reduction in fat content and fibrosis of the liver [58,59]. With these data, it is hoped that this drug will finally receive FDA approval. A second drug, obeticholic acid—a farnesoid X receptor (FXR) agonist—did not receive FDA approval despite positive phase III data [60].

With the lack of specific treatment options for MASLD, more unspecific treatment options primarily aimed at weight loss have become popular among overweight patients and patients with fatty liver disease. GLP-1 antagonists such as Semaglutide were demonstrated and approved for weight loss [61]. After FDA approval, the drug gained high popularity and was even temporarily sold out. Although GLP-1 may aid in weight loss efforts, it must be stressed that, according to the current data from a randomized phase II trial in patients with MASH and compensated cirrhosis, Semaglutide did not significantly improve fibrosis or achieve MASH resolution versus placebo [62].

In the absence of therapeutic drug options, bariatric surgery currently remains one of the main treatment options. Several surgical strategies were established, of which Roux-Y Gastric Bypass (RYGB) and Sleeve Gastrectomy are presently the most commonly applied procedures. No matter what procedure is used, bariatric surgery leads to effective weight loss in most overweight and fatty liver disease patients, but also reduces the cardiovascular risk and, ultimately, even patient mortality [63].

Overall, given the high prevalence of patients with MASLD, currently estimated to affect approximately 30% of the general population around the globe, its considerable impact on liver failure and carcinogenesis, improvements in the awareness and prevention of fatty liver disease have become a key issue in public health in most countries. With the current lack of available drugs, bariatric surgery should be considered in patients with advanced MASH and the failure of sufficient weight loss, as, in general, a weight loss of approximately 10% was demonstrated to be associated with a significant improvement in fatty liver disease.

6. Secondary Prevention of Esophageal Bleeding in Patients with Liver Cirrhosis

Even after the development of liver cirrhosis, preventative measurements may be of high clinical benefit for patients. Esophageal variceal bleeding because of portal hypertension may be prevented, or at least significantly reduced, through non-selective beta-blocker (NSBB) treatment.

The use of NSBBs in the primary prevention of esophageal bleeding is well established. By reducing cardiac output and splanchnic vasoconstriction resulting in a decrease in portal collateral blood flow [64], NSBBs were repetitively shown to be effective in both the primary and secondary prevention of esophageal bleeding [65–67]. In recent years, carvedilol was favored over propranolol [68], as an intrinsic anti-a1 adrenergic effect was demonstrated to cause intrahepatic vasodilatation and an additional decrease in portal pressure [64]. Thus, the most recent Baveno VII guidelines recommend carvedilol as the preferred NSBB in compensated cirrhosis, being more effective at reducing HVPG, providing better tolerability, and improving survival in compensated patients with clinically significant portal hypertension [69]. The use of NSBB in decompensated cirrhosis remains controversial. Based on initial data and trials discouraging NSBB use in decompensated cirrhosis, they are currently not recommended in many guidelines. However, several recent studies have questioned that dogma and provided evidence that a cautious use of NSBB may also be feasible and potentially effective in decompensated cirrhosis [70–73].

Alternatively, endoscopic band ligation (EBL) is currently thought to be equally effective in the primary prevention of first esophageal bleeding in patients with (high-risk) varices, although the data basis for their evaluation is somewhat heterogeneous. First, EBL had to be compared against NSBB treatment as they were already established by the time EBL became available. Several randomized controlled trials have shown a benefit of EBL in preventing first variceal hemorrhage. However, this effect was not visible in larger trials with more than 100 patients and longer follow-ups. Summarizing the available evidence, a Cochrane analysis found a beneficial effect of EBL on the primary prevention of upper GI bleeding in patients with esophageal varices. However, this effect did not impact mortality [74]. As it is currently assumed that EBL and NSBBs are equally effective, it is important to point out that NSBBs, in particular carvedilol, have a lower risk of serious complications compared to EBL [75,76].

Having established a potential role of EBL in the prevention of variceal bleeding, the use of both NSBBs and EBL is clinical routine in many hepatological centers around the globe. This leads to the obvious question of whether a combination of both treatment options would further increase the efficacy of bleeding prevention. Current data remain inconclusive, with some studies supporting a reduced probability of first bleeding [77], others stating a lower recurrence of varices if propranolol is added to EBL [78], and additional ones with data concluding that a combination does not add any benefit for the patient [79].

7. Prevention of Liver Cancer—HCC Surveillance

As previously stated, patients with liver cirrhosis have a significantly higher risk of developing HCC. Depending on the underlying disease, up to 6.5% of patients develop HCC each year [80]. Other risk constellations associated with an increased HCC rate are chronic hepatitis B and advanced fibrosis and cirrhosis. This significantly increased incidence justifies the regular surveillance of these patients using ultrasound or other imaging such as computed tomography (CT) or magnetic resonance imaging (MRI) [81].

Most recently, a large meta-analysis including almost 150,000 patients showed that patients who receive regular surveillance (ultrasound every 6 months) have a longer overall survival period [82]. This is due to the earlier detection of tumors in early tumor stages, among other factors [82]. Thus, more patients can be referred to curative therapy. An independent study also recently showed that shortening the surveillance interval to 3 months did not improve survival in these patients. In terms of the sensitivity and specificity of early HCC detection, conventional ultrasound and CT were comparable, while MRI performed better [83]. However, as MRI resources remain limited in many areas and ultrasound is ubiquitously available, ultrasound currently remains the imaging method of choice. In this respect, surveillance using ultrasound or other imaging methods has found its way into the recommendations of various gastroenterology or hepatology societies [81,84].

In contrast, a benefit of biomarkers in the early detection of liver cancer is still highly controversial. Heterogeneous data and the resulting meta-analyses have so far not shown an advantage for the use of AFP/AFP-L3 in the early detection of HCC [85,86]. Combinations of several markers including AFP, such as the GALAD (gender, age, AFP-L3, AFP, des-gamma-carboxy prothrombin) factor, are currently being discussed and undergoing extended clinical evaluation [85,87]. A combination of imaging with laboratory chemical markers such as AFP has also not been shown to have an advantage or additional benefit [88].

8. Preventive Substances—Metformin, Aspirin, Coffee, Statins

Numerous publications have suggested the preventive effects of diverse drugs and nutrients on the decompensation of liver cirrhosis and the development of HCC. Although none of these preventive options are clearly recommended in clinical practice guidelines, cumulating data, mostly from retrospective data analyses, at least warrant further evaluation of their potential use. Among the most prominent examples is metformin, which is encouraged to be evaluated in patients with liver cirrhosis for HCC prevention by German HCC guidelines [81]. Also, accumulating evidence on coffee consumption resulted in a recommendation to encourage patients with chronic liver diseases to drink coffee in order to decrease liver-related mortality and HCC development by the EASL clinical practice guideline [84]. Other substances such as statins, acetylsalicylic acid (ASA), or vitamin D supplementation were also under evaluation and were reported to show efficacy in larger retrospective analyses. We recently summarized the accumulating evidence elsewhere [89]. Finally, decreased anti-oxidant capabilities were repetitively discussed for chronic liver diseases, and the supplementation of vitamin E has been suggested by some authors as a possible "scavenger" of oxidative stress products. However, the role of vitamin E supplementation remains controversial [90], particularly since some reports discussed an association with an increased risk of prostate cancer [91,92]. However, the successful implementation of some of these strategies may potentially lead to additional improvement in the prevention of HCC development in patients with liver cirrhosis.

9. Further Improving Prevention in Hepatology—Awareness, Risk Stratification, and Big Data

The development of several effective prevention concepts is certainly among the biggest success stories in hepatology. As previously stated, with effective vaccination strategies, new therapeutic options, nutritional interventions, and even preventative surgical procedures, chronic progression to liver cirrhosis and its associated complications can be avoided for many liver diseases. Most importantly, this saves the patient considerable suffering and also saves society significant costs in the treatment of these liver diseases and secondary complications or even liver transplantation. Despite the availability of these effective preventive tools, it is evident and undisputed that there is still significant room for improvement.

Essential to further improvements in prevention in hepatology must be to increase awareness of liver diseases, preventive possibilities, and the usefulness of (regular) testing.

To illustrate this issue, we launched a public website for patients to inquire about their risk for liver disease. Among more than 117,000 participants, 50.7% were uncertain about their liver enzyme status. In addition, approx. 30% were unsure about their hepatitis B vaccination status [93]. Obviously, even for such a potentially essential preventative intervention like hepatitis B vaccination, awareness could be significantly improved.

Simultaneously, it has recently been shown by the German Hepatitis C Registry and others that if we are, in fact, to regularly test more patients (i.e., regular follow-ups), we will indeed improve liver health. In Germany, health insurance companies just recently agreed that testing for hepatitis B and C would be part of regular check-up tests at the age of 35. The effectiveness of these measurements was evaluated by the German Hepatitis C Registry. Out of the 13,000 patients tested, 52 people had previously unknown anti-HCV antibodies and 8 were even HCV-RNA-positive. Thus, the number of patients that needed to be screened was 262; narrowing the HCV screening to risk factors such as (previous) drug abuse, blood transfusion before 1992, immigration to Germany, and elevated ALT further reduced the number needed to be screened to 111 [94].

Broad screening approaches and the detailed analysis of the resulting Big Data in healthcare could further improve prevention in hepatology. As an example, we recently demonstrated that influenza vaccination in patients with alcoholic liver cirrhosis may lead to a significant improvement in survival using the large public electronic health record collection of the international OHDSI (Observational Health Data Sciences and Informatics) consortium [95]. However, we found surprisingly few data available on vaccinations other than hepatitis A and B. In a similar approach, these methods were also used to further validate the beneficial effects of metformin and aspirin, but also the link to poorer prognosis in patients receiving catecholamines in patients with alcoholic liver disease [96]. Given the rapidly increasing and available health-related data for chronic liver disease, the joint efforts of clinicians and medical informaticists could unravel multiple further preventive aspects, particularly for co-medications and co-morbidities in patients with chronic liver disease.

With the increasing identification of patients with elevated liver values at risk for chronic liver disease, it will be important to refer patients with an elevated risk of chronic deterioration, a lack of prevention or treatment options in general practice, or an acute course of the disease to specialized centers. Given the enormous number of patients and considering that the prevalence of fatty liver disease alone is nearly 30% of the adult population worldwide, the stratification of patients is becoming increasingly important. In addition, close networking between general practitioners and specialists will be essential in order to enable the optimal use of the health system's limited resources [97].

Since most of the health consequences of liver disease result from progressive chronic disease and from progressive fibrosis/cirrhosis, determining the degree of fibrosis using liver biopsy or elastography has emerged as an effective selection criterion. However, since these are not ubiquitously available and—particularly biopsy—involve considerable logistical effort, serological markers have increasingly emerged and been validated in recent years, particularly regarding liver-related events and mortality.

The Fibrosis-4 (FIB-4) Index for Liver Fibrosis is currently preferred in most clinical evaluations, particularly because of the easy availability of the included parameters (i.e., age, platelets, and transaminases) [98]. The further development of integrated care concepts must now involve the validation of these care concepts by incorporating serological or elastography markers into the stratified care of liver diseases [97]. If such validation is successful, it will certainly lead to a much more efficient use of the health system's available resources, ultimately also enabling an even broader rollout of the already available, highly successful tool to prevent and halt the progression of liver disease.

Finally, novel developments in MASLD, hepatitis B, and hepatitis D treatment may, in the near future, offer even more options for the effective prevention of liver cirrhosis and its associated complications. Resmetirom has demonstrated positive effects on fat content and fibrosis in MASLD and may soon become the first FDA-approved drug to treat the disease and prevent its progression to cirrhosis [58,59]. Furthermore, recent progress in early clinical trials and novel approaches for a cure of hepatitis B could eventually translate into effective treatment strategies and a reduction in the global hepatitis B prevalence [99]. Subsequently, this would lead to a significant reduction in associated liver cirrhosis. Finally, the successful introduction of Bulevirtide for hepatitis D treatment in some countries may also help to prevent the development of liver cirrhosis in hepatitis B/D co-infected patients. A global introduction and approval may further aid in the efforts made towards limiting disease progression [100].

10. Conclusions

The development of preventive strategies and treatments in hepatology over the past decades truly is a success story. However, despite this medical progress, the prevalence of chronic liver diseases is increasing, and currently as much as 30% of the global adult population is assumed to suffer from elevated liver enzymes and chronic liver disease. Thus, continuous testing in primary care and awareness campaigns to motivate patients to be tested are crucial for further improvements in prevention in hepatology. Primary prevention through universal vaccination was proven highly effective for hepatitis B and is of high impact, particularly in high-prevalence areas. With the large number of patients suffering from fatty liver disease, stratification will be necessary for secondary prevention, as those with viral, metabolic, or autoimmune diseases or with a higher fibrosis grade need more specialized treatment in order to prevent liver cirrhosis and its complications. In contrast, patients with low fibrosis and a high likelihood of fatty liver disease may very well undergo an attempt of weight loss under the guidance of their primary care physician. Therefore, the FIB-4 was established as a simple marker for estimating fibrosis load. Finally, HCCs are detected earlier through consistent surveillance using ultrasound and patients are treated curatively more frequently.

In conclusion, vaccination and the early identification of patients for further surveillance and early treatment, as well as effective patient stratification, may further improve prevention in hepatology, as effective preventive options are already available for many diseases.

Author Contributions: A.-M.M.-R., M.-C.N. and A.T. drafted the manuscript. J.D., M.G., E.R. and G.G. provided expert insights and discusssions. All authors have read and agreed to the published version of the manuscript.

Funding: A.T. received a grant from the Sino-German Center for Research Promotion (GZ-1546), the State Ministry of Baden-Wuerttemberg for Sciences, Research and Arts supporting the Clinical Cooperation Unit Healthy Metabolism at the Center for Preventive Medicine and Digital Health (CCU Healthy Metabolism), the Baden-Wuerttemberg Center for Digital Early Disease Detection and Prevention (BW-ZDFP) and the Federal MinistryEducation and Research (BMBF, Q-HCC, 01KD22214).

Institutional Review Board Statement: Not applicable.

Informed Consent Statement: Not applicable.

Data Availability Statement: Not applicable.

Conflicts of Interest: The authors declare no conflict of interest.

References

1. Asrani, S.K.; Devarbhavi, H.; Eaton, J.; Kamath, P.S. Burden of liver diseases in the world. *J. Hepatol.* **2019**, *70*, 151–171. [CrossRef] [PubMed]
2. Ginès, P.; Krag, A.; Abraldes, J.G.; Solà, E.; Fabrellas, N.; Kamath, P.S. Liver cirrhosis. *Lancet* **2021**, *398*, 1359–1376. [CrossRef] [PubMed]
3. GBD 2019 Diseases and Injuries Collaborators. Global burden of 369 diseases and injuries in 204 countries and territories, 1990–2019: A systematic analysis for the Global Burden of Disease Study 2019. *Lancet* **2020**, *396*, 1204–1222. [CrossRef] [PubMed]
4. Kaplan, D.E.; Chapko, M.K.; Mehta, R.; Dai, F.; Skanderson, M.; Aytaman, A.; Baytarian, M.; D'Addeo, K.; Fox, R.; Hunt, K.; et al. Healthcare Costs Related to Treatment of Hepatocellular Carcinoma among Veterans with Cirrhosis in the United States. *Clin. Gastroenterol. Hepatol.* **2018**, *16*, 106–114.e5. [CrossRef]

5. Marcellin, P.; Kutala, B.K. Liver diseases: A major, neglected global public health problem requiring urgent actions and large-scale screening. *Liver Int.* **2018**, *38*, 2–6. [CrossRef] [PubMed]
6. Lee, D.U.; Hastie, D.J.; Lee, K.J.; Addonizio, E.; Fan, G.H.; Chou, H.; Jung, D.; Lee, K.; Lominadze, Z. The trends in cost associated with liver transplantation in the US: Analysis of weighted hospital data. *Liver Transpl.* **2023**, *29*, 626–643. [CrossRef]
7. D'Amico, G.; Morabito, A.; D'Amico, M.; Pasta, L.; Malizia, G.; Rebora, P.; Valsecchi, M.G. Clinical states of cirrhosis and competing risks. *J. Hepatol.* **2018**, *68*, 563–576. [CrossRef] [PubMed]
8. Roehlen, N.; Crouchet, E.; Baumert, T.F. Liver Fibrosis: Mechanistic Concepts and Therapeutic Perspectives. *Cells* **2020**, *9*, 875. [CrossRef]
9. Karlsen, T.H.; Sheron, N.; Zelber-Sagi, S.; Carrieri, P.; Dusheiko, G.; Bugianesi, E.; Pryke, R.; Hutchinson, S.J.; Sangro, B.; Martin, N.K.; et al. The EASL-Lancet Liver Commission: Protecting the next generation of Europeans against liver disease complications and premature mortality. *Lancet* **2022**, *399*, 61–116. [CrossRef]
10. Buechter, M.; Dorn, D.; Möhlendick, B.; Siffert, W.; Baba, H.A.; Gerken, G.; Kahraman, A. Characteristics and Long-Term Outcome of 535 Patients with Autoimmune Hepatitis-The 20-Year Experience of a High-Volume Tertiary Center. *J. Clin. Med.* **2023**, *12*, 4192. [CrossRef]
11. Buechter, M.; Gerken, G. Liver Function-How to Screen and to Diagnose: Insights from Personal Experiences, Controlled Clinical Studies and Future Perspectives. *J. Pers. Med.* **2022**, *12*, 1657. [CrossRef] [PubMed]
12. di Filippo Villa, D.; Navas, M.-C. Vertical Transmission of Hepatitis B Virus—An Update. *Microorganisms* **2023**, *11*, 1140. [CrossRef] [PubMed]
13. Funk, A.L.; Lu, Y.; Yoshida, K.; Zhao, T.; Boucheron, P.; van Holten, J.; Chou, R.; Bulterys, M.; Shimakawa, Y. Efficacy and safety of antiviral prophylaxis during pregnancy to prevent mother-to-child transmission of hepatitis B virus: A systematic review and meta-analysis. *Lancet Infect. Dis.* **2021**, *21*, 70–84. [CrossRef] [PubMed]
14. Yuen, M.F.; Lim, S.G.; Plesniak, R.; Tsuji, K.; Janssen, H.L.A.; Pojoga, C.; Gadano, A.; Popescu, C.P.; Stepanova, T.; Asselah, T.; et al. Efficacy and Safety of Bepirovirsen in Chronic Hepatitis B Infection. *N. Engl. J. Med.* **2022**, *387*, 1957–1968. [CrossRef] [PubMed]
15. Chang, M.-H. Prevention of Hepatitis B Virus Infection and Liver Cancer. In *Viruses and Human Cancer: From Basic Science to Clinical Prevention*; Chang, M.H., Jeang, K.-T., Eds.; Springer: Berlin/Heidelberg, Germany, 2014; pp. 75–95. [CrossRef]
16. Rivière, L.; Ducroux, A.; Buendia, M.A. The Oncogenic Role of Hepatitis B Virus. In *Viruses and Human Cancer: From Basic Science to Clinical Prevention*; Chang, M.H., Jeang, K.-T., Eds.; Springer: Berlin/Heidelberg, Germany, 2014; pp. 59–74. [CrossRef]
17. Woller, N.; Kühnel, F. Virus Infection, Inflammation and Prevention of Cancer. In *Viruses and Human Cancer: From Basic Science to Clinical Prevention*; Chang, M.H., Jeang, K.-T., Eds.; Springer: Berlin/Heidelberg, Germany, 2014; pp. 33–58. [CrossRef]
18. Yang, H.I.; Lu, S.N.; Liaw, Y.F.; You, S.L.; Sun, C.A.; Wang, L.Y.; Hsiao, C.K.; Chen, P.J.; Chen, D.S.; Chen, C.J.; et al. Hepatitis B e Antigen and the Risk of Hepatocellular Carcinoma. *N. Engl. J. Med.* **2002**, *347*, 168–174. [CrossRef]
19. Chen, C.J.; Yang, H.I.; Su, J.; Jen, C.L.; You, S.L.; Lu, S.N.; Huang, G.T.; Iloeje, U.H.; REVEAL-HBV Study Group. Risk of Hepatocellular Carcinoma across a Biological Gradient of Serum Hepatitis B Virus DNA Level. *JAMA* **2006**, *295*, 65–73. [CrossRef]
20. Nelson, N.P.; Easterbrook, P.J.; McMahon, B.J. Epidemiology of Hepatitis B Virus Infection and Impact of Vaccination on Disease. *Clin. Liver Dis.* **2016**, *20*, 607–628. [CrossRef]
21. Chang, M.H.; You, S.L.; Chen, C.J.; Liu, C.J.; Lee, C.M.; Lin, S.M.; Chu, H.C.; Wu, T.C.; Yang, S.S.; Kuo, H.S.; et al. Decreased Incidence of Hepatocellular Carcinoma in Hepatitis B Vaccinees: A 20-Year Follow-up Study. *JNCI J. Natl. Cancer Inst.* **2009**, *101*, 1348–1355. [CrossRef]
22. Mohanty, P.; Jena, P.; Patnaik, L. Vaccination against Hepatitis B: A Scoping Review. *Asian Pac. J. Cancer Prev.* **2020**, *21*, 3453–3459. [CrossRef]
23. Pattyn, J.; Hendrickx, G.; Vorsters, A.; Van Damme, P. Hepatitis B Vaccines. *J. Infect. Dis.* **2021**, *224*, S343–S351. [CrossRef]
24. Ríos-Hincapié, C.Y.; Murad-Rivera, R.; Tohme, R.A.; Ropero, A.M.; Gómez, B.; Librado Cardona, D.; Forest Duque, B.N.; Cuellar, D.; Cardenas, I.; Krow-Lucal, E.; et al. Progress towards the elimination of hepatitis B in children in Colombia: A novel two-phase study approach. *J. Viral Hepat.* **2022**, *29*, 737–747. [CrossRef] [PubMed]
25. Vogel, A.; Meyer, T.; Sapisochin, G.; Salem, R.; Saborowski, A. Hepatocellular carcinoma. *Lancet* **2022**, *400*, 1345–1362. [CrossRef] [PubMed]
26. Choi, J.; Lim, Y.-S. Secondary prevention of hepatitis B virus-related hepatocellular carcinoma with current antiviral therapies. *Kaohsiung J. Med. Sci.* **2021**, *37*, 262–267. [CrossRef] [PubMed]
27. Choi, W.M.; Yip, T.C.; Wong, G.L.; Kim, W.R.; Yee, L.J.; Brooks-Rooney, C.; Curteis, T.; Cant, H.; Chen, C.H.; Chen, C.Y.; et al. Hepatocellular carcinoma risk in patients with chronic hepatitis B receiving tenofovir- vs. entecavir-based regimens: Individual patient data meta-analysis. *J. Hepatol.* **2023**, *78*, 534–542. [CrossRef]
28. Chen, H.L.; Lee, C.N.; Chang, C.H.; Ni, Y.H.; Shyu, M.K.; Chen, S.M.; Hu, J.J.; Lin, H.H.; Zhao, L.L.; Mu, S.C.; et al. Efficacy of maternal tenofovir disoproxil fumarate in interrupting mother-to-infant transmission of hepatitis B virus. *Hepatology* **2015**, *62*, 375–386. [CrossRef]
29. Di Marco, L.; La Mantia, C.; Di Marco, V. Hepatitis C: Standard of Treatment and What to Do for Global Elimination. *Viruses* **2022**, *14*, 505. [CrossRef]
30. Polaris Observatory Collaborators. Global prevalence, cascade of care, and prophylaxis coverage of hepatitis B in 2022: A modelling study. *Lancet Gastroenterol. Hepatol.* **2023**, *8*, 879–907. [CrossRef]

31. Chan, C.P.; Uemura, H.; Kwan, T.H.; Wong, N.S.; Oka, S.; Chan, D.P.C.; Lee, S.S. Review on the molecular epidemiology of sexually acquired hepatitis C virus infection in the Asia-Pacific region. *J. Int. AIDS Soc.* **2020**, *23*, e25618. [CrossRef]
32. McGlynn, K.A.; Petrick, J.L.; El-Serag, H.B. Epidemiology of Hepatocellular Carcinoma. *Hepatology* **2021**, *73*, 4–13. [CrossRef]
33. Pisano, M.B.; Giadans, C.G.; Flichman, D.M.; Ré, V.E.; Preciado, M.V.; Valva, P. Viral hepatitis update: Progress and perspectives. *World J. Gastroenterol.* **2021**, *27*, 4018–4044. [CrossRef]
34. Wang, S.; Toy, M.; Hang Pham, T.T.; So, S. Causes and trends in liver disease and hepatocellular carcinoma among men and women who received liver transplants in the U.S., 2010–2019. *PLoS ONE* **2020**, *15*, e0239393. [CrossRef] [PubMed]
35. Westbrook, R.H.; Dusheiko, G. Natural history of hepatitis C. *J. Hepatol.* **2014**, *61*, S58–S68. [CrossRef] [PubMed]
36. Rein, D.B.; Wittenborn, J.S.; Weinbaum, C.M.; Sabin, M.; Smith, B.D.; Lesesne, S.B. Forecasting the morbidity and mortality associated with prevalent cases of pre-cirrhotic chronic hepatitis C in the United States. *Dig. Liver Dis.* **2011**, *43*, 66–72. [CrossRef] [PubMed]
37. Polaris Observatory HCV Collaborators. Global change in hepatitis C virus prevalence and cascade of care between 2015 and 2020: A modelling study. *Lancet Gastroenterol. Hepatol.* **2022**, *7*, 396–415. [CrossRef]
38. Davidov, Y.; Kleinbaum, Y.; Inbar, Y.; Cohen-Ezra, O.; Veitsman, E.; Weiss, P.; Berdichevski, T.; Katsherginsky, S.; Hassid, A.; Tsaraf, K.; et al. Noninvasive Assessment of Fibrosis Regression after Direct-Acting Antiviral Treatment in Hepatitis C Virus Patients. *Isr. Med. Assoc. J. IMAJ* **2021**, *23*, 794–800. Available online: https://www.ima.org.il/MedicineIMAJ/viewarticle.aspx?year=2021&month=12&page=794 (accessed on 7 July 2023).
39. Cuesta-Sancho, S.; Márquez-Coello, M.; Illanes-Álvarez, F.; Márquez-Ruiz, D.; Arizcorreta, A.; Galán-Sánchez, F.; Montiel, N.; Rodriguez-Iglesias, M.; Girón-González, J.A. Hepatitis C: Problems to extinction and residual hepatic and extrahepatic lesions after sustained virological response. *World J. Hepatol.* **2022**, *14*, 62–79. [CrossRef]
40. Elsharkawy, A.; Samir, R.; El-Kassas, M. Fibrosis regression following hepatitis C antiviral therapy. *World J. Hepatol.* **2022**, *14*, 1120–1130. [CrossRef]
41. Kim, D.H.; Choi, J.-I. Current status of image-based surveillance in hepatocellular carcinoma. *Ultrasonography* **2021**, *40*, 45–56. [CrossRef]
42. Yang, J.D.; Hainaut, P.; Gores, G.J.; Amadou, A.; Plymoth, A.; Roberts, L.R. A global view of hepatocellular carcinoma: Trends, risk, prevention and management. *Nat. Rev. Gastroenterol. Hepatol.* **2019**, *16*, 589–604. [CrossRef]
43. Younossi, Z.M.; Wong, G.; Anstee, Q.M.; Henry, L. The Global Burden of Liver Disease. *Clin. Gastroenterol. Hepatol.* **2023**, *21*, 1978–1991. [CrossRef]
44. Schuppan, D.; Schattenberg, J.M. Non-alcoholic steatohepatitis: Pathogenesis and novel therapeutic approaches. *J. Gastroenterol. Hepatol.* **2013**, *28* (Suppl. S1), 68–76. [CrossRef] [PubMed]
45. Sundaram, V.; Jalan, R.; Shah, P.; Singal, A.K.; Patel, A.A.; Wu, T.; Noureddin, M.; Mahmud, N.; Wong, R.J. Acute on Chronic Liver Failure from Nonalcoholic Fatty Liver Disease: A Growing and Aging Cohort with Rising Mortality. *Hepatology* **2021**, *73*, 1932–1944. [CrossRef] [PubMed]
46. Adams, L.A.; Anstee, Q.M.; Tilg, H.; Targher, G. Non-alcoholic fatty liver disease and its relationship with cardiovascular disease and other extrahepatic diseases. *Gut* **2017**, *66*, 1138–1153. [CrossRef] [PubMed]
47. Ekstedt, M.; Hagström, H.; Nasr, P.; Fredrikson, M.; Stål, P.; Kechagias, S.; Hultcrantz, R. Fibrosis stage is the strongest predictor for disease-specific mortality in NAFLD after up to 33 years of follow-up. *Hepatology* **2015**, *61*, 1547–1554. [CrossRef]
48. Angulo, P.; Kleiner, D.E.; Dam-Larsen, S.; Adams, L.A.; Bjornsson, E.S.; Charatcharoenwitthaya, P.; Mills, P.R.; Keach, J.C.; Lafferty, H.D.; Stahler, A.; et al. Liver Fibrosis, but No Other Histologic Features, Is Associated with Long-Term Outcomes of Patients with Nonalcoholic Fatty Liver Disease. *Gastroenterology* **2015**, *149*, 389–397.e10. [CrossRef]
49. Mirbagheri, S.A.; Rashidi, A.; Abdi, S.; Saedi, D.; Abouzari, M. Liver: An alarm for the heart? *Liver Int.* **2007**, *27*, 891–894. [CrossRef]
50. Assy, N.; Djibre, A.; Farah, R.; Grosovski, M.; Marmor, A. Presence of coronary plaques in patients with nonalcoholic fatty liver disease. *Radiology* **2010**, *254*, 393–400. [CrossRef]
51. Wang, M.; Zhou, B.G.; Zhang, Y.; Ren, X.F.; Li, L.; Li, B.; Ai, Y.W. Association between Non-alcoholic Fatty Liver Disease and Risk of Stroke: A Systematic Review and Meta-Analysis. *Front. Cardiovasc. Med.* **2022**, *9*, 812030. [CrossRef]
52. Younossi, Z.M.; Henry, L. Epidemiology of non-alcoholic fatty liver disease and hepatocellular carcinoma. *JHEP Rep.* **2021**, *3*, 100305. [CrossRef]
53. Glass, L.M.; Dickson, R.C.; Anderson, J.C.; Suriawinata, A.A.; Putra, J.; Berk, B.S.; Toor, A. Total body weight loss of $\geq$10% is associated with improved hepatic fibrosis in patients with nonalcoholic steatohepatitis. *Dig. Dis. Sci.* **2015**, *60*, 1024–1030. [CrossRef]
54. Brunner, K.T.; Henneberg, C.J.; Wilechansky, R.M.; Long, M.T. Nonalcoholic Fatty Liver Disease and Obesity Treatment. *Curr. Obes. Rep.* **2019**, *8*, 220–228. [CrossRef] [PubMed]
55. Hallsworth, K.; Adams, L.A. Lifestyle modification in NAFLD/NASH: Facts and figures. *JHEP Rep.* **2019**, *1*, 468–479. [CrossRef] [PubMed]
56. Anderson, J.W.; Konz, E.C.; Frederich, R.C.; Wood, C.L. Long-term weight-loss maintenance: A meta-analysis of US studies. *Am. J. Clin. Nutr.* **2001**, *74*, 579–584. [CrossRef] [PubMed]
57. Hall, K.D.; Kahan, S. Maintenance of Lost Weight and Long-Term Management of Obesity. *Med. Clin. N. Am.* **2018**, *102*, 183–197. [CrossRef] [PubMed]

58. Harrison, S.A.; Ratziu, V.; Anstee, Q.M.; Noureddin, M.; Sanyal, A.J.; Schattenberg, J.M.; Bedossa, P.; Bashir, M.R.; Schneider, D.; Taub, R.; et al. Design of the phase 3 MAESTRO clinical program to evaluate resmetirom for the treatment of nonalcoholic steatohepatitis. *Aliment. Pharmacol. Ther.* **2023**, *59*, 51–63. [CrossRef] [PubMed]
59. Thomas, H. EASL Congress 2023. *Lancet Gastroenterol. Hepatol.* **2023**, *8*, 700. [CrossRef] [PubMed]
60. Intercept. Press Release: Intercept Receives Complete Response Letter from FDA for Obeticholic Acid as a Treatment for Pre-Cirrhotic Fibrosis due to NASH. Available online: https://ir.interceptpharma.com/news-releases/news-release-details/intercept-receives-complete-response-letter-fda-obeticholic-0 (accessed on 6 August 2023).
61. FDA. FDA Approves New Drug Treatment for Chronic Weight Management, First Since 2014. Available online: https://www.fda.gov/news-events/press-announcements/fda-approves-new-drug-treatment-chronic-weight-management-first-2014 (accessed on 6 August 2023).
62. Loomba, R.; Abdelmalek, M.F.; Armstrong, M.J.; Jara, M.; Kjær, M.S.; Krarup, N.; Lawitz, E.; Ratziu, V.; Sanyal, A.J.; Schattenberg, J.M.; et al. Semaglutide 2·4 mg once weekly in patients with non-alcoholic steatohepatitis-related cirrhosis: A randomised, placebo-controlled phase 2 trial. *Lancet Gastroenterol. Hepatol.* **2023**, *8*, 511–522. [CrossRef] [PubMed]
63. van Veldhuisen, S.L.; Gorter, T.M.; van Woerden, G.; de Boer, R.A.; Rienstra, M.; Hazebroek, E.J.; van Veldhuisen, D.J. Bariatric surgery and cardiovascular disease: A systematic review and meta-analysis. *Eur. Heart J.* **2022**, *43*, 1955–1969. [CrossRef]
64. Rodrigues, S.G.; Mendoza, Y.P.; Bosch, J. Beta-blockers in cirrhosis: Evidence-based indications and limitations. *JHEP Rep.* **2020**, *2*, 100063. [CrossRef]
65. Garcia-Tsao, G.; Bosch, J. Management of varices and variceal hemorrhage in cirrhosis. *N. Engl. J. Med.* **2010**, *362*, 823–832. [CrossRef]
66. Pascal, J.P.; Cales, P. Propranolol in the prevention of first upper gastrointestinal tract hemorrhage in patients with cirrhosis of the liver and esophageal varices. *N. Engl. J. Med.* **1987**, *317*, 856–861. [CrossRef] [PubMed]
67. Lebrec, D.; Poynard, T.; Hillon, P.; Benhamou, J.P. Propranolol for prevention of recurrent gastrointestinal bleeding in patients with cirrhosis: A controlled study. *N. Engl. J. Med.* **1981**, *305*, 1371–1374. [CrossRef] [PubMed]
68. Bañares, R.; Moitinho, E.; Piqueras, B.; Casado, M.; García-Pagán, J.C.; de Diego, A.; Bosch, J. Carvedilol, a new nonselective beta-blocker with intrinsic anti- Alpha1-adrenergic activity, has a greater portal hypotensive effect than propranolol in patients with cirrhosis. *Hepatology* **1999**, *30*, 79–83. [CrossRef] [PubMed]
69. de Franchis, R.; Bosch, J.; Garcia-Tsao, G.; Reiberger, T.; Ripoll, C.; Baveno VII Faculty. Baveno VII—Renewing consensus in portal hypertension. *J. Hepatol.* **2022**, *76*, 959–974. [CrossRef] [PubMed]
70. Bang, U.C.; Benfield, T.; Hyldstrup, L.; Jensen, J.-E.B.; Bendtsen, F. Effect of propranolol on survival in patients with decompensated cirrhosis: A nationwide study based Danish patient registers. *Liver Int.* **2016**, *36*, 1304–1312. [CrossRef] [PubMed]
71. Bossen, L.; Krag, A.; Vilstrup, H.; Watson, H.; Jepsen, P. Nonselective β-blockers do not affect mortality in cirrhosis patients with ascites: Post Hoc analysis of three randomized controlled trials with 1198 patients. *Hepatology* **2016**, *63*, 1968–1976. [CrossRef] [PubMed]
72. Ngwa, T.; Orman, E.; Gomez, E.V.; Vuppalanchi, R.; Kubal, C.; Chalasani, N.; Ghabril, M. Non-selective β-blockers are associated with improved survival in patients with ascites listed for liver transplantation. *Gut* **2015**, *64*, 1111–1119.
73. Sinha, R.; Lockman, K.A.; Mallawaarachchi, N.; Robertson, M.; Plevris, J.N.; Hayes, P.C. Carvedilol use is associated with improved survival in patients with liver cirrhosis and ascites. *J. Hepatol.* **2017**, *67*, 40–46. [CrossRef] [PubMed]
74. Gluud, L.L.; Krag, A. Banding ligation versus beta-blockers for primary prevention in oesophageal varices in adults. *Cochrane Database Syst. Rev.* **2012**, *8*, CD004544. [CrossRef]
75. Dardari, L.; Taha, M.; Dahat, P.; Toriola, S.; Satnarine, T.; Zohara, Z.; Adelekun, A.; Seffah, K.D.; Salib, K.; Arcia Franchini, A.P. Efficacy of carvedilol versus propranolol versus variceal band ligation for primary prevention of variceal bleeding. *Hepatol. Int.* **2018**, *12*, 75–82.
76. Sharma, M.; Singh, S.; Desai, V.; Shah, V.H.; Kamath, P.S.; Murad, M.H.; Simonetto, D.A. Comparison of Therapies for Primary Prevention of Esophageal Variceal Bleeding: A Systematic Review and Network Meta-Analysis. *Hepatology* **2019**, *69*, 1657–1675. [CrossRef] [PubMed]
77. Pepe, V.; Angeli, P.; Di Pascoli, M. Combined Pharmacological and Endoscopic Treatment for Worsening Gastroesophageal Varices in Patients with Cirrhosis. *Clin. Exp. Gastroenterol.* **2022**, *15*, 59–65. [CrossRef] [PubMed]
78. Sarin, S.K.; Wadhawan, M.; Agarwal, S.R.; Tyagi, P.; Sharma, B.C. Endoscopic variceal ligation plus propranolol versus endoscopic variceal ligation alone in primary prophylaxis of variceal bleeding. *Am. J. Gastroenterol.* **2005**, *100*, 797–804. [CrossRef] [PubMed]
79. Pfisterer, N.; Dexheimer, C.; Fuchs, E.M.; Bucsics, T.; Schwabl, P.; Mandorfer, M.; Gessl, I.; Sandrieser, L.; Baumann, L.; Riedl, F.; et al. Betablockers do not increase efficacy of band ligation in primary prophylaxis but they improve survival in secondary prophylaxis of variceal bleeding. *Aliment. Pharmacol. Ther.* **2018**, *47*, 966–979. [CrossRef]
80. Kanwal, F.; Kramer, J.R.; Asch, S.M.; Cao, Y.; Li, L.; El-Serag, H.B. Long-Term Risk of Hepatocellular Carcinoma in HCV Patients Treated with Direct Acting Antiviral Agents. *Hepatology* **2020**, *71*, 44. [CrossRef]
81. Voesch, S.; Bitzer, M.; Malek, N. Clinical relevance of the new S3 guideline on hepatocellular carcinoma and biliary tract cancer for practitioners. *Radiologe* **2022**, *62*, 200–204. [CrossRef]
82. Singal, A.G.; Zhang, E.; Narasimman, M.; Rich, N.E.; Waljee, A.K.; Hoshida, Y.; Yang, J.D.; Reig, M.; Cabibbo, G.; Nahon, P.; et al. HCC surveillance improves early detection, curative treatment receipt, and survival in patients with cirrhosis: A meta-analysis. *J. Hepatol.* **2022**, *77*, 128–139. [CrossRef]

83. Ayoub, W.S.; Steggerda, J.; Yang, J.D.; Kuo, A.; Sundaram, V.; Lu, S.C. Current status of hepatocellular carcinoma detection: Screening strategies and novel biomarkers. *Ther. Adv. Med. Oncol.* **2019**, *11*, 1758835919869120. [CrossRef]
84. European Association for the Study of the Liver. EASL Clinical Practice Guidelines: Management of hepatocellular carcinoma. *J. Hepatol.* **2018**, *69*, 182–236. [CrossRef]
85. Wang, X.; Zhang, Y.; Yang, N.; He, H.; Tao, X.; Kou, C.; Jiang, J. Evaluation of the Combined Application of AFP, AFP-L3%, and DCP for Hepatocellular Carcinoma Diagnosis: A Meta-Analysis. *Biomed. Res. Int.* **2020**, *2020*, 5087643. [CrossRef]
86. Zhou, J.-M.; Wang, T.; Zhang, K.-H. AFP-L3 for the diagnosis of early hepatocellular carcinoma: A meta-analysis. *Medicine* **2021**, *100*, e27673. [CrossRef] [PubMed]
87. Yang, J.D.; Addissie, B.D.; Mara, K.C.; Harmsen, W.S.; Dai, J.; Zhang, N.; Wongjarupong, N.; Ali, H.M.; Ali, H.A.; Hassan, F.A.; et al. GALAD Score for Hepatocellular Carcinoma Detection in Comparison with Liver Ultrasound and Proposal of GALADUS Score. *Cancer Epidemiol. Biomark. Prev.* **2019**, *28*, 531–538. [CrossRef] [PubMed]
88. Zangneh, H.F.; Cerocchi, O.; Khalili, K.; El-Karim, L.A.; Vecchio, M.; Winick, J.; Mori, Y.; Yamada, H.; Janssen, H.; Hansen, B. Prospective randomized controlled trial of biomarkers for early detection of hepatocellular carcinoma. *J. Hepatol.* **2022**, *77*, S3. [CrossRef]
89. Kusnik, A.; Hunter, N.; Rasbach, E.; Miethke, T.; Reissfelder, C.; Ebert, M.P.; Teufel, A. Co-Medication and Nutrition in Hepatocellular Carcinoma: Potentially Preventative Strategies in Hepatocellular Carcinoma. *Dig. Dis.* **2021**, *39*, 526–533. [CrossRef] [PubMed]
90. Di Sario, A.; Candelaresi, C.; Omenetti, A.; Benedetti, A. Vitamin E in chronic liver diseases and liver fibrosis. *Vitam. Horm.* **2007**, *76*, 551–573. [PubMed]
91. Klein, E.A.; Thompson IMJr Tangen, C.M.; Crowley, J.J.; Lucia, M.S.; Goodman, P.J.; Minasian, L.M.; Ford, L.G.; Parnes, H.L.; Gaziano, J.M.; Karp, D.D.; et al. Vitamin E and the risk of prostate cancer: The Selenium and Vitamin E Cancer Prevention Trial (SELECT). *JAMA* **2011**, *306*, 1549–1556. [CrossRef]
92. Vivarelli, F.; Canistro, D.; Cirillo, S.; Papi, A.; Spisni, E.; Vornoli, A.; Croce, C.M.D.; Longo, V.; Franchi, P.; Filippi, S.; et al. Co-carcinogenic effects of vitamin E in prostate. *Sci. Rep.* **2019**, *9*, 11636. [CrossRef]
93. Kautz, A.; Dorner, R.; Antoni, C.; Ebert, M.; Teufel, A. Self-testing for liver disease—Response to an online liver test questionnaire. *Scand. J. Gastroenterol.* **2020**, *55*, 67–73. [CrossRef]
94. Petroff, D.; Wolffram, I.; Bätz, O.; Jedrysiak, K.; Kramer, J.; Berg, T.; Wiegand, J. Confirmation of guideline-defined hepatitis C screening strategies within the 'Check-Up35+' examination in the primary care setting. *Liver Int.* **2023**, *43*, 785–793. [CrossRef]
95. Itzel, T.; Falconer, T.; Daza, J.; Roig, A.; Park, J.; Cheong, J.Y.; Park, R.W.; Wiest, I.; Ebert, M.; Hripcsak, G.; et al. Letter to the editor: Vaccination against upper respiratory infections is a matter of survival in alcoholic liver disease. *Gut* **2023**, *72*, 208–209. [CrossRef]
96. Itzel, T.; Falconer, T.; Roig, A.; Daza, J.; Park, J.; Cheong, J.Y.; Park, R.W.; Wiest, I.; Ebert, M.P.; Hripcsak, G.; et al. Efficacy of Co-Medications in Patients with Alcoholic Liver Disease. *Dig. Dis.* **2023**, *41*, 780–788. [CrossRef] [PubMed]
97. Teufel, A.; Geier, A.; Sarrazin, C.; Schattenberg, J.M.; Kautz, A.; Dorner, R.; Kramer, J.; Jerysiak, K.; Baars, T.; Hönscheid, B.; et al. Intersectoral management of patients with abnormal liver enzymes and non-alcoholic fatty liver disease (NAFLD). *Z. Gastroenterol.* **2023**, *61*, 1028–1036. [PubMed]
98. Vallet-Pichard, A.; Mallet, V.; Nalpas, B.; Verkarre, V.; Nalpas, A.; Dhalluin-Venier, V.; Fontaine, H.; Pol, S. FIB-4: An inexpensive and accurate marker of fibrosis in HCV infection. comparison with liver biopsy and fibrotest. *Hepatology* **2007**, *46*, 32–36. [CrossRef] [PubMed]
99. Spyrou, E.; Smith, C.I.; Ghany, M.G. Hepatitis B: Current Status of Therapy and Future Therapies. *Gastroenterol. Clin. N. Am.* **2020**, *49*, 215–238. [CrossRef]
100. Lampertico, P.; Degasperi, E.; Sandmann, L.; Wedemeyer, H.; Delta Cure 2022 Working Group. Hepatitis D virus infection: Pathophysiology, epidemiology and treatment. Report from the first international delta cure meeting 2022. *JHEP Rep.* **2023**, *5*, 100818. [CrossRef]

Journal of *Personalized Medicine*

MDPI

Article

Longterm Outcome of Therapeutic Vaccination with a Third Generation Pre-S/S HBV Vaccine (PreHevbrio^R) of Chronically HBV Infected Patients

Hedwig Roggendorf [1,*], Daniel Shouval [2], Michael Roggendorf [3] and Guido Gerken [4]

1 Institute of Molecular Immunology, University Hospital TUM, 81675 Munich, Germany
2 Liver Unit, Hadassah Medical Center, POB 12000, Jerusalem 91120, Israel; shouval@hadassah.org.il
3 Institute of Virology, Technical University of Munich/Helmholtz Zentrum, 81675 Munich, Germany; michael.roggendorf@tum.de
4 Department of Gastroenterology, Helios Klinikum Niedernberg, 42551 Velbert, Germany; guido.gerken@helios-gesundheit.de
* Correspondence: hedwig.roggendorf@tum.de

Abstract: Several antiviral treatment regimens for chronic hepatitis B (CHB) virus infection have been shown to be effective in suppressing viral load and reducing the risk of hepatocellular injury and its complications. It has been hypothesized that high levels of circulating HBV surface antigen(s) may lead to immune tolerance against HBV and contribute to chronic carriership. Conversely, low-level HBsAg may create a window for the reconstitution of an HBV-specific immune response through vaccination and control of infection. Previous studies in non-responders to yeast-derived HBV vaccines, using a third-generation pre-S/S vaccine, have led to up to 95% anti-HBs seroconversion. This report evaluates the long-term outcome after experimental vaccination with a pre-S/S HBV vaccine intended as a therapeutic intervention in chronic HBV carriers. Four low-level HBsAg carriers (<500 IU/mL) were vaccinated three to seven times with 20 μg PreHevbrio^R. Three out of four carriers eliminated HBsAg completely and seroconverted to anti-HBs. One patient seroconverted to anti-HBs but remained with a borderline HBsAg titer (10 IU/mL). Serum anti-HBs levels following repeated vaccination varied between 27 and >1000 IU/L, respectively. Long-term observation (>6 years) showed that after discontinuing NUC treatment for at least two years, HBsAg and HBV DNA remained negative with anti-HBs positive titers ranging between 80 and >1000 IU/L. Based on our preliminary observations, there is a rationale to further evaluate the role of this vaccine as a therapeutic agent.

Keywords: chronic hepatitis B; NUC treatment; low-level HBsAg carriers; therapeutic vaccination; third generation pre-S/S vaccine; long-term observation

Citation: Roggendorf, H.; Shouval, D.; Roggendorf, M.; Gerken, G. Longterm Outcome of Therapeutic Vaccination with a Third Generation Pre-S/S HBV Vaccine (PreHevbrio^R) of Chronically HBV Infected Patients. *J. Pers. Med.* **2024**, *14*, 364. https://doi.org/10.3390/jpm14040364

Academic Editor: Jesús Troya

Received: 31 January 2024
Revised: 13 March 2024
Accepted: 28 March 2024
Published: 29 March 2024

1. Introduction and Rational

Currently, chronic HBV infection affects an estimated 300 million people worldwide, leading to about 1 million deaths per year caused by complications of liver disease and hepatocellular carcinoma (HCC) [1]. For more than 30 years, hepatitis B vaccines have efficiently decreased the incidence worldwide, also in countries known to be heavily affected (endemic) for HB, such as China [2]. Yet, a large pool of individuals with chronic infections remain at risk of developing liver cirrhosis and hepatocellular carcinoma. Those with chronic HBV infections serve as the main reservoir for viral spread. In the United States, fewer than one-third of individuals with chronic HBV are aware of their infection status, increasing the likelihood of transmission to susceptible populations and putting close family members, household contacts, and sexual partners at risk of infection [1].

An effective HBV-specific T-cell response is regarded as a key determinant in resolving acute infection. The human immune system has the potential to resolve an acute HBV infection in about 90% of infected individuals. However, in patients who fail to resolve acute infection and develop chronic hepatitis B, an effective T-cell response against the virus is functionally impaired or absent. Furthermore, a humoral anti-HBs immune response is required for protection against HBV infection due to exposure or re-exposure in patients who recover from acute hepatitis B, as well as in patients successfully immunized against HBV.

Current available therapies for chronic hepatitis B include the short-term administration of pegylated interferon-alfa (PEG-IFNα) or long-term nucleoside analogues (NUC) treatment. Fortunately, the current generation of NUCs, such as Entecavir (ETV) and Tenofovir (TDF), effectively suppress viral replication with a high barrier to resistance in the majority of patients [3]. However, NUCs cannot reconstitute immunological control and completely eliminate HBV in patients. The infection of woodchucks with woodchuck hepatitis virus (WHV) is an important preclinical model for chronic hepatitis B since it has a very similar pathogenesis as HBV infection in humans. This includes a high rate of chronic infection in early life, frequent integration in the genome of hepatocytes, cccDNA in the replication cycle of the virus and development of HCC. Studies in patients and woodchucks indicate a high frequency of HBV DNA or WHV DNA integration into the host genome during chronic infection, respectively [4]. Both integrated viral DNA and free episomal are important sources for the continuous production of large amounts of envelope proteins (HBsAg or WHsAg), which remain unaffected by NUC therapy [5]. HBsAg, present in high levels, seems to induce an immune inhibitory effect on both adaptive and innate immune functions [6,7]. The secretion of the small surface antigen, especially in the early phase of infection, has been suggested to interfere with the generation of an effective immune response against HBV through the induction of immune tolerance toward viral antigens [8].

This tolerance can promote an enhanced viral load and paradoxically reduce the intensity of hepatocellular injury [9].

There are rare cases of chronically HBV-infected patients who spontaneously overcome HBV-specific immune tolerance and clear HBsAg [10]. Additionally, HBsAg clearance rates have been reported in the years following the discontinuation of NUC therapy [11,12], with such cessation proposed as a strategy to induce enhanced immune control. The highest HBsAg seroclearance rate in this context reached 39%, 4–5 years after discontinuation of NUC therapy [13]. However, a recent study by Jeng et al. evaluating the cessation of NUC monotherapy in HBeAg-negative and HBeAg-positive patients determined the annual HBsAg seroclearance rate to be only 1.78% [14]. A low baseline HBsAg concentration was the only statistically significant correlate for HBsAg seroclearance, indicating that a high HBsAg level may be an important factor in maintaining immunotolerance in chronic hepatitis B carriers.

This immune tolerance may explain, at least in part, why attempts at therapeutic vaccination with conventional HBsAg vaccines have failed in several clinical trials [15]. In pre-clinical trials in the mouse model, it has been shown that vaccination of HBV transgenic mice with a high level of HBsAg are associated with a low level of B- and T-cell response to HBsAg or the core protein and do not reduce HBsAg or induce anti-HBs [16].

Nevertheless, low-level HBsAg in HBV carriers may create a window to reconstitute an HBV-specific immune response through so-called therapeutic vaccination in a compassionate access program setting.

It was therefore hypothesized that patients under NUC therapy, who have a low HBsAg load, may be good candidates for an effective therapeutic intervention through vaccination against HBV.

1.1. Development of Vaccines for Protection against Hepatitis B Virus Infection

First-generation, plasma-derived hepatitis B vaccines containing HBsAg harvested from patients with chronic Hepatitis B were developed in the late 1970s. In the mid-1980s, second-generation recombinant DNA hepatitis B vaccines were constructed in yeast transfected with HBV-DNA sequences coding for the small hepatitis B virus surface protein

(SHBs). These vaccines have gradually replaced the first-generation plasma-derived [17] vaccines and are currently used for universal vaccination of newborns and adults in more than 170 countries worldwide [18].

According to WHO guidelines, successful seroprotection against HBV infection after vaccination is defined by an anti-HBs titer of $\geq$ 10 IU/L (following immunization with at least three vaccine doses). The same threshold anti-HBs level applies to the definition of protection against hepatitis B after resolution of "wild-type" HBV infection. While the majority of vaccinees develop protection against HBV, 5–10% do not respond to the currently used conventional vaccines. These non-responders have anti-HBs levels < 10 IU/L after three or more injections with the conventional vaccine and remain susceptible to HBV infection. Thus, protection of such non-responders to conventional vaccination against HBV remains an important goal for specific risk groups (e.g., medical personnel involved in exposure-prone procedures, babies born to HBsAg positive mothers, or spouses of HBsAg carriers). The recently developed US FDA/EMA approved pre-S1/pre-S2/S HBV vaccine (PreHevbrio^{R}) has been shown to bypass resistance to vaccination in such non-responders to conventional vaccines containing only the small surface antigen [17,19–21].

1.2. Development of Therapeutic Vaccines for Control of Chronic Hepatitis B

Therapy with NUCs does not completely eliminate HBV, and there is a need to explore additional therapeutic regimens. With a few exceptions, vaccines are traditionally used for prevention of infection and not for treatment of established infections. An early attempt at therapeutic vaccination with an HBV vaccine containing the small HBsAg was indeed unsuccessful [15], possibly due to a failure to generate an effective cellular immune response against HBV.

We hypothesized that a new third-generation highly immunogenic HBV vaccine, Sci-B-Vac™, recently renamed PreHevbrio^{R}, containing all three envelope proteins, may harbor a therapeutic potential against HBV, possibly leading to a functional cure of HBV infection [20,21]. Indeed, restoring an antiviral B-cell and T-cell response not only in non-responders to the vaccine but also in patients with persistent HBV infection, remains a major challenge.

Consequently, starting in 2010, we initiated an experimental "Compassionate Access Program" in patients with chronic hepatitis B (CHB) with low-level HBsAg (<500 IU/mL) to assess the safety and efficacy of pre-S/S (PreHevbrio^{R}) vaccination in chronic carriers [22]. This therapeutic vaccination was not structured as a clinical trial.

Our early results reveal that a functional serologic cure was induced in a small number of HBsAg-positive patients treated with NUCs and immunized repeatedly with the pre-S1/pre-S2/S vaccine. We hypothesized that ongoing vaccinations using the pre-S/S vaccine in combination with NUC may lead to a functional cure of HBV. Consequently, we initiated a long-term follow-up of such patients.

The desired clinical outcome of our treatment protocol is the loss of HBsAg and induction of seroconversion to anti-HBs, leading to a putative "functional cure" as a result of suppression of viral replication with an aim to induce a sterilizing cure [23], eliminating hepatocytes containing episomal covalently closed circular HBV DNA (cccDNA).

2. Materials and Methods

2.1. Patients

An anonymous data request was sent to two virologic diagnostic laboratories to determine the frequency of low-level carriers of HBsAg who might be good candidates for this therapeutic vaccination protocol [22].

Four patients with chronic, low-level HBsAg were included in this pilot "Compassionate Access Program" using the third-generation pre-S/S HBV vaccine (PreHevbrio^{R}). According to the Declaration of Helsinki § 37: "In the treatment of an individual patient, where proven interventions do not exist or other known interventions have been ineffective, the physician, after seeking expert advice, with informed consent from the patient or a legally authorised representative, may use an unproven intervention if in the physician's judgement it offers hope of saving life, re-establishing health or alleviating suffering. This intervention should subsequently be made the object of research, designed to evaluate its safety and efficacy. In all cases, new information must be recorded and, where appropriate, made publicly available".

We followed this guideline in our clinical observation. All patients to be vaccinated with the pre-S/S vaccine (PreHevbrio^R) gave written consent to this procedure.

These patients were identified through the surveillance system of the registry at the University Hospital Essen and the Helios Klinikum Niedernberg. After screening our cohort of patients, only low-level carriers were chosen to give them this vaccine as compassionate use. There were no long-term HBsAg measurements prior to start of vaccination and HBV DNA below detection limit (HBV DNA "negative" as it is written in clinical diagnostic reports) was a prerequisite prior to vaccination.

Among them, two were females and two were males, with ages ranging between 43 and 68 years. Baseline characteristics (age, gender, HBV-DNA, HbsAg, anti-HBs status, NUC therapy, e.g., Baraclude, Viread, Entecavir, Tenofovir) of the four vaccinated patients with low-level HBsAg before and after vaccination are shown in Table 1. All patients were HBeAg-negative and anti-HBc-positive.

Table 1. Summary of therapeutic vaccination in four patients.

Patient	HBV-DNA Quant. IU/L (CobasX800)	HBs-Ag IU/mL (CMIA)	Anti-HBs IU/L (ECLIA)
Patient 1 Male, 46 y			
Prior to Vaccination	Negative	22	Negative
Number of Vaccinations: 4 with PreHevbrio^R (10 µg)			
Number of Vaccinations: 3 with PreHevbrio^R (20 µg)			
Post-Vaccination (48 months)	Negative	Negative	1000
Termination of NUC Treatment (24 Months after Vaccination): 2021			
Patient 2 Male, 68 y			
Prior to Vaccination	Negative	264	Negative
Number of Vaccinations: 4 with PreHevbrio^R (20 µg)			
Post-Vaccination (60 months)	Negative	10	112
Termination of NUC Treatment (30 Months after Vaccination): 2014			
Patient 3 Female, 43 y			
Prior to Vaccination	Negative	21	Negative
Number of Vaccinations: 5 with PreHevbrio^R (20 µg)			
Post-vaccination (48 months)	Negative	Negative	58
Termination of NUC Treatment (24 Months after Vaccination): 2020			
Patient 4 Female, 61 y			
Prior to Vaccination	Negative	350	Negative
Number of Vaccinations: 3 with PreHevbrio^R (20 µg)			
Post-Vaccination (7 months)	Negative	Negative	27
NUC Treatment: Ongoing			

The following inclusion criteria were used for recruitment: HBsAg-positive carrier status for more than five years; quantitative HBsAg level currently below 500 IU/mL; NUC treatment (e.g., Baraclude, Viread, Entecavir, Tenofovir) for at least two years; HBV DNA-negative; HBeAg-negative; anti-HBs-negative.

As controls, two high-level carriers of HBsAg (27,000 IU/mL and 8552 IU/mL, respectively) were also vaccinated. These patients were treated with NUCs (Entecavir) and had HBV-DNA below the detection limit. Baseline characteristics (age, gender, HBV-DNA, HBsAg, anti-HBs status, NUC therapy: Entecavir, Tenofovir) of these two vaccinated control patients (C 1, C 2) with high-level HBsAg before and after vaccination are provided in Table 2.

Table 2. Summary of therapeutic vaccination in two control patients.

Patient	HBV-DNA Quant. IU/L (CobasX800)	HBs-Ag IU/mL (CMIA)	Anti-HBs IU/L (ECLIA)
Patient C1 Male, 68 y			
Prior to Vaccination	Negative	24.000	Negative
Number of Vaccinations: 10 with PreHevbrio^{R} (20 µg)			
Post-Vaccination (40 months)	negative	24.000	Negative
Termination of NUC Treatment (32 Months after Vaccination): 2019			
Patient C2 Female, 65 y			
Prior to Vaccination	Negative	8442	Negative
Number of Vaccinations: 3 with PreHevbrio^{R} (20 µg)			
Post-Vaccination (6 months)	Negative	7687	>1.000
NUC Treatment: Ongoing			

2.2. *Pre-S1/Pre-S2/S HBV Vaccine*

The third-generation vaccine pre-S1/pre-S2/S HBV (PreHevbrio^{R}) is an aluminum hydroxide adjuvanted recombinant hepatitis B vaccine [18], currently manufactured by VBI Vaccines Ltd., Cambridge, MA, USA. This pre-S1/pre-S2/S vaccine is produced in mammalian Chinese hamster ovary (CHO) cells transfected with appropriate HBV sequences that code for the three HBV envelope proteins: the small S hepatitis B surface antigen (SHBs), the middle pre-S2 (MHBs), and the large pre-S1 envelope protein (LHBs). The purified HBsAg particles, secreted by the transfected CHO cells, mainly consist of SHBs (75–77% p24, gp27), MHBs (17–21% gp33, gp36), and LHBs (3–7% p39, gp42).

This pre-S1/pre-S2/S vaccine was approved by the FDA in November 2021.The European Medicines Agency (EMA) authorized this vaccine for use in the EU (including Germany) since 18.5.2022. Authorization details: EMA Product Nr: EMEA/H/C/005466 Reference Number: EMA/129611/2022.

2.3. *Vaccination Protocol and Monitoring*

In this "Compassionate Access Program," four patients who met the inclusion criteria were initially vaccinated according to the protocol on Day 0, Day 30, and Day 90. As there is no experience at which time intervals and how frequent therapeutic vaccination is needed to reach functional cure status, we used different schedules and numbers for vaccination. The numbers of vaccinations for each patient are given in Figures 1 and 2. All patients were presented individually because they started therapy in different years. In Patient 1, we started with a single dose of 10 µg, but he showed no response in terms of anti-HBsAg. Therefore, he was vaccinated more frequently and consecutively with 20 µg. After that, we administered 20 µg of PreHevbrio^{R} initially in the other patients, which showed better response.

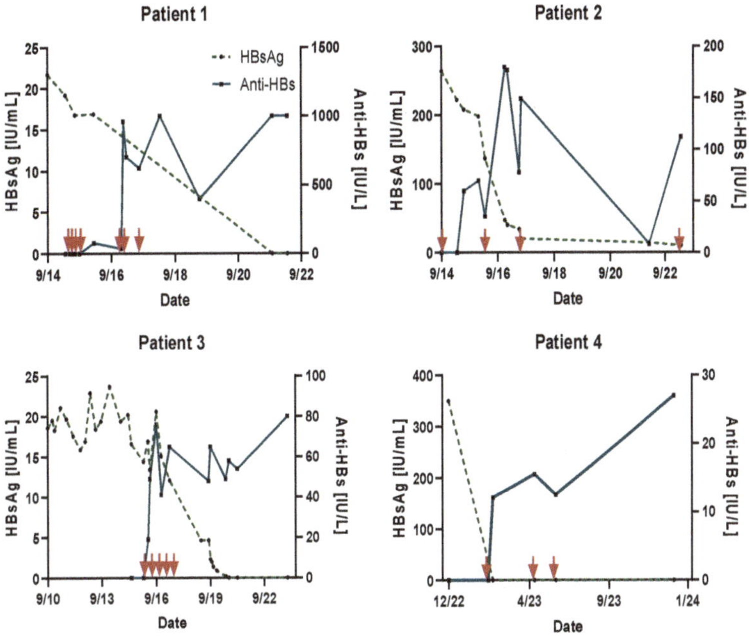

Figure 1. Kinetics of HBsAg and anti-HBs in vaccinated chronic low-level (HBsAg < 500 IU/mL) HBV carriers over time. Red arrows indicate the timing of vaccination/re-vaccination. The x-axis represents the month/year.

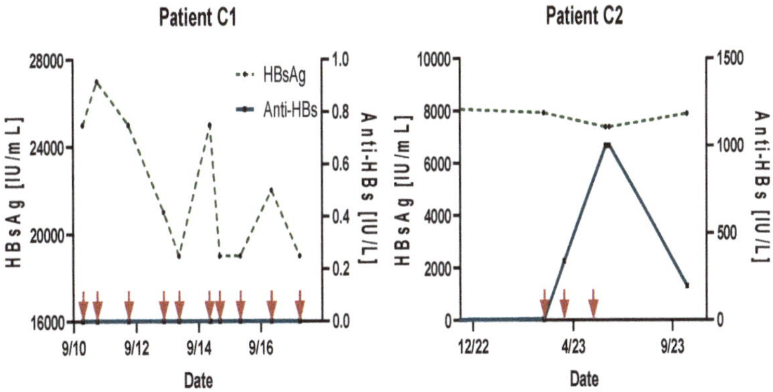

Figure 2. Kinetics of HBsAg and anti-HBs in vaccinated chronic high-level (HBsAg > 5000 IU/mL) HBV carriers over time. Red arrows represent time of vaccination/re-vaccination. The x-axis represents the month/year.

Anti-HBs levels (ECLIA) were determined 4 weeks after each vaccination. If a low immune response was observed after three vaccine doses (defined as an anti-HBs level < 100 IU/L), additional vaccine shots were administered. HBsAg (CMIA) and HBV DNA levels (CobasX800) were determined before vaccination. Monitoring of transaminases (GOT, GPT, GGT) and bilirubin was performed simultaneously (Tables 3 and 4).

Table 3. Summary of transaminases development in four patients.

Patient	GOT/GPT IU/L 10–50	GGT IU/L < 66
Patient 1 Male, 46 y		
Prior to Vaccination	41/60	45
Post-Vaccination (48 months)	39/47	43
Termination of NUC Treatment (24 Months after Vaccination): 2021		
Patient 2 Male, 68 y		
Prior to Vaccination	27/38	24
Post-Vaccination (60 months)	23/28	13
Termination of NUC Treatment (30 Months after Vaccination): 2014		
Patient 3 Female, 43 y		
Prior to Vaccination	24/16	12
Post-Vaccination (48 months)	21/14	10
Termination of NUC Treatment (24 Months after Vaccination): 2020		
Patient 4 Female, 61 y		
Prior to Vaccination	18/20	18
Post-Vaccination (7 months)	25/27	22
NUC Treatment: Ongoing		

Table 4. Summary of transaminases development in two control patients.

Patient	GOT/GPT IU/L 10–50	GGT IU/L < 66
Patient C1 Male, 68 y		
Prior to Vaccination	24/18	34
Post-Vaccination (2 months)	28/22	56
Termination of NUC Treatment (32 Months after Vaccination): 2019		
Patient C2 Female, 65 y		
Prior to Vaccination	23/17	19
Post-Vaccination (1.5 months)	24/12	12
NUC Treatment: Ongoing		

3. Results

Four patients, who were low-level carriers of HBsAg with levels less than 500 IU/mL (ranging from 18 IU/mL to 350 IU/mL), underwent vaccination with the PreHevbrio^R vaccine. They received between three and ten doses of 20 µg each, following pre-treatment with nucleos(t)ide analogues (NUCs). The inclusion criteria for these patients included: ongoing antiviral NUC therapy for more than 2 years; undetectable HBV-DNA; negative HBeAg; HBsAg levels under 500 IU/mL; and normal liver enzymes (GOT/GPT) below 50 U/L.

All four HBV carriers, with undetectable HBV DNA and under NUC treatment, who received three or more doses of the vaccine, seroconverted to anti-HBs at varying levels, as detailed in Table 1. The administration of the vaccine was well-tolerated, with no patients reporting serious adverse events (AEs), either locally or systemically, at any point during the study.

Patient 1, a 46-year-old male, was HBsAg-positive (22 IU/mL) prior to vaccination. After receiving seven doses over two years, he completely eliminated HBsAg and seroconverted to anti-HBs with levels greater than 1000 IU/L (see Figure 1). NUC treatment with Tenofovir was discontinued two years after the last vaccination. The total observation period for this patient was eight years.

Patient 2, a 68-year-old male, was HBsAg-positive (264 IU/mL) before vaccination. After three doses over 2.5 years, he seroconverted to anti-HBs (180 IU/L), and his HBsAg levels dropped to borderline reactive levels (10 IU/mL). His NUC treatment with Tenofovir was stopped. Over a monitoring period of six years, his anti-HBs level decreased to 8 IU/L. Following a booster shot, his anti-HBs levels increased to 112 IU/L two weeks after vaccination (see Figure 1). The total observation period for this patient was nine years.

Patient 3, a 43-year-old female, had an HBsAg level of 18 IU/mL before vaccination. After five doses over 1.5 years, she cleared HBsAg and seroconverted to anti-HBs with a level of 80 IU/L (see Figure 1). Her NUC treatment with Tenofovir was discontinued two years post-vaccination. The total observation period for this patient was eight years.

Patient 4, a female with an initial HBsAg level of 350 IU/mL, seroconverted to anti-HBs (27 IU/L) after three doses over four months and completely eliminated HBsAg (see Figure 1). She has been followed for one year so far, with ongoing NUC treatment with Entecavir. The total observation period for this patient has been one year.

3.1. *Monitoring Unvaccinated Carriers*

Given that a small number of NUC-treated patients may spontaneously clear HBsAg [24], we monitored four such unvaccinated low-level HBsAg carriers under NUC treatment for over two years. No significant change in HBsAg concentration was observed during this period.

3.2. *Follow-Up of Vaccinated Patients with or without Ongoing NUC Treatment*

Patients with chronic hepatitis B who discontinue treatment with NUCs often experience a relapse of HBV replication, characterized by the reappearance of HBV DNA and increased HBsAg levels in the serum [3]. Consequently, we monitored our vaccinated patients who discontinued NUC treatment for two to three years after their last vaccination.

Patient 1 remained HBsAg- and HBV DNA-negative six years after ceasing NUC treatment, with an anti-HBs titer of >1000 IU/L documented five years post-vaccination.

Patient 2 dropped his HBsAg level to a borderline level of 14 IU/mL one year post-vaccination and his anti-HBs Titer was 180 IU/L. NUC treatment was stopped two years after his last vaccination.

Patient 3 HBsAg level dropped to a borderline 14 IU/mL one year post-vaccination, with an anti-HBs titer of 180 IU/L. NUC treatment was halted two years after the final vaccination.

Patient 4 lost HBsAg but maintained low anti-HBs titers; thus, NUC treatment continues. Having recently completed the baseline vaccination series, further monitoring and potential re-vaccination are planned.

3.3. *Vaccination of High-Level HBsAg Carrier Controls*

Prior research has demonstrated that high concentrations of HBsAg can significantly suppress the immune response [24]. To verify ongoing immune tolerance during and after vaccination in patients with high levels of HBsAg (Table 2), we vaccinated a male chronic HBV carrier (Patient C1) with an extremely high HBsAg concentration of 27,000 IU/mL. Despite receiving ten doses of 20 μg PreHevbrio^{R} over seven years, there was no notable reduction in HBsAg nor seroconversion to anti-HBs (Figure 2).

Another patient (C2), a 65-year-old female initially HBsAg-positive (8552 IU/mL) before vaccination, maintained similar HBsAg levels (7896 IU/mL) after vaccination. This patient eventually seroconverted to anti-HBs > 1000 IU/L, which decreased to 197 IU/L within three months. Due to her high HBsAg levels, NUC treatment is ongoing.

4. Discussion

In this pilot observation, we followed four HBV carriers with low-level HBsAg (<500 IU/L) and undetectable HBV DNA, who were under NUC treatment, for up to 6 years after an attempt of therapeutic vaccination with the third-generation vaccine PreHevbrio^{R}. Remarkably, all four carriers seroconverted to anti-HBs, showing increasing antibody concentrations. This indicates a breakthrough in immune tolerance against HBsAg, which dropped below detection levels in patients 1, 3, and 4, while remaining at very low levels (10 IU/mL) in patient 2. The persistence of circulating HBsAg in patient 2 may be partly explained by the continuous secretion of HBsAg, potentially encoded by integrated HBV DNA—a phenomenon indistinguishable from HBsAg synthesized by episomal HBV-DNA.

The long-term observations of patients 1, 2, and 3 suggest the potential establishment of a functional cure of HBV infection over six years of follow-up. Patient 4, while HBsAg-negative, exhibited low anti-HBs levels and is scheduled for further booster shots and regular anti-HBs monitoring. Ongoing surveillance of these patients will continue to monitor for HBV reactivation or the development of liver cirrhosis or hepatocellular carcinoma (HCC). Should the anti-HBs titer drop below 100 IU/L, additional booster doses will be considered.

Patients C1 and C2, who had initially high HBsAg titers (27,000 IU/mL and 8552 IU/mL, respectively) and did not meet the inclusion criteria for the "Compassionate Access Program," were vaccinated with up to ten doses of PreHevbrio. Unlike the low-level HBsAg carriers, both patients showed no HBsAg reduction or clearance. However, a substantial anti-HBs titer (>1000 IU/L) was observed in Control Patient C2 over two months, despite no change in the high HBsAg level.

These outcomes suggest that the low HBsAg concentration in some patients may have a less tolerizing effect on the immune response compared to high-level carriers [25].

The results of this pilot study indicate that low-HBsAg levels in HBV carriers may be a prerequisite for therapeutic intervention through immunization with a preS/S vaccine. To validate this hypothesis, a significantly larger cohort of HBV carrier candidates will be necessary. We already conducted a preliminary survey to assess the number of candidates for such a trial for therapeutic intervention through vaccination: HBsAg levels were tested in two cohorts of patients, using data obtained at the Technical University Hospital in Munich ($n = 351$) and the University Hospital Duisburg-Essen ($n = 1131$). We identified that 30% of patients had HBsAg levels below 500 IU/mL [22], making them potential candidates for a larger therapeutic vaccination trial with PreHevbrio combined with NUC treatment.

The cumulative experience from several studies highlights that NUC monotherapy significantly reduces HBV DNA but often fails to decrease HBsAg levels in chronic HBV (CHB) [3,14]. Thus, exploring new strategies for suppressing HBsAg production, such as treatment with iRNA (Interference RNA) compounds or nucleic acid polymers (NAPs), is warranted [26–29]. NAPs therapy has shown promise in two clinical studies, leading to a 2–7 log reduction of HBsAg concentration in 12 patients [30,31].

In addition to a B-cell response in patient 1 [22], who exhibited the highest antibody response, we observed an HBV-specific T-cell response. However, T-cell responses in the remaining patients were below the cutoff. Despite using highly sensitive assays, the HBV-specific T-cell numbers may have been too low for detection. Further research is needed to confirm that HBV-specific CD4 and CD8 T cells contribute to sterilizing immunity against HBV, enhanced by NUC treatment in combination with a pre-S/S HBV vaccine.

The methodology employed in the current compassionate access program naturally has its limitations, including the inability to evaluate the impact of the vaccination protocol on HBV-cccDNA. Another important point of discussion is the role immunosenescence regarding responsiveness to vaccinations in general but especially to HBV vaccines [32]. Nonetheless, our findings suggest that vaccination with a pre-S/S HBV vaccine may contribute not only to a better vaccine efficacy but even more important to the functional cure of persistent HBV infection and warrants evaluation in future clinical trials alongside new antiviral agents aimed at disrupting cccDNA. Finally, we sought to determine if therapeutic vaccination of HBV carriers with a pre-S/S vaccine could result in long-lasting elimination of HBsAg and seroconversion to anti-HBs even after ceasing nucleoside therapy. Three patients (1, 2, and 3) discontinued Entecavir or Tenofovir treatment after their last vaccine dose. Two to five years post-NUC therapy, three patients remained anti-HBs positive, indicating a sustained B cell response without NUC therapy. The anti-HBs antibodies induced through preS/S vaccination may appear due to overcome the immune tolerance and neutralize minor amounts of circulating HBV, preventing de novo infection of previously uninfected hepatocytes. Further studies in a larger cohort of vaccinated patients with persistent HBV are needed to confirm the induction of a functional cure through long-term monitoring of viral replication, anti-HBs levels, and evaluating a specific T cell response.

Short overview:

- The results of the present observation revealed an excellent anti-HBs seroconversion response in three out of four HBsAg carriers immunized repeatedly with a PreS/S vaccine.
- Furthermore, these data suggest that low-level HBV carriership seems to be a prerequisite for successful therapeutic vaccination resulting in functional cure.
- No patient in this long-term observation (1 to 6 years) had a relapse or deterioration of the clinical status.
- These findings provide strong support for the development of a larger prospective study focused on low-level HBV carriers, utilizing a pre-S/S HBV vaccine in combination with NUC therapy.

Author Contributions: Conceptualization, H.R. and D.S.; methodology, M.R.; investigation, G.G.; resources, G.G.; data curation, H.R.; writing—original draft preparation, H.R.; writing—review and editing, M.R.; supervision, D.S. All authors have read and agreed to the published version of the manuscript.

Funding: This research received no external funding.

Institutional Review Board Statement: Ethical review and approval were waived for this individual Compassionate Access Program due to individual treatment of patients with chronic hepatitis B outside a clinical study on an off-label basis. These patients who have no other options of treatment to eliminate HBsAg were vaccinated with a vaccine, which was approved for prophylactic use so far. This helps to make this vaccine, which is still under development and consideration as a therapeutic treatment, available to individual patients in this special setting.

Informed Consent Statement: Informed consent was obtained from all subjects for this individual Compassionate Access Program.

Data Availability Statement: Data is unavailable due to privacy.

Conflicts of Interest: The authors declare no conflict of interest.

References

1. WHO. Available online: https://who.int/news-room/fact-sheets/detail/hepatitis-b (accessed on 15 March 2024).
2. Liu, Z.; Li, M.; Hutton, D.W.; Wagner, A.L.; Yao, Y.; Zhu, W.; Cao, L.; Tang, S.; Pan, J.; Wang, Y.; et al. Impact of the national hepatitis B immunization program in China: A modeling study. *Infect. Dis. Poverty* **2022**, *11*, 106. [CrossRef] [PubMed]
3. Chien, R.N.; Liaw, Y.F. Current Trend in Antiviral Therapy for Chronic Hepatitis B. *Viruses* **2022**, *14*, 434. [CrossRef]
4. Brechot, C.; Pourcel, C.; Louise, A.; Rain, B.; Tiollais, P. Presence of integrated hepatitis B virus DNA sequences in cellular DNA of human hepatocellular carcinoma. *Nature* **1980**, *286*, 533–535. [CrossRef] [PubMed]
5. Grudda, T.; Hwang, H.S.; Taddese, M.; Quinn, J.; Sulkowski, M.S.; Sterling, R.K.; Balagopal, A.; Thio, C.L. Integrated hepatitis B virus DNA maintains surface antigen production during antiviral treatment. *Clin. Investig.* **2022**, *132*, e161818. [CrossRef] [PubMed]
6. Kondo, Y.; Ninomiya, M.; Kakazu, E.; Kimura, O.; Shimosegawa, T. Hepatitis B surface antigen could contribute to the immunopathogenesis of hepatitis B virus infection. *ISRN Gastroenterol.* **2013**, *2013*, 935295. [CrossRef] [PubMed]
7. Tsai, K.N.; Kuo, C.F.; Ou, J.J. Mechanisms of Hepatitis B Virus Persistence. *Trends Microbiol.* **2018**, *26*, 33–42. [CrossRef] [PubMed]

8. Menne, S.; Roneker, C.A.; Roggendorf, M.; Gerin, J.L.; Cote, P.J.; Tennant, B.C. Deficiencies in the acute-phase cell-mediated immune response to viral antigens are associated with development of chronic woodchuck hepatitis virus infection following neonatal inoculation. *J. Virol.* **2002**, *76*, 1769–1780. [CrossRef]
9. Thomson, A.W.; Knolle, P.A. Antigen-presenting cell function in the tolerogenic liver environment. *Nat. Rev. Immunol.* **2010**, *10*, 753–766. [CrossRef]
10. Alawad, A.; Auh, S.; Suarez, D.; Ghany, M. Durability of Spontaneous and Treatment-Related Loss of Hepatitis B s Antigen. *Clin. Gastroenterol. Hepatol.* **2020**, *18*, 700–709.e3. [CrossRef]
11. Berg, T.; Simon, K.G.; Mauss, S.; Schott, E.; Heyne, R.; Klass, D.M.; Eisenbach, C.; Welzel, T.M.; Zachoval, R.; Felten, G.; et al. FINITE CHB study investigators [First investigation in stopping TDF Long-term response after stopping tenofovir disoproxil fumarate in non-cirrhotic HBeAg-negative patients treatment after long-term virological suppression in HBeAg-negative chronic hepatitis B]. *J. Hepatol.* **2017**, *67*, 918–924. [CrossRef]
12. Wong, G.L.H.; Gane, E.; Lok, A.S.F. How to achieve functional cure of HBV: Stopping NUCs, adding interferon or new drug development? *J. Hepatol.* **2022**, *76*, 1249–1262. [CrossRef] [PubMed]
13. Hadziyannis, S.J.; Sevastianos, V.; Rapti, I.; Vassilopoulos, D.; Hadziyannis, E. Sustained responses and loss of HBsAg in HBeAg-negative patients with chronic hepatitis B who stop long-term treatment with adefovir. *Gastroenterology* **2012**, *143*, 629–636. [CrossRef]
14. Jeng, W.J.; Chen, Y.C.; Chien, R.N.; Sheen, I.S.; Liaw, Y.F. Incidence and predictors of hepatitis B surface antigen seroclearance after cessation of nucleos(t)ide analogue therapy in hepatitis B e antigen-negative chronic hepatitis B. *Hepatology* **2018**, *68*, 425–434. [CrossRef] [PubMed]
15. Pol, S.; Nalpas, B.; Driss, F.; Michel, M.L.; Tiollais, P.; Denis, J.; Brécho, C. Multicenter study group. Efficacy and limitations of a specific immunotherapy in chronic hepatitis B. *J. Hepatol.* **2001**, *34*, 917–921. [CrossRef]
16. Michler, T.; Kosinska, A.; Festag, J.; Bunse, T.; Su, J.; Ringelhan, M.; Imhof, H.; Grimm, D.; Steiger, K.; Mogler, C.; et al. Knockdown of Virus Antigen Expression Increases Therapeutic Vaccine Efficacy in High-Titer Hepatitis B Virus Carrier Mice. *Gastroenterology* **2020**, *158*, 1762–1775.e9. [CrossRef] [PubMed]
17. Vesikari, T.; Finn, A.; van Damme, P.; Leroux-Roels, I.; Leroux-Roels, G.; Segall, N.; Toma, A.; Vallieres, G.; Aronson, R.; Reich, D.; et al. CONSTANT Study Group Immunogenicity and Safety of a 3-Antigen Hepatitis B Vaccine vs a Single-Antigen Hepatitis B Vaccine: A Phase 3 Randomized Clinical Trial. *JAMA Netw. Open* **2021**, *4*, e2128652. [CrossRef] [PubMed]
18. Shouval, D. Hepatitis B vaccines. *J. Hepatol.* **2003**, *39* (Suppl. S1), S70–S76. [CrossRef] [PubMed]
19. Shouval, D.; Roggendorf, H.; Roggendorf, M. Enhanced immune response to hepatitis B vaccination through immunization with a Pre-S1/Pre-S2/S Vaccine. *Med. Microbiol. Immun.* **2015**, *204*, 57–68. [CrossRef] [PubMed]
20. Krawczyk, A.; Ludwig, C.; Jochum, C.; Fiedler, M.; Heinemann, F.M.; Shouval, D.; Roggendorf, M.; Roggendorf, H.; Lindemann, M. Induction of a robust T- and B-cell immune response in non- and low-responders to conventional vaccination against hepatitis B by using a third generation PreS/S vaccine. *Vaccine* **2014**, *32*, 5077–5082. [CrossRef]
21. Rendi-Wagner, P.; Shouval, D.; Genton, B.; Lurie, Y.; Rümke, H.; Boland, G.; Cerny, A.; Heim, M.; Bach, D.; Schroeder, M.; et al. Comparative immunogenicity of a PreS/S hepatitis B vaccine in non- and low responders to conventional vaccine. *Vaccine* **2006**, *24*, 2781–2789. [CrossRef]
22. Roggendorf, H.; Krawczyk, A.; Lindemann, M.; Shouval, D.; Michler, T.; Roggendorf, M.; Gerken, G. Induction of Functional Control in Chronic Hepatitis B Patients with Low Level HBsAg Using a Combination of a PreS1/S2/S HBV Vaccine (Sci-B-VacTM) and a Nucleoside Analogue. *J. Infect. Dis. Ther.* **2019**, *7*, 390. [CrossRef]
23. Lok, A.S.; Zoulim, F.; Dusheiko, G.; Ghany, M.G. Hepatitis B cure: From discovery to regulatory approval. *J. Hepatol.* **2017**, *67*, 847–861. [CrossRef] [PubMed]
24. Knolle, P.A.; Thimme, R. Hepatic immune regulation and its involvement in viral hepatitis infection. *Gastroenterology* **2014**, *146*, 1193–1207. [CrossRef]
25. Dembek, C.; Protzer, U.; Roggendorf, M. Overcoming immunetolerance in chronic hepatitis B by therapeutic vaccination. *Curr. Opin. Virol.* **2018**, *30*, 58–67. [CrossRef]
26. Gish, R.G.; Yuen, M.-F.; Chan, H.L.Y.; Given, B.D.; Lai, C.-L.; Locarnini, S.A.; Lau, J.Y.; Wooddell, C.I.; Schluep, T.; Lewis, D.L. Synthetic RNAi triggers and their use in chronic hepatitis B therapies with curative intent. *Antivir. Res.* **2015**, *121*, 97–108. [CrossRef] [PubMed]
27. Yuen, M.F.; Lim, S.G.; Plesniak, R.; Tsuji, K.; Janssen, H.L.A.; Pojoga, C.; Gadano, A.; Popescu, C.P.; Stepanova, T.; Asselah, T.; et al. Efficacy and Safety of Bepirovirsen in Chronic Hepatitis B Infection. *N. Engl. J. Med.* **2022**, *387*, 1957–1968. [CrossRef] [PubMed]
28. Kasianchuk, N.; Dobrowolska, K.; Harkava, S.; Bretcan, A.; Zarębska-Michaluk, D.; Jaroszewicz, J.; Flisiak, R.; Rzymski, P. Gene-Editing and RNA Interference in Treating Hepatitis B: A Review. *Viruses* **2023**, *15*, 2395. [CrossRef] [PubMed]
29. Van Gulck, E.; Conceição-Neto, N.; Aerts, L.; Pierson, W.; Verschueren, L.; Vleeschouwer, M.; Krishna, V.; Nájera, I.; Pauwels, F. Retreatment with HBV siRNA Results in Additional Reduction in HBV Antigenemia and Immune Stimulation in the AAV-HBV Mouse Model. *Viruses* **2024**, *16*, 347. [CrossRef]
30. Al-Mahtab, M.; Bazinet, M.; Vaillant, A. Safety and Efficacy of Nucleic Acid Polymers in Monotherapy and Combined with Immunotherapy in Treatment-Naive Bangladeshi Patients with HBeAg+ Chronic Hepatitis B Infection. *PLoS ONE* **2016**, *11*, e0156667. [CrossRef]

31. Bazinet, M.; Pantea, V.; Cebotarescu, V.; Cojuhari, L.; Jimbei, P.; Albrecht, J.; Schmid, P.; Le Gal, F.; Gordien, E.; Krawczyk, A.; et al. Safety and efficacy of REP 2139 and pegylated interferon alfa-2a for treatment-naive patients with chronic hepatitis B virus and hepatitis D virus co-infection (REP 301 and REP 301-LTF): A non-randomised, openlabel, phase 2 trial. *Lancet Gastroenterol. Hepatol.* **2017**, *2*, 877–889. [CrossRef]
32. Liang, P.; Zu, J.; Zhuang, G. A Literature Review of Mathematical Models of Hepatitis B Virus Transmission Applied to Immunization Strategies From 1994 to 2015. *J. Epidemiol.* **2018**, *28*, 221–229. [CrossRef] [PubMed]

Journal of *Personalized Medicine*

MDPI

Review

Acute-On-Chronic Liver Failure: Current Interventional Treatment Options and Future Challenges

Markus Kimmann [1,*] and Jonel Trebicka [1,2,*]

[1] Department of Internal Medicine B, University of Münster, 48149 Münster, Germany
[2] European Foundation for the Study of Chronic Liver Failure, EFCLIF, 08021 Barcelona, Spain
* Correspondence: markus.kimmann@ukmuenster.de (M.K.); jonel.trebicka@ukmuenster.de (J.T.)

Abstract: Acute-on-chronic liver failure (ACLF) is a frequent complication in patients with liver cirrhosis that has high short-term mortality. It is characterized by acute decompensation (AD) of liver cirrhosis, intra- and extrahepatic organ failure, and severe systemic inflammation (SI). In the recent past, several studies have investigated the management of this group of patients. Identification and treatment of precipitants of decompensation and ACLF play an important role, and management of the respective intra- and extrahepatic organ failures is essential. However, no specific treatment for ACLF has been established to date, and the only curative treatment option currently available for these patients is liver transplantation (LT). It has been shown that ACLF patients are at severe risk of waitlist mortality, and post-LT survival rates are high, making ACLF patients suitable candidates for LT. However, only a limited number of patients are eligible for LT due to related contraindications such as uncontrolled infections. In this case, bridging strategies (e.g., extracorporeal organ support systems) are required. Further therapeutic approaches have recently been developed and evaluated. Thus, this review focuses on current management and potential future treatment options.

Keywords: hepatology; liver cirrhosis; liver transplantation; decompensated cirrhosis; acute-on-chronic liver failure

Citation: Kimmann, M.; Trebicka, J. Acute-On-Chronic Liver Failure: Current Interventional Treatment Options and Future Challenges. *J. Pers. Med.* **2023**, *13*, 1052. https://doi.org/10.3390/jpm13071052

Academic Editor: Guido Gerken

Received: 12 May 2023
Revised: 20 June 2023
Accepted: 25 June 2023
Published: 26 June 2023

1. Introduction

Liver cirrhosis is associated with high morbidity and mortality, and represents a considerable public healthcare burden worldwide [1–3]. A recent study by Gu et al. investigated the epidemiology of cirrhosis in Germany. It was shown that cirrhosis was diagnosed in 0.94% of all patients admitted to hospital. Remarkably, 54.8% of these patients were diagnosed with cirrhosis or cirrhosis-related complications as a comorbidity, while the primary reasons for hospital admission were other diagnoses. Alcoholic liver cirrhosis accounted for 52% of admissions with cirrhosis [1]. Patients with liver cirrhosis are at risk of acute decompensation (AD), which is defined by the occurrence of cirrhosis-related complications and hospitalization, and it is associated with increased mortality [4]. The prevalence of different complications of cirrhosis has changed over time, with a decrease in bleeding complications and an increase in the prevalence of portal vein thrombosis (PVT), infections, hepatic encephalopathy (HE), ascites, hepatorenal syndrome, and hepatocellular carcinoma [1]. Recent studies have investigated the clinical course of AD and identified distinct clinical phenotypes ranging from stable decompensated cirrhosis (SDC) to unstable decompensated cirrhosis (UDC) to pre-ACLF patients [5,6]. The latter develop acute-on-chronic liver failure (ACLF), which is especially driven by severe systemic inflammation (SI) [7,8] within 90 days. According to the EASL definition, ACLF is defined by intra- and extrahepatic organ failure in patients with acutely decompensated liver cirrhosis, and is associated with a 28-day mortality rate in about 30% of cases [9]. Recent studies have demonstrated a high prevalence of ACLF in patients hospitalized due to AD worldwide [10]. Different precipitants for AD have been identified, i.e., bacterial infections, severe alcoholic hepatitis, bleeding with shock, and drug-induced toxic encephalopathy [11]. Currently, management of these patients consists of identification and treatment of the precipitant as well as the respective intra- and extrahepatic organ failures. However, no specific therapy exists for this group of patients, and to date the only curative treatment is liver transplantation (LT). On the one hand, there is a scarcity of donor organs because of strong competition for patients on the waiting list, while on the other hand, in terms of eligibility, ACLF patients may present with contraindications for LT, i.e., uncontrolled bacterial infection. Extracorporeal liver support systems (ECLS) such as albumin dialysis have been and are currently being evaluated, and new approaches and experimental therapeutic strategies are being tested.

2. Precipitants and Definition of AD and ACLF

AD in patients with liver cirrhosis is defined by the occurrence of cirrhosis-related complications such as acute gastrointestinal bleeding, development of ascites, HE, and bacterial infection, which lead to hospital admission [6]. Development of AD based on precipitants, i.e., bacterial infection, alcohol-induced hepatitis, gastrointestinal bleeding, and drug induced toxic encephalopathy, marks a crucial time point in the clinical course of patients with liver cirrhosis [12]. Interestingly, the PREDICT study found that more than 96% of patients with precipitants showed proven bacterial infection and/or severe alcoholic hepatitis. However, no precipitant was identified in 39% of patients with AD who developed ACLF. Bacterial translocation might play a role in this group of patients. Furthermore, it was shown that the number of events, not the type of event, seemed to determine prognosis in these patients [11]. A recent meta-analysis showed that bacterial infections were the most common precipitants worldwide [10]. However, there were regional differences, with alcohol-induced hepatitis being the most common precipitant in East Asia and North America. In particular, the first episode of AD marks a shift towards end-stage cirrhosis leading to a significant reduction in terms of median survival time. Around 30% of patients with AD develop intra- or extrahepatic organ failures and suffer from severe systemic inflammatory response [13]. However, due to divergent definitions of ACLF, cautious interpretation is needed when comparing international studies and data. These definitions and characteristics are summarized in Table 1.

The Chronic Liver Failure Acute-on-Chronic Liver Failure in Cirrhosis (CANONIC) study was able to determine major risk factors for mortality in patients presenting with AD, and defined ACLF as a distinct syndrome in patients with decompensated cirrhosis [9]. Based on these findings, the Chronic Liver Failure–Sequential Organ Failure Assessment (CLIF-SOFA) score and the Chronic Liver Failure Consortium Organ Failure (CLIF-C OF) score were developed to assess ACLF in patients. According to the European Association for the Study of the Liver Chronic Liver Failure (EASL-CLIF) Consortium definition, ACLF is present in patients with acutely decompensated liver cirrhosis with single kidney failure (serum creatinine $\geq$ 2 mg/dL) or single organ failure combined with kidney dysfunction (serum creatinine range 1.5–1.9 mg/dL) and/or mild-to-moderate HE or presence of two or more organ failures [14].

The North American Consortium for the Study of End-Stage Liver Disease (NACSELD) defines ACLF as a condition in patients with acutely decompensated cirrhosis with or without prior episode(s) of decompensation and two or more organ system failures (maximum of four organ failures) that is associated with increased mortality within three months in the absence of treatment of the underlying liver disease, liver support, or LT. The organ failures that are taken into account include kidney (need for renal replacement therapy), brain (HE grade III or IV according to the West Haven Criteria), circulation (shock, mean arterial pressure < 60 mmHg), and respiration (need for mechanical ventilation) [15].

Table 1. Different definitions and characteristics of Acute-on-chronic liver failure.

	Asian Pacific Association for the Study of the Liver (APASL) ACLF Research Consortium (AARC) [16]	**European Association for the Study of the Liver-Chronic Liver Failure (EASL-CLIF) Consortium [9]**	**North American Consortium for the Study of End-Stage Liver Disease (NASCELD) [15]**
Patients considered in the definition	Acute liver deterioration in patients with diagnosed or undiagnosed chronic liver disease (including liver cirrhosis) Both compensated cirrhosis and non-cirrhotic chronic liver disease (chronic hepatitis with fibrosis, or fibrosis due to other reasons, non-alcoholic fatty liver disease, related chronic hepatic injury) qualify as chronic liver disease	Patients with an acute decompensation of liver cirrhosis with or without prior episode(s) of decompensation	Patients with an acute decompensation of liver cirrhosis with or without prior episode(s) of decompensation
Precipitating disorders	Extrahepatic (bacterial infection), intrahepatic (HBV reactivation), or both	Extrahepatic (infection, gastrointestinal bleeding), intrahepatic (alcoholic hepatitis), or both	Extrahepatic (infection)
Definition	ACLF definition is based on the presence of liver dysfunction. Extrahepatic organ failures may develop but are not included in the definition	ACLF definition is based on the existence of the failure of 1 or more of the 6 major organ considered in the CLIF-C Organ Failure scale (organ systems considered: Liver, kidney, brain, coagulation, circulation, respiration)	ACLF definition is based on the existence of 2 organ system failures or more (maximum 4) according to the NASCELD definition (organ systems considered: kidney, brain, circulation and respiration)
Stratification of ACLF	Jaundice (total bilirubin levels of 5 mg/dL or more) and coagulopathy (INR of 1.5 or more, or prothrombin activity of less than 40%) as a result of an acute hepatic insult which is complicated within 4 weeks by ascites, encephalopathy, or both. The severity of ACLF is assessed using the AARC score	ACLF is divided into 3 grades with increasing severity and mortality. ACLF grade 1 includes: - patients with single kidney failure - patients with single liver, coagulation, circulatory or lung failure that is associated with creatinine levels ranging from 1.5 mg/dL to 1.9 mg/dL or hepatic encephalopathy grade 1 or grade 2, or both - patients with single brain failure with creatinine levels ranging from 1.5 mg/dL to 1.9 mg/dL ACLF grade 2 includes: - patients with 2 organ failures ACLF grade 3 includes: - patients with 3 organ failures or more had ACLF grade 3	The stratification of patients is based according to the number of organ failures 2, 3, or all 4 organ failures, respectively
Approximate short-term mortality according to the stratification of ACLF	By 28 days: Grade 1: 13% Grade 2: 45% Grade 3: 86%	By 28 days: Grade 1: 23% Grade 2: 31% Grade 3: 74%	By 30 days: 2 organ failures: 18 to 43% 3 organ failures: 45 to 68% 4 organ failures: 77%

The Asian Pacific Association for the Study of the Liver (APASL) defines ACLF as an acute hepatic insult with jaundice (defined as serum bilirubin $\geq$ 5mg/dL) and coagulopathy (defined by international normalized ratio (INR) $\geq$ 1.5) complicated within four weeks by clinically manifestation of ascites and/or HE in patients with or without diagnosis of chronic liver disease and cirrhosis and a high 28-day mortality; organ failures are not included in the definition. The grade of ACLF is assessed using the AARC scoring system, which includes bilirubin, HE grade, INR, lactate, and creatinine [16].

3. Clinical Courses and Pathophysiology of AD

The CANONIC and PREDICT studies provided data on the role of SI as a crucial determinant in the development of cirrhosis-related complications and organ failure in addition to the well-known role of portal hypertension (PHT) and its complications, supporting the systemic inflammation hypothesis [8]. Furthermore, it has been shown that SI is associated with disease severity as well as with patient survival rates.

The PREDICT study was able to show that AD is characterized by different clinical phenotypes. These phenotypes depict heterogenous clinical conditions with distinct pathophysiology and a different prognosis. Therefore, the authors of the PREDICT study suggested a novel classification into three patient groups to better identify and differentiate these distinct courses of AD.

Most patients admitted with AD belong to the group of patients with SDC, who show low SI, are more likely to be recompensated quickly, show fewer cirrhosis-associated complications, and have the lowest 1-year mortality risk of the three groups.

The second clinical course of AD suggested by the PREDICT study is UDC. These patients mainly suffer from PHT-driven complications, present with a high prevalence of bacterial infections, i.e., spontaneous bacterial peritonitis, and have a higher risk of further decompensation and significantly decreased survival rates. In terms of pathophysiology, severe PHT seems much more relevant than in the other courses of AD, while SI is present in these patients as well.

Lastly, there are patients with AD who are characterized by development of ACLF within 90 days, constituting the group of pre-ACLF patients. These patients show severe progression of SI, possibly a key factor in development of intra- and extrahepatic organ failures, and show the highest short-term mortality among the three groups [6].

Overall, both PHT and SI are important in the pathophysiology of AD, with different characteristics depending on the clinical course of AD. The results of the PREDICT study suggest that the relevance of SI increases over the clinical course of AD and is most prominent in pre-ACLF and ACLF patients, while PHT is especially relevant in UDC patients [6] (Figure 1).

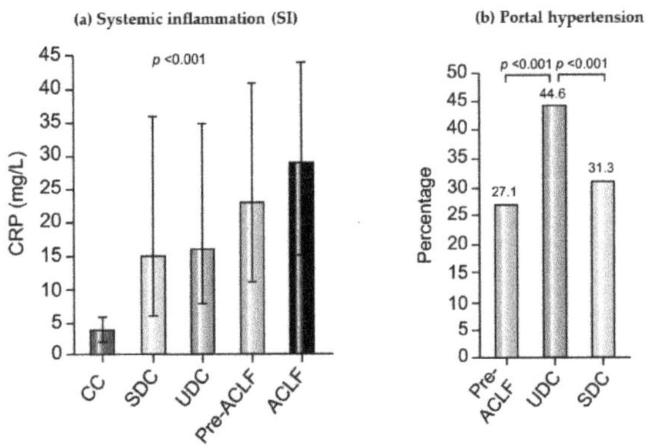

Figure 1. The role of Systemic inflammation (SI) and Portal Hypertension in different stages of cirrhosis according to the results of the PREDICT study. (**a**) Plasma levels of CRP as a marker of SI in

patients with compensated cirrhosis (CC, no prior history of AD), SDC, UDC, pre-ACLF, and ACLF. *p* values were obtained using Kruskal–Wallis test. (**b**) Percentage of patients presenting at least one surrogate of severe portal hypertension during the 6-month observational period of the PREDICT study in the Pre-ACLF, UDC, and SDC groups. *p* values were obtained using chi-square test. ACLF, acute-on-chronic liver failure; AD, acute decompensation; CC, compensated cirrhosis; CRP, C-reactive protein; SDC, stable decompensated cirrhosis; UDC, unstable decompensated cirrhosis. Figure modified from Trebicka et al., 2020 [6].

4. Management of AD and ACLF

4.1. Prediction of Decompensation

The early detection of patients at risk of decompensation is essential to allow early treatment and optimize their prognosis. The ANTICIPATE study highlighted the diagnostic value of liver stiffness measurements (LSM) in combination with platelet count for identifying clinically significant portal hypertension [17]. The Baveno VII consensus further highlighted the diagnostic value of LSM and stated the "rule of five" for LSM by the transient elastography (10–15–20–25 kPa) approach to denote higher relative risk of decompensation and liver-related death [18]. Additionally, a recent international multicenter cohort study provided an efficient and simple algorithm (M10LS20 algorithm) for risk stratification of patients with chronic liver disease that was externally validated. This algorithm consists of the Model of End Stage Liver Disease (MELD) score and LSM by shear wave elastography. A combined cutoff of a MELD score $\geq$ 10 and a liver stiffness of $\geq$20 kPa was able to identify those patients with a poor prognosis who have high mortality and risk of development or worsening of decompensation [19].

4.2. Prevention of AD and ACLF

Several therapeutic options have been evaluated for prevention of decompensation and for the management of patients with liver cirrhosis.

The role of non-selective beta blockers (NSBB) has been the topic of research in the recent past. In the PREDESCI study on patients with compensated cirrhosis and clinically significant portal hypertension the use of NSBB was associated with increased decompensation-free survival, mainly by reducing the incidence of ascites [20].

Aspirin should not be discouraged in patients with an approved indication, as it may reduce the risk of hepatocellular carcinoma, death, and liver-related complications [18,21]. The same applies for anticoagulation in patients with an approved indication, as it may improve the prognosis of these patients [18,22].

The use of statins in patients with liver cirrhosis and an indication for statins has been shown to improve survival, and may even decrease portal pressure [23]. Furthermore, Mahmud et al. were able to demonstrate that statin use and the duration of therapy significantly reduced the risk of ACLF [24]. However, statins should be used at a lower dose (simvastatin at max. 20 mg/d) in patients with Child–Pugh B and C cirrhosis due to a higher rate of adverse events, and patients should be followed for muscle and liver toxicity [25]. It remains unclear whether patients with decompensated liver cirrhosis without indication for statins benefit from this medication, and more data is necessary. Currently, we are awaiting the results of a phase III multicenter double-blind placebo-controlled randomized clinical trial (NCT03780673) investigating the use of simvastatin plus rifaximin in patients with decompensated cirrhosis to prevent ACLF.

Antibiotic prophylaxis is a keystone in preventing bacterial infections, and especially in preventing spontaneous bacterial peritonitis (SBP), which are both common precipitants of AD and ACLF. Thus, primary antibiotic prophylaxis is recommended in patients with gastrointestinal bleeding and Child–Pugh C cirrhosis with low protein ascites, as they are at high risk of SBP development. In patients with a history of previous SBP, secondary antibiotic prophylaxis is indicated [18,26].

However, while rifaximin is not indicated for primary or secondary prophylaxis of SBP, it is indicated for secondary prophylaxis of HE, and should be considered in patients undergoing elective transjugular intrahepatic portosystemic shunt (TIPS) implantation who have a history of overt HE [18].

The role of albumin administration, which has been discussed as a disease modifying agent, is a topic of current research. In this context, short-term and long-term administration of albumin must be distinguished.

In patients with spontaneous bacterial peritonitis (SBP) and specific risk factors (bilirubin > 4 mg/dL or creatinine > 1 mg/dL), albumin infusion has been shown to decrease the risk of acute kidney injury, and has been associated with improved prognosis [27].

According to the Baveno VII consortium, there is an indication for short-term albumin administration in patients with SBP, acute kidney injury (AKI), large-volume paracentesis (>5 L), and, in combination with terlipressin, in the treatment of hepatorenal syndrome (HRS) [18]. Furthermore, a recent study by Arora et al. demonstrated that albumin infusion decreases the incidence of paracentesis-induced circulatory dysfunction and mortality in ACLF patients receiving paracentesis < 5 L [28].

The role of long-term albumin administration is more controversial. The multicenter open-label ATTIRE study by China et al. found that targeted albumin infusion in patients hospitalized with liver cirrhosis and serum albumin levels < 30 g/L did not decrease the risk of onset of infection, acute kidney injury, or death [29].

However, the multicenter open-label ANSWER study investigated the effect of long-term albumin administration (40 g twice weekly for two weeks and then 40 g weekly for up to 18 months) in patients with decompensated cirrhosis and uncomplicated ascites, and showed increased overall survival in the group of patients with long-term albumin administration. Currently, no clear recommendation regarding long-term albumin administration can be provided due to insufficient data.

In this regard, the Prevention of Mortality with Long-Term Administration of Human Albumin in Subjects With Decompensated Cirrhosis and Ascites (PRECIOSA) trial (NCT03451292), which is currently recruiting, will hopefully improve the evidence base regarding long-term administration of albumin. Furthermore, a study on personalized long-term albumin treatment in patients with decompensated cirrhosis and ascites (Albtrial) (NCT05056220) guided by the MICROB-PREDICT biomarker is expected to provide more evidence.

4.3. *Treatment of the Precipitants of AD and ACLF*

As a result of the lack of currently available specific treatment options for AD and ACLF, management consists of identifying, preventing, and treating precipitants of ACLF as well as management of the respective intra- and extrahepatic organ failures [26,30].

If a precipitant is ascertainable, early detection and adequate treatment are both crucial.

The PREDICT study demonstrated that the most frequent precipitants in Europe are bacterial infections and severe alcoholic hepatitis, and that the number of precipitants is relevant, the latter being significantly associated with surrogates for SI and increased 90-day mortality [11].

Liver cirrhosis is associated with dysfunctions of various components of the immune system, which together have been described as cirrhosis-associated immune dysfunction [31]. Patients with liver cirrhosis have an increased risk of developing bacterial infections, which are the most common precipitants of ACLF in Europe [11]. If a bacterial infection is confirmed, it is essential to begin timely and adequate antibiotic treatment, which has been shown to decrease the incidence of ACLF development in patients with AD and to decrease mortality in patients with ACLF [32,33]. The most common bacterial infection in this context is SBP [34]. Due to the increasing prevalence of multidrug-resistant organisms (MDRO), the high prevalence of these organisms in patients with ACLF, and their lower infection resolution rates, antibiotic treatment of these patients is challenging [32]. Interestingly, antibiotic prophylaxis of SBP with norfloxacin was not associated with an increased incidence of MDRO [35]. However, it is crucial to initially start a broad empirical antibiotic treatment that takes into account the local resistance spectrum as well as the recent medical history of the patient (e.g., recent interventions such as ascites drainage) and to subsequently adjust the therapy according to microbiological results [35,36].

A recent study by Fernandez et al. investigated the effect of albumin administration in patients with advanced cirrhosis and non-SBP infections. While there was no difference in terms of in-hospital mortality, in the group of patients who received albumin there was a higher rate of ACLF resolution and a lower proportion of nosocomial infections despite the fact that the group of patients who received albumin were more seriously ill at baseline [32]. Thus, albumin administration in this subgroup of patients might be beneficial.

Another common trigger for ACLF is severe alcoholic hepatitis, which is linked to massive inflammation. The STOPAH trial investigated therapeutic approaches (pentoxifylline and prednisolone) for these patients and found no significant survival benefit for either drug. However, there was a benefit found in the 28-day survival rate in the group of patients who received prednisolone [32], possibly due to the higher likelihood of infections being developed during the additional observation period. According to the relevant EASL guideline, prednisolone 40 mg/day should be administered in patients with a Glasgow Alcoholic Hepatitis score (GAHS) of ≥9 [37]. GAHS was developed to identify patients at risk of death in case of alcoholic hepatitis [38]. After seven days of prednisolone therapy, treatment response is assessed by the Lille model for alcoholic hepatitis, which is a risk stratification tool for patients with alcoholic hepatitis receiving steroids for seven days [39]. In patients with a Lille score < 0.45, prednisolone administration should be continued for a total of 28 days, while in patients with a Lille score > 0.45 prednisolone should be discontinued. Patients with severe alcoholic hepatitis who receive prednisolone therapy have significantly increased mortality in terms of infection. Therefore, if an infection is already present, prednisolone therapy should not be administered when severe alcoholic hepatitis is present [37]. Furthermore, a French multicenter study suggests that in severe alcoholic hepatitis the combination of N-acetylcysteine with prednisolone has a better 1-month survival and a lower incidence of renal failure and infection compared with prednisolone therapy alone. However, 6-month survival was not shown to be different [35].

It has to be taken into account that the number of organ failures at admission, which itself indicates the severity and grade of ACLF, is negatively correlated with response rates to prednisolone therapy [40]. In most allocation systems, patients with alcohol-related liver disease need to be abstinent for at least six months in order to be considered for LT. However, Mathurin et al. evaluated the option of LT in patients with a first episode of severe alcoholic hepatitis not responding to medical therapy who failed to be abstinent for at least six months, demonstrating that in selected patients early transplantation can improve survival [41]. Another recent multicenter non-randomized study confirmed the survival benefit related to early liver transplantation for severe alcoholic hepatitis, showing similar 2-year survival rates for the early and standard (six months of alcohol abstinence) transplantation groups. However, non-inferiority in terms of rate of alcohol relapse post-transplant between early liver transplantation and standard transplantation could not be concluded [42]. Overall, this topic remains controversial.

Another trigger of ACLF is upper gastrointestinal bleeding, which is a potentially acute life-threatening event [43]. The management of upper gastrointestinal bleeding, whether variceal or nonvariceal, first consists of monitoring the patient and providing hemodynamic and respiratory stabilization. Volume replacement with crystalloid fluids and, if necessary, catecholamine therapy and transfusion (restrictively from Hemoglobin < 7 g/dL) play a major role in the management of these patients [26,44]. Administration of an initial bolus of proton-pump inhibitor (PPI) seems reasonable, as up to 50% of upper gastrointestinal bleedings in patients with liver cirrhosis are not varicose. However, if purely variceal bleeding is confirmed there is no indication for long-term PPI administration. Because the size of postligation ulcers is reduced by PPI therapy, short-term administration should be considered [26]. In suspected variceal bleeding, intravenous therapy with vasoconstrictors of the splanchnic area (terlipressin, somatostatin, and octreotide) should be initiated immediately and even before endoscopy. In confirmed variceal bleeding this therapy should be continued for a total of five days to prevent recurrent bleeding. Terlipressin should be administered initially as a bolus and ideally continuously during the course of therapy [26,45]. Furthermore, immediate antibiotic therapy (e.g., third generation cephalosporines), usually for seven days, should be administered, as it improves bleeding control and survival and is associated with a reduced rate of recurrent bleeding [46]. Interestingly, bacterial infections are present in about 50% of patients with acute esophageal variceal bleeding, and are often precipitants of the bleeding [46]. Endoscopic diagnosis and therapy should be performed within the first 12 h after hemodynamic stabilization. However, up to 15% of patients develop recurrent bleeding [26]. Regarding indication for protective endotracheal intubation, there are no safe intubation criteria except for coma and suspected airway obstruction. The European Association of Gastrointestinal Endoscopy (ESGE) suggests endotracheal intubation before endoscopy in patients with active hematemesis, encephalopathy, or agitation [47].

In high-risk patients (Child–Pugh B and active bleeding on screening endoscopy or Child–Pugh C < 14 points), early TIPS (within the first 24–72 h) can be considered primarily, while in the case of recurrent bleeding it should be evaluated secondarily [18]. A recent multicenter international observational study identified ACLF at admission as an independent predictor of rebleeding and mortality in patients with acute variceal bleeding. Furthermore, preemptive TIPS placement was associated with improved survival (42-day and 1-year survival) in patients with ACLF and acute variceal bleeding, which indicates the important role of portal hypertension in these patients [48].

4.4. Organ Liver Support in ACLF

ACLF is associated with intra- and extrahepatic organ failures, and the supportive treatment of these respective organ failures plays an important part in the management of these patients. A frequent syndrome that occurs in decompensated cirrhosis is HRS-AKI. It is defined by the International Ascites Club as AKI in patients with cirrhosis, acute liver failure, or acute-on-chronic liver failure which does not show full or partial response after at least two days of diuretic withdrawal and volume expansion with albumin (1 g/kg of body weight per day to a maximum of 100 g/day) in the absence of shock, treatment with nephrotoxic drugs, and absence of parenchymal kidney disease [49]. Vasoconstrictors (e.g., terlipressin) and albumin are considered first line therapy for patients with HRS-AKI [26].

Renal replacement alone is useful in the short term for acute renal failure requiring dialysis; however, it has not been shown to be suitable as a medium- or long-term option in HRS [50]. In the context of liver failure and ACLF, both toxic hydrophilic substances which could be removed from the circulation by conventional dialysis and non-hydrophilic substances accumulate in the body [51]. Thus, ECLS systems have been developed that can eliminate albumin-bound substances. In the past, two large randomized controlled trials (HELIOS and RELIEF) which evaluated the influence of two different ECLS systems (MARS® and Prometheus®) in ACLF patients were unable to show a significant benefit in overall survival in the overall cohort [45,46]. However, at least in the HELIOS trial, the subgroup of patients with a MELD score > 30 showed improved survival [52]. A recent meta-analysis of 25 randomized controlled trials of ELCS systems showed at least a moderate certainty regarding reduction of mortality (RR 0.84, 95%CI 0.74–0.96) [53]. Furthermore, another meta-analysis of individual patient data investigated the use of MARS® and treatment intensity in ACLF, finding survival benefits in the group of patients who received high-intensity therapy (more than four MARS® sessions), especially in the first ten days [54]. Another recent meta-analysis which included 16 randomized controlled trials concluded that of all support systems for ACLF patients, plasma exchange might be the best current treatment option [55]. Currently, the APACHE trial, a large randomized controlled trial, is being performed to provide more evidence on plasma exchange. A general recommendation for the use of any ECLS in ACLF patients cannot be provided due to the lack of sufficient evidence. However, in selected cases or in the context of clinical trials ECLS systems should be considered as a therapeutic option, especially as a bridging strategy to improve short-term survival in patients who are potentially eligible for LT [25]. The results of current and future trials will identify patients who could benefit from ECLS.

4.5. LT in AD and ACLF

Although prevention and timely adequate treatment of precipitants of ACLF is essential even when supportive treatment of the respective intra- and extrahepatic organ failures is performed, the only curative and potentially life-saving therapeutic option for patients with AD and ACLF remains LT.

Recent data have generated a consensus that patients with ACLF, especially grades 1 and 2, should be listed for LT and benefit from timely evaluation for LT. This concept is supported by the fact that even ACLF patients who recover from their ACLF episode are at high risk of future decompensation and ensuing ACLF development, and suffer from high mortality [48]. Recently, it has been shown by Sundaram et al. that ACLF patients, especially ACLF grade 3 patients, have high waitlist mortality. At the same time, 1-year survival rates in ACLF patients who received LT were not significantly different from patients without ACLF in this study [56]. However, there are uncertainties regarding timing and selection of patients for LT, especially in case of ACLF grade 3 patients. For example, in patients with ACLF grade 3 and PVT, mortality was significantly higher than in ACLF grade 3 patients without PVT, indicating that LT should be approached cautiously in this subgroup of patients [56]. Interestingly, a recent publication by Zhang et al. demonstrated that in ACLF grade 3 patients earlier transplantation improves survival even if the organ is suboptimal. In particular, in the first week the use of borderline organs is significantly more advantageous than waiting for the patient to achieve a lower ACLF grade or to regain organ function. This effect was particularly noticeable in patients older than 60 years or with 4–6 organ failures [57].

The PREDICT study was able to demonstrate that UDC and pre-ACLF patients have higher short-term mortality than patients with SDC. These patients are at high risk of progression to ACLF and development of organ failure, which is associated with potentially rapid clinical deterioration and even higher short-term mortality [6].

The current scoring tools used for liver transplant allocation, such as MELD [58], MELD-sodium (MELD-Na) [59], and Child–Pugh score [60], may not adequately reflect the high short-term mortality of ACLF patients. These models have been widely discussed in the recent past, and it is known that different clinical conditions (e.g., frailty, sarcopenia, recurrent HE) are not adequately reflected by the underlying scores. Regarding ACLF, the scores have no surrogate for SI, i.e., CRP, ferritin, or white blood cell count (which is the main pathophysiological driver of ACLF progression and has been found to be strongly correlated with mortality rate in these patients) [7,61,62]. Additionally, neither the MELD score nor the MELD-Na score includes markers for portal hypertension, which is the main driver in UDC and is a relevant factor in the development of complications such as acute variceal bleeding or development of ascites and SBP in these patients. Limitations of the current risk stratification allocation policies are further underlined by a recent study of Sundaram et al. which demonstrated that patients with ACLF grade 3 and MELD-Na < 25 have higher mortality than patients with MELD-Na > 35 without ACLF [56]. Hernaez et al. were able to demonstrate that the MELD-Na score distinctly underestimates the 90-day mortality of ACLF patients [63].

In an awareness of these limitations that are especially relevant for patient with ACLF, the CLIF-C-ACLF score was specifically more than adequate in rating the mortality risk in ACLF patients. This score consists of the number of organ failures, which are reflected by the CLIF-OF score, age and the white blood cell count, which is used as the parameter indicating the severity of SI [64]. Each of these parameters has been investigated as a predictor of mortality in ACLF patients. The predictive accuracy of the CLIF-C ACLF score, especially in terms of short-term mortality in ACLF patients, has been affirmed in the recent past, and has been shown to be superior to other prognostic models in ACLF patients [64,65]. Furthermore, a CLIF-C ACLF score of 64 points or above is regarded as the threshold of futility of care, and is used as a tool to identify patients for whom supportive care has to be critically discussed if LT is not a valid option [66]. A recent publication by Schulz et al. found that pulmonary impairment independently determined mortality in critically ill patients with ACLF. They proposed the CLIF-C ACLF-R score, which is a modified CLIF-C ACLF score that uses a calibration variable to adjust for the presence or absence of mechanical ventilation or pulmonary failure. After further external validation, this simple modification could be used in clinical practice to improve the stratification of these patients [67]. Another recent publication by Weiss et al. investigated the role of metabolites reflecting SI, mitochondrial dysfunction, and sympathetic system activation in predicting short term mortality in patients with ACLF, and invented the CLIF-C MET score. However, cost-effectiveness analysis and prospective validation of these markers and scores remains necessary [68].

Overall, these findings advocate for the need to discuss the necessity of improving and modifying the current allocation systems to better reflect waitlist mortality, and especially to adequately reflect the prognosis of ACLF patients.

5. Outlook for Possible Future Therapeutic Options for ACLF Management of AD and ACLF

To date, no drug has been approved as a specific treatment for ACLF. However, different therapeutic options are evaluated here.

Granulocyte-colony stimulating factor (G-CSF) was considered a novel therapy for patients with ACLF, and showed promising results in small single center studies. However, a multicenter randomized phase-II trial by Engelmann et al. revealed that it improved neither patient survival rates nor organ function and that it failed to reduce the rate of complications [69]. Yet, based on promising results in experimental mouse models (inhibition of inflammation and promoting hepatocyte regeneration) [70], the combination of G-CSF and TAK-242 (an inhibitor of Toll-like receptor-4) in ACLF patients is scheduled to be investigated in a randomized trial (EU-funded A-TANGO project).

Omega-3 fatty acids as a treatment option for ACLF were investigated in a small open-label randomized controlled trial of 90 patients. The patients were randomized into three groups (1, regular diet; 2, regular diet plus 50 mL/d of intralipid 20% with omega-6 fatty acids for five days; and 3, regular diet plus 100 mL/d of 10% omega-3 fatty acids for five days). The study concluded that omega-3 infusion is safe and effective in reducing SI in ACLF. Even 28-day LT-free survival was significantly higher in the group of patients who received omega-3 [71]. However, the evidence is limited as yet, and more studies, ideally multicenter randomized trials, are needed in order to evaluate whether a general recommendation for patients with ACLF can be provided.

A recent study by Moreau et al. was able to demonstrate that in ACLF patients SI is associated with blood metabolite accumulation as well as profound alterations in major metabolic pathways. The inhibition of mitochondrial energy production might especially contribute to the development of intra- and extrahepatic organ failure [72]. Approaches with liposome-supported peritoneal dialysis, especially for the extraction of ammonia and other potentially harmful metabolites, have been the topic of research in the past [73].

A poster presentation by Uschner et al. at AASLD in 2021 demonstrated the results of the phase-I-b clinical trial on VS-01, which is a novel intraperitoneal pH-gradient liposomal infusion drug that has been shown to enhance clearance of ammonia and other potentially harmful metabolites. It showed the safety and tolerability of the intraperitoneal application of VS-01. Furthermore, in this small group of patients it demonstrated promising clinical efficacy, with the results supporting future development of VS-01 for patients with ACLF and organ failures [74]. A trial on the application of VS-01 in ACLF in a larger cohort to evaluate its clinical efficacy is expected in the near future.

Furthermore, the intravenous application of human allogeneic liver-derived progenitor cells (HALPC) (HepaStem®; Promethera Biosciences, Mont-Saint-Guibert, Belgium) is currently being investigated for patients with AD and ACLF. Nevens et al. investigated the use of HALPC in 24 patients (nine AD, 15 ACLF), and were able to demonstrate a reduction of markers of SI and altered liver function in the surviving patients. The 28-day and 3-month survival rates were 83% and 71%, respectively, and no patient had ACLF at month three [75]. Currently, the DHELIVER study, an interventional double blind randomized and placebo-controlled phase-II-b study, is being performed to investigate the use of HALPC in patients recently diagnosed with ACLF grade 1 or 2 in a larger cohort (NCT04229901).

The results of these and further studies will show whether the newly developed substances named above can offer feasible treatment options for patients with ACLF.

Author Contributions: M.K. and J.T.: Conceptualization, literature research, writing—original draft preparation, writing—review and editing. All authors have read and agreed to the published version of the manuscript.

Funding: J.T. was supported by the German Research Foundation (DFG) project ID 403224013—SFB 1382 (A09), by the German Federal Ministry of Education and Research (BMBF) for the DEEP-HCC project and by the Hessian Ministry of Higher Education, Research and the Arts (HMWK) for the ENABLE and ACLF-I cluster projects. The MICROB-PREDICT (project ID 825694), DECISION (project ID 847949), GALAXY (project ID 668031), LIVERHOPE (project ID 731875), and IHMCSA (project ID 964590) projects have received funding from the European Union's Horizon 2020 research and innovation program. The manuscript reflects only the authors' views, and the European Commission is not responsible for any use that may be made of the information it contains. The funders had no role in the design of the study, in the collection, analyses, or interpretation of data, in the writing of the manuscript or in the decision to publish the results.

Institutional Review Board Statement: Not applicable.

Informed Consent Statement: Not applicable.

Data Availability Statement: Not applicable.

Conflicts of Interest: J.T. has received speaking and/or consulting fees from Versantis, Gore, Boehringer-Ingelheim, Falk, Grifols, Genfit and CSL Behring. The funders had no role in the design of the study, in the collection, analyses, or interpretation of data, in the writing of the manuscript or in the decision to publish the results.

References

1. Gu, W.; Hortlik, H.; Erasmus, H.-P.; Schaaf, L.; Zeleke, Y.; Uschner, F.E.; Ferstl, P.; Schulz, M.; Peiffer, K.-H.; Queck, A.; et al. Trends and the Course of Liver Cirrhosis and Its Complications in Germany: Nationwide Population-Based Study (2005 to 2018). *Lancet Reg. Health Eur.* **2022**, *12*, 100240. [CrossRef] [PubMed]
2. Jepsen, P.; Younossi, Z.M. The Global Burden of Cirrhosis: A Review of Disability-Adjusted Life-Years Lost and Unmet Needs. *J. Hepatol.* **2021**, *75*, S3–S13. [CrossRef]
3. Ginès, P.; Krag, A.; Abraldes, J.G.; Solà, E.; Fabrellas, N.; Kamath, P.S. Liver Cirrhosis. *Lancet* **2021**, *398*, 1359–1376. [CrossRef] [PubMed]
4. D'Amico, G.; Garcia-Tsao, G.; Pagliaro, L. Natural History and Prognostic Indicators of Survival in Cirrhosis: A Systematic Review of 118 Studies. *J. Hepatol.* **2006**, *44*, 217–231. [CrossRef]
5. Arroyo, V.; Moreau, R.; Jalan, R. Acute-on-Chronic Liver Failure. *N. Engl. J. Med.* **2020**, *382*, 2137–2145. [CrossRef] [PubMed]
6. Trebicka, J.; Fernandez, J.; Papp, M.; Caraceni, P.; Laleman, W.; Gambino, C.; Giovo, I.; Uschner, F.E.; Jimenez, C.; Mookerjee, R.; et al. The PREDICT Study Uncovers Three Clinical Courses of Acutely Decompensated Cirrhosis That Have Distinct Pathophysiology. *J. Hepatol.* **2020**, *73*, 842–854. [CrossRef]
7. Clària, J.; Stauber, R.E.; Coenraad, M.J.; Moreau, R.; Jalan, R.; Pavesi, M.; Amorós, À.; Titos, E.; Alcaraz-Quiles, J.; Oettl, K.; et al. Systemic Inflammation in Decompensated Cirrhosis: Characterization and Role in Acute-on-Chronic Liver Failure. *Hepatology* **2016**, *64*, 1249–1264. [CrossRef]
8. Arroyo, V.; Angeli, P.; Moreau, R.; Jalan, R.; Clària, J.; Trebicka, J.; Fernández, J.; Gustot, T.; Caraceni, P.; Bernardi, M.; et al. The Systemic Inflammation Hypothesis: Towards a New Paradigm of Acute Decompensation and Multiorgan Failure in Cirrhosis. *J. Hepatol.* **2021**, *74*, 670–685. [CrossRef]
9. Moreau, R.; Jalan, R.; Gines, P.; Pavesi, M.; Angeli, P.; Cordoba, J.; Durand, F.; Gustot, T.; Saliba, F.; Domenicali, M.; et al. Acute-on-Chronic Liver Failure Is a Distinct Syndrome That Develops in Patients with Acute Decompensation of Cirrhosis. *Gastroenterology* **2013**, *144*, 1426–1437.e9. [CrossRef]
10. Mezzano, G.; Juanola, A.; Cardenas, A.; Mezey, E.; Hamilton, J.P.; Pose, E.; Graupera, I.; Ginès, P.; Solà, E.; Hernaez, R. Global Burden of Disease: Acute-on-Chronic Liver Failure, a Systematic Review and Meta-Analysis. *Gut* **2022**, *71*, 148–155. [CrossRef]

11. Trebicka, J.; Fernandez, J.; Papp, M.; Caraceni, P.; Laleman, W.; Gambino, C.; Giovo, I.; Uschner, F.E.; Jansen, C.; Jimenez, C.; et al. PREDICT Identifies Precipitating Events Associated with the Clinical Course of Acutely Decompensated Cirrhosis. *J. Hepatol.* **2021**, *74*, 1097–1108. [CrossRef]
12. Ferstl, P.; Trebicka, J. Acute Decompensation and Acute-on-Chronic Liver Failure. *Clin. Liver Dis.* **2021**, *25*, 419–430. [CrossRef]
13. Jalan, R.; Williams, R. Acute-on-Chronic Liver Failure: Pathophysiological Basis of Therapeutic Options. *Blood Purif.* **2002**, *20*, 252–261. [CrossRef]
14. Jalan, R.; Saliba, F.; Pavesi, M.; Amoros, A.; Moreau, R.; Ginès, P.; Levesque, E.; Durand, F.; Angeli, P.; Caraceni, P.; et al. Development and Validation of a Prognostic Score to Predict Mortality in Patients with Acute-on-Chronic Liver Failure. *J. Hepatol.* **2014**, *61*, 1038–1047. [CrossRef]
15. Bajaj, J.S.; O'Leary, J.G.; Reddy, K.R.; Wong, F.; Biggins, S.W.; Patton, H.; Fallon, M.B.; Garcia-Tsao, G.; Maliakkal, B.; Malik, R.; et al. Survival in Infection-Related Acute-on-Chronic Liver Failure Is Defined by Extrahepatic Organ Failures. *Hepatology* **2014**, *60*, 250–256. [CrossRef] [PubMed]
16. Sarin, S.K.; Kedarisetty, C.K.; Abbas, Z.; Amarapurkar, D.; Bihari, C.; Chan, A.C.; Chawla, Y.K.; Dokmeci, A.K.; Garg, H.; Ghazinyan, H.; et al. Acute-on-Chronic Liver Failure: Consensus Recommendations of the Asian Pacific Association for the Study of the Liver (APASL) 2014. *Hepatol. Int.* **2014**, *8*, 453–471. [CrossRef] [PubMed]
17. Abraldes, J.G.; Bureau, C.; Stefanescu, H.; Augustin, S.; Ney, M.; Blasco, H.; Procopet, B.; Bosch, J.; Genesca, J.; Berzigotti, A.; et al. Noninvasive Tools and Risk of Clinically Significant Portal Hypertension and Varices in Compensated Cirrhosis: The "Anticipate" Study. *Hepatology* **2016**, *64*, 2173–2184. [CrossRef]
18. de Franchis, R.; Bosch, J.; Garcia-Tsao, G.; Reiberger, T.; Ripoll, C. Baveno VII Faculty Baveno VII—Renewing Consensus in Portal Hypertension. *J. Hepatol.* **2022**, *76*, 959–974. [CrossRef] [PubMed]
19. Trebicka, J.; Gu, W.; de Ledinghen, V.; Aubé, C.; Krag, A.; Praktiknjo, M.; Castera, L.; Dumortier, J.; Bauer, D.J.M.; Friedrich-Rust, M.; et al. Two-Dimensional Shear Wave Elastography Predicts Survival in Advanced Chronic Liver Disease. *Gut* **2022**, *71*, 402–414. [CrossRef] [PubMed]
20. Villanueva, C.; Albillos, A.; Genescà, J.; Garcia-Pagan, J.C.; Calleja, J.L.; Aracil, C.; Bañares, R.; Morillas, R.M.; Poca, M.; Peñas, B.; et al. β Blockers to Prevent Decompensation of Cirrhosis in Patients with Clinically Significant Portal Hypertension (PREDESCI): A Randomised, Double-Blind, Placebo-Controlled, Multicentre Trial. *Lancet Lond. Engl.* **2019**, *393*, 1597–1608. [CrossRef]
21. Association of Aspirin with Hepatocellular Carcinoma and Liver-Related Mortality | NEJM. Available online: https://www.nejm.org/doi/full/10.1056/NEJMoa1912035 (accessed on 30 April 2023).
22. Protopapas, A.A.; Savopoulos, C.; Skoura, L.; Goulis, I. Anticoagulation in Patients with Liver Cirrhosis: Friend or Foe? *Dig. Dis. Sci.* **2023**, *68*, 2237–2246. [CrossRef] [PubMed]
23. Pose, E.; Trebicka, J.; Mookerjee, R.P.; Angeli, P.; Ginès, P. Statins: Old Drugs as New Therapy for Liver Diseases? *J. Hepatol.* **2019**, *70*, 194–202. [CrossRef]
24. Mahmud, N.; Chapin, S.; Goldberg, D.S.; Reddy, K.R.; Taddei, T.H.; Kaplan, D.E. Statin Exposure Is Associated with Reduced Development of Acute-on-Chronic Liver Failure in a Veterans Affairs Cohort. *J. Hepatol.* **2022**, *76*, 1100–1108. [CrossRef] [PubMed]
25. Pose, E.; Napoleone, L.; Amin, A.; Campion, D.; Jimenez, C.; Piano, S.; Roux, O.; Uschner, F.E.; de Wit, K.; Zaccherini, G.; et al. Safety of Two Different Doses of Simvastatin plus Rifaximin in Decompensated Cirrhosis (LIVERHOPE-SAFETY): A Randomised, Double-Blind, Placebo-Controlled, Phase 2 Trial. *Lancet Gastroenterol. Hepatol.* **2020**, *5*, 31–41. [CrossRef]
26. Angeli, P.; Bernardi, M.; Villanueva, C.; Francoz, C.; Mookerjee, R.P.; Trebicka, J.; Krag, A.; Laleman, W.; Gines, P. EASL Clinical Practice Guidelines for the Management of Patients with Decompensated Cirrhosis. *J. Hepatol.* **2018**, *69*, 406–460. [CrossRef]
27. Sort, P.; Navasa, M.; Arroyo, V.; Aldeguer, X.; Planas, R.; Ruiz-del-Arbol, L.; Castells, L.; Vargas, V.; Soriano, G.; Guevara, M.; et al. Effect of Intravenous Albumin on Renal Impairment and Mortality in Patients with Cirrhosis and Spontaneous Bacterial Peritonitis. *N. Engl. J. Med.* **1999**, *341*, 403–409. [CrossRef]
28. Arora, V.; Vijayaraghavan, R.; Maiwall, R.; Sahney, A.; Thomas, S.S.; Ali, R.; Jain, P.; Kumar, G.; Sarin, S.K. Paracentesis-Induced Circulatory Dysfunction With Modest-Volume Paracentesis Is Partly Ameliorated by Albumin Infusion in Acute-on-Chronic Liver Failure. *Hepatology* **2020**, *72*, 1043–1055. [CrossRef] [PubMed]
29. China, L.; Freemantle, N.; Forrest, E.; Kallis, Y.; Ryder, S.D.; Wright, G.; Portal, A.J.; Becares Salles, N.; Gilroy, D.W.; O'Brien, A.; et al. A Randomized Trial of Albumin Infusions in Hospitalized Patients with Cirrhosis. *N. Engl. J. Med.* **2021**, *384*, 808–817. [CrossRef]
30. Nanchal, R.; Subramanian, R.; Karvellas, C.J.; Hollenberg, S.M.; Peppard, W.J.; Singbartl, K.; Truwit, J.; Al-Khafaji, A.H.; Killian, A.J.; Alquraini, M.; et al. Guidelines for the Management of Adult Acute and Acute-on-Chronic Liver Failure in the ICU: Cardiovascular, Endocrine, Hematologic, Pulmonary, and Renal Considerations. *Crit. Care Med.* **2020**, *48*, e173. [CrossRef]
31. Albillos, A.; Lario, M.; Álvarez-Mon, M. Cirrhosis-Associated Immune Dysfunction: Distinctive Features and Clinical Relevance. *J. Hepatol.* **2014**, *61*, 1385–1396. [CrossRef]
32. Fernández, J.; Piano, S.; Bartoletti, M.; Wey, E.Q. Management of Bacterial and Fungal Infections in Cirrhosis: The MDRO Challenge. *J. Hepatol.* **2021**, *75* (Suppl. S1), S101–S117. [CrossRef] [PubMed]
33. Fernández, J.; Acevedo, J.; Wiest, R.; Gustot, T.; Amoros, A.; Deulofeu, C.; Reverter, E.; Martínez, J.; Saliba, F.; Jalan, R.; et al. Bacterial and Fungal Infections in Acute-on-Chronic Liver Failure: Prevalence, Characteristics and Impact on Prognosis. *Gut* **2018**, *67*, 1870–1880. [CrossRef] [PubMed]

34. Fernández, J.; Navasa, M.; Gómez, J.; Colmenero, J.; Vila, J.; Arroyo, V.; Rodés, J. Bacterial Infections in Cirrhosis: Epidemiological Changes with Invasive Procedures and Norfloxacin Prophylaxis. *Hepatology* **2002**, *35*, 140–148. [CrossRef]
35. Piano, S.; Singh, V.; Caraceni, P.; Maiwall, R.; Alessandria, C.; Fernandez, J.; Soares, E.C.; Kim, D.J.; Kim, S.E.; Marino, M.; et al. Epidemiology and Effects of Bacterial Infections in Patients With Cirrhosis Worldwide. *Gastroenterology* **2019**, *156*, 1368–1380.e10. [CrossRef] [PubMed]
36. Singer, M.; Deutschman, C.S.; Seymour, C.W.; Shankar-Hari, M.; Annane, D.; Bauer, M.; Bellomo, R.; Bernard, G.R.; Chiche, J.-D.; Coopersmith, C.M.; et al. The Third International Consensus Definitions for Sepsis and Septic Shock (Sepsis-3). *JAMA* **2016**, *315*, 801–810. [CrossRef] [PubMed]
37. Thursz, M.; Gual, A.; Lackner, C.; Mathurin, P.; Moreno, C.; Spahr, L.; Sterneck, M.; Cortez-Pinto, H. EASL Clinical Practice Guidelines: Management of Alcohol-Related Liver Disease. *J. Hepatol.* **2018**, *69*, 154–181. [CrossRef] [PubMed]
38. Forrest, E.H.; Evans, C.D.J.; Stewart, S.; Phillips, M.; Oo, Y.H.; McAvoy, N.C.; Fisher, N.C.; Singhal, S.; Brind, A.; Haydon, G.; et al. Analysis of Factors Predictive of Mortality in Alcoholic Hepatitis and Derivation and Validation of the Glasgow Alcoholic Hepatitis Score. *Gut* **2005**, *54*, 1174–1179. [CrossRef]
39. Louvet, A.; Naveau, S.; Abdelnour, M.; Ramond, M.-J.; Diaz, E.; Fartoux, L.; Dharancy, S.; Texier, F.; Hollebecque, A.; Serfaty, L.; et al. The Lille Model: A New Tool for Therapeutic Strategy in Patients with Severe Alcoholic Hepatitis Treated with Steroids. *Hepatology* **2007**, *45*, 1348–1354. [CrossRef]
40. Sersté, T.; Cornillie, A.; Njimi, H.; Pavesi, M.; Arroyo, V.; Putignano, A.; Weichselbaum, L.; Deltenre, P.; Degré, D.; Trépo, E.; et al. The Prognostic Value of Acute-on-Chronic Liver Failure during the Course of Severe Alcoholic Hepatitis. *J. Hepatol.* **2018**, *69*, 318–324. [CrossRef]
41. Mathurin, P.; Moreno, C.; Samuel, D.; Dumortier, J.; Salleron, J.; Durand, F.; Castel, H.; Duhamel, A.; Pageaux, G.-P.; Leroy, V.; et al. Early Liver Transplantation for Severe Alcoholic Hepatitis. *N. Engl. J. Med.* **2011**, *365*, 1790–1800. [CrossRef]
42. Louvet, A.; Labreuche, J.; Moreno, C.; Vanlemmens, C.; Moirand, R.; Féray, C.; Dumortier, J.; Pageaux, G.-P.; Bureau, C.; Chermak, F.; et al. Early Liver Transplantation for Severe Alcohol-Related Hepatitis Not Responding to Medical Treatment: A Prospective Controlled Study. *Lancet Gastroenterol. Hepatol.* **2022**, *7*, 416–425. [CrossRef] [PubMed]
43. Carbonell, N.; Pauwels, A.; Serfaty, L.; Fourdan, O.; Lévy, V.G.; Poupon, R. Improved Survival after Variceal Bleeding in Patients with Cirrhosis over the Past Two Decades. *Hepatology* **2004**, *40*, 652–659. [CrossRef] [PubMed]
44. Villanueva, C.; Colomo, A.; Bosch, A.; Concepción, M.; Hernandez-Gea, V.; Aracil, C.; Graupera, I.; Poca, M.; Alvarez-Urturi, C.; Gordillo, J.; et al. Transfusion Strategies for Acute Upper Gastrointestinal Bleeding. *N. Engl. J. Med.* **2013**, *368*, 11–21. [CrossRef]
45. Seo, Y.S.; Park, S.Y.; Kim, M.Y.; Kim, J.H.; Park, J.Y.; Yim, H.J.; Jang, B.K.; Kim, H.S.; Hahn, T.; Kim, B.I.; et al. Lack of Difference among Terlipressin, Somatostatin, and Octreotide in the Control of Acute Gastroesophageal Variceal Hemorrhage. *Hepatology* **2014**, *60*, 954–963. [CrossRef]
46. Bernard, B.; Grangé, J.D.; Khac, E.N.; Amiot, X.; Opolon, P.; Poynard, T. Antibiotic Prophylaxis for the Prevention of Bacterial Infections in Cirrhotic Patients with Gastrointestinal Bleeding: A Meta-Analysis. *Hepatology* **1999**, *29*, 1655–1661. [CrossRef]
47. Karstensen, J.G.; Ebigbo, A.; Aabakken, L.; Dinis-Ribeiro, M.; Gralnek, I.; Le Moine, O.; Vilmann, P.; Ijoma, U.; Anigbo, G.; Afihene, M.; et al. Nonvariceal Upper Gastrointestinal Hemorrhage: European Society of Gastrointestinal Endoscopy (ESGE) Cascade Guideline. *Endosc. Int. Open* **2018**, *6*, E1256–E1263. [CrossRef] [PubMed]
48. Trebicka, J.; Gu, W.; Ibáñez-Samaniego, L.; Hernández-Gea, V.; Pitarch, C.; Garcia, E.; Procopet, B.; Giráldez, Á.; Amitrano, L.; Villanueva, C.; et al. Rebleeding and Mortality Risk Are Increased by ACLF but Reduced by Pre-Emptive TIPS. *J. Hepatol.* **2020**, *73*, 1082–1091. [CrossRef]
49. Angeli, P.; Garcia-Tsao, G.; Nadim, M.K.; Parikh, C.R. News in Pathophysiology, Definition and Classification of Hepatorenal Syndrome: A Step beyond the International Club of Ascites (ICA) Consensus Document. *J. Hepatol.* **2019**, *71*, 811–822. [CrossRef]
50. Witzke, O.; Baumann, M.; Patschan, D.; Patschan, S.; Mitchell, A.; Treichel, U.; Gerken, G.; Philipp, T.; Kribben, A. Which Patients Benefit from Hemodialysis Therapy in Hepatorenal Syndrome? *J. Gastroenterol. Hepatol.* **2004**, *19*, 1369–1373. [CrossRef]
51. Durand, F.; Roux, O.; Weiss, E.; Francoz, C. Acute-on-Chronic Liver Failure: Where Do We Stand? *Liver Int.* **2021**, *41*, 128–136. [CrossRef]
52. Kribben, A.; Gerken, G.; Haag, S.; Herget-Rosenthal, S.; Treichel, U.; Betz, C.; Sarrazin, C.; Hoste, E.; Vlierberghe, H.V.; Escorsell, À.; et al. Effects of Fractionated Plasma Separation and Adsorption on Survival in Patients With Acute-on-Chronic Liver Failure. *Gastroenterology* **2012**, *142*, 782–789.e3. [CrossRef] [PubMed]
53. Alshamsi, F.; Alshammari, K.; Belley-Cote, E.; Dionne, J.; Albrahim, T.; Albudoor, B.; Ismail, M.; Al-judaibi, B.; Baw, B.; Subramanian, R.M.; et al. Extracorporeal Liver Support in Patients with Liver Failure: A Systematic Review and Meta-Analysis of Randomized Trials. *Intensive Care Med.* **2020**, *46*, 1–16. [CrossRef]
54. Bañares, R.; Ibáñez-Samaniego, L.; Torner, J.M.; Pavesi, M.; Olmedo, C.; Catalina, M.V.; Albillos, A.; Larsen, F.S.; Nevens, F.; Hassanein, T.; et al. Meta-Analysis of Individual Patient Data of Albumin Dialysis in Acute-on-Chronic Liver Failure: Focus on Treatment Intensity. *Ther. Adv. Gastroenterol.* **2019**, *12*, 1756284819879565. [CrossRef]
55. Ocskay, K.; Kanjo, A.; Gede, N.; Szakács, Z.; Pár, G.; Erőss, B.; Stange, J.; Mitzner, S.; Hegyi, P.; Molnár, Z. Uncertainty in the Impact of Liver Support Systems in Acute-on-Chronic Liver Failure: A Systematic Review and Network Meta-Analysis. *Ann. Intensive Care* **2021**, *11*, 10. [CrossRef] [PubMed]

56. Sundaram, V.; Jalan, R.; Wu, T.; Volk, M.L.; Asrani, S.K.; Klein, A.S.; Wong, R.J. Factors Associated with Survival of Patients With Severe Acute-On-Chronic Liver Failure Before and After Liver Transplantation. *Gastroenterology* **2019**, *156*, 1381–1391.e3. [CrossRef]
57. Zhang, S.; Suen, S.-C.; Gong, C.L.; Pham, J.; Trebicka, J.; Duvoux, C.; Klein, A.S.; Wu, T.; Jalan, R.; Sundaram, V. Early Transplantation Maximizes Survival in Severe Acute-on-Chronic Liver Failure: Results of a Markov Decision Process Model. *JHEP Rep.* **2021**, *3*, 100367. [CrossRef]
58. Kamath, P.S.; Kim, W.R. The Model for End-Stage Liver Disease (MELD). *Hepatology* **2007**, *45*, 797–805. [CrossRef] [PubMed]
59. Biggins, S.W.; Kim, W.R.; Terrault, N.A.; Saab, S.; Balan, V.; Schiano, T.; Benson, J.; Therneau, T.; Kremers, W.; Wiesner, R.; et al. Evidence-Based Incorporation of Serum Sodium Concentration Into MELD. *Gastroenterology* **2006**, *130*, 1652–1660. [CrossRef]
60. Christensen, E.; Schlichting, P.; Fauerholdt, L.; Gluud, C.; Andersen, P.K.; Juhl, E.; Poulsen, H.; Tygstrup, N.; Diseases, T.C.S.G.F.L. Prognostic Value of Child-Turcotte Criteria in Medically Treated Cirrhosis. *Hepatology* **1984**, *4*, 430–435. [CrossRef]
61. Laleman, W.; Claria, J.; Van der Merwe, S.; Moreau, R.; Trebicka, J. Systemic Inflammation and Acute-on-Chronic Liver Failure: Too Much, Not Enough. *Can. J. Gastroenterol. Hepatol.* **2018**, *2018*, e1027152. [CrossRef]
62. Trebicka, J.; Amoros, A.; Pitarch, C.; Titos, E.; Alcaraz-Quiles, J.; Schierwagen, R.; Deulofeu, C.; Fernandez-Gomez, J.; Piano, S.; Caraceni, P.; et al. Addressing Profiles of Systemic Inflammation Across the Different Clinical Phenotypes of Acutely Decompensated Cirrhosis. *Front. Immunol.* **2019**, *10*, 476. [CrossRef]
63. Hernaez, R.; Liu, Y.; Kramer, J.R.; Rana, A.; El-Serag, H.B.; Kanwal, F. Model for End-Stage Liver Disease-Sodium Underestimates 90-Day Mortality Risk in Patients with Acute-on-Chronic Liver Failure. *J. Hepatol.* **2020**, *73*, 1425–1433. [CrossRef] [PubMed]
64. Jalan, R.; Pavesi, M.; Saliba, F.; Amorós, A.; Fernandez, J.; Holland-Fischer, P.; Sawhney, R.; Mookerjee, R.; Caraceni, P.; Moreau, R.; et al. The CLIF Consortium Acute Decompensation Score (CLIF-C ADs) for Prognosis of Hospitalised Cirrhotic Patients without Acute-on-Chronic Liver Failure. *J. Hepatol.* **2015**, *62*, 831–840. [CrossRef] [PubMed]
65. Barosa, R.; Roque Ramos, L.; Patita, M.; Nunes, G.; Fonseca, J. CLIF-C ACLF Score Is a Better Mortality Predictor than MELD, MELD-Na and CTP in Patients with Acute on Chronic Liver Failure Admitted to the Ward. *Rev. Esp. Enferm. Dig.* **2017**, *109*, 399–405. [CrossRef] [PubMed]
66. Engelmann, C.; Thomsen, K.L.; Zakeri, N.; Sheikh, M.; Agarwal, B.; Jalan, R.; Mookerjee, R.P. Validation of CLIF-C ACLF Score to Define a Threshold for Futility of Intensive Care Support for Patients with Acute-on-Chronic Liver Failure. *Crit. Care* **2018**, *22*, 254. [CrossRef]
67. Schulz, M.S.; Mengers, J.; Gu, W.; Drolz, A.; Ferstl, P.G.; Amoros, A.; Uschner, F.E.; Ackermann, N.; Guttenberg, G.; Queck, A.; et al. Pulmonary Impairment Independently Determines Mortality in Critically Ill Patients with Acute-on-Chronic Liver Failure. *Liver Int.* **2023**, *43*, 180–193. [CrossRef]
68. Weiss, E.; de la Peña-Ramirez, C.; Aguilar, F.; Lozano, J.-J.; Sánchez-Garrido, C.; Sierra, P.; Martin, P.I.-B.; Diaz, J.M.; Fenaille, F.; Castelli, F.A.; et al. Sympathetic Nervous Activation, Mitochondrial Dysfunction and Outcome in Acutely Decompensated Cirrhosis: The Metabolomic Prognostic Models (CLIF-C MET). *Gut* **2023**. [CrossRef]
69. Engelmann, C.; Herber, A.; Franke, A.; Bruns, T.; Reuken, P.; Schiefke, I.; Zipprich, A.; Zeuzem, S.; Goeser, T.; Canbay, A.; et al. Granulocyte-Colony Stimulating Factor (G-CSF) to Treat Acute-on-Chronic Liver Failure: A Multicenter Randomized Trial (GRAFT Study). *J. Hepatol.* **2021**, *75*, 1346–1354. [CrossRef]
70. Engelmann, C.; Habtesion, A.; Hassan, M.; Kerbert, A.J.; Hammerich, L.; Novelli, S.; Fidaleo, M.; Philips, A.; Davies, N.; Ferreira-Gonzalez, S.; et al. Combination of G-CSF and a TLR4 Inhibitor Reduce Inflammation and Promote Regeneration in a Mouse Model of ACLF. *J. Hepatol.* **2022**, *77*, 1325–1338. [CrossRef]
71. Kulkarni, A.V.; Anand, L.; Vyas, A.K.; Premkumar, M.; Choudhury, A.K.; Trehanpati, N.; Benjamin, J.; Kumar, G.; Joshi, Y.K.; Sarin, S.K. Omega-3 Fatty Acid Lipid Emulsions Are Safe and Effective in Reducing Endotoxemia and Sepsis in Acute-on-Chronic Liver Failure: An Open-Label Randomized Controlled Trial. *J. Gastroenterol. Hepatol.* **2021**, *36*, 1953–1961. [CrossRef]
72. Moreau, R.; Clària, J.; Aguilar, F.; Fenaille, F.; Lozano, J.J.; Junot, C.; Colsch, B.; Caraceni, P.; Trebicka, J.; Pavesi, M.; et al. Blood Metabolomics Uncovers Inflammation-Associated Mitochondrial Dysfunction as a Potential Mechanism Underlying ACLF. *J. Hepatol.* **2020**, *72*, 688–701. [CrossRef] [PubMed]
73. Hart, K.; Harvey, M.; Tang, M.; Wu, Z.; Cave, G. Liposomes to Augment Dialysis in Preclinical Models: A Structured Review. *Pharmaceutics* **2021**, *13*, 395. [CrossRef] [PubMed]
74. Uschner, F.E.; Schulz, M.; Tyc, O.; Ferstl, P.; Stoffers, P.; Erasmus, H.-P.; Masseli, J.; Peiffer, K.-H.; Finkelmeier, F.; Pathil, A.; et al. Safety and Preliminary Efficacy and Pharmacokinetics of Intraperitoneal VS-01 Infusions in Patients with Decompensated Liver Cirrhosis: A First-in-Human, Open-Label, Phase 1b Clinical Trial. *Hepatology* **2021**, *74*, 139A. [CrossRef]
75. Nevens, F.; Gustot, T.; Laterre, P.-F.; Lasser, L.L.; Haralampiev, L.E.; Vargas, V.; Lyubomirova, D.; Albillos, A.; Najimi, M.; Michel, S.; et al. A Phase II Study of Human Allogeneic Liver-Derived Progenitor Cell Therapy for Acute-on-Chronic Liver Failure and Acute Decompensation. *JHEP Rep. Innov. Hepatol.* **2021**, *3*, 100291. [CrossRef] [PubMed]

Journal of *Personalized Medicine*

MDPI

Perspective

Liver-Gut-Interaction: Role of Microbiome Transplantation in the Future Treatment of Metabolic Disease

Vanessa Stadlbauer [1,2]

[1] Department of Gastroenterology and Hepatology, Medical University of Graz, 8036 Graz, Austria; vanessa.stadlbauer@medunigraz.at

[2] Center for Biomarker Research in Medicine (CBmed), 8010 Graz, Austria

Abstract: The association between shifts in gut microbiome composition and metabolic disorders is a well-recognized phenomenon. Clinical studies and experimental data suggest a causal relationship, making the gut microbiome an attractive therapeutic goal. Fecal microbiome transplantation (FMT) is a method to alter a person's microbiome composition. Although this method allowed for the establishment of proof of concept for using microbiome modulation to treat metabolic disorders, the method is not yet ready for broad application. It is a resource-intensive method that also carries some procedural risks and whose effects are not always reproducible. This review summarizes the current knowledge on FMT to treat metabolic diseases and gives an outlook on open research questions. Further research is undoubtedly required to find applications that are less resource-intensive, such as oral encapsulated formulations, and have strong and predictable results. Furthermore, a clear commitment from all stakeholders is necessary to move forward in the direction of developing live microbial agents, next-generation probiotics, and targeted dietary interventions.

Keywords: microbiome; obesity; metabolic syndrome; fecal microbiota transplantation

check for updates

Citation: Stadlbauer, V. Liver-Gut-Interaction: Role of Microbiome Transplantation in the Future Treatment of Metabolic Disease. *J. Pers. Med.* **2023**, *13*, 220. https://doi.org/10.3390/jpm13020220

Academic Editor: Guido Gerken

Received: 6 January 2023
Revised: 20 January 2023
Accepted: 25 January 2023
Published: 27 January 2023

1. Introduction

Obesity and the metabolic syndrome lead to a considerable disease burden due to complications, such as cardiovascular disorders, cancer, dementia, and fertility disorders, and are also associated with socio-economic disadvantages [1]. Obesity and its relation to shifts in the microbiome composition were among the first described associations when culture-independent techniques to study the complex ecosystem of the gut microbiome were developed about 20 years ago [2]. Starting from a description of shifts in microbiome composition at the phylum level, a large body of literature has evolved that describes the bidirectional interaction between the gut microbiome and human metabolism [3]. Human association studies clearly demonstrated an association between the gut microbiome composition and obesity. The causal relationship was first suggested by the finding that the phylum Bacteroides was reduced in obesity and that weight loss through diet led to an increase [4,5]. Data from in vitro systems and animal models suggest that diet or drugs (e.g., antibiotics) alter the microbiome, and this dysbiosis is mechanistically involved in disrupting molecular metabolism and signaling through bacterial metabolites (e.g., bile acids), which impacts energy intake and leads to metabolic disorders. For example, a high-fat and high-glucose diet lead to increased gut permeability, translocation of bacterial products, a low grade inflammatory response, and insulin resistance, indicating that the gut-liver axis is involved in the metabolic sequelae of the western diet [6,7]. Certain metabolites, such as bile acids, that are bidirectionally interacting with the microbiome have been identified as playing a mechanistic role in the development of metabolic syndrome via the gut microbiome. In obesity, altering bile acid composition via reduction of microbial diversity through an antibiotic leads to increased insulin resistance [8]. Furthermore, host genetics play a role since immune control of the microbiome maintains beneficial microbial populations that constrain lipid metabolism to prevent metabolic syndrome [9]. The strongest evidence for a causal relationship between the gut microbiome and obesity is derived from animal experiments in germ-free mice, which received fecal transplantation from human twins discordant in obesity. The mice transplanted with the microbiome of the obese twin developed obesity, while the mice who received the microbiome of the lean twin stayed lean [10]. As a result of this strong link between the gut microbiome and metabolic disorders, new therapeutic modalities targeting the altered commensal bacteria as a means of treating metabolic syndrome have gained a lot of interest.

2. Fecal Microbiota Transplantation (FMT) for Metabolic Syndrome

FMT, the transfer of fecal matter from one individual to another with the aim to improve/restore the composition of the gut microbiome and thereby treat a disease, has gained much attention in the scientific field and also in the general public. As a method that in principle dates back to the fourth century in China, its modern application took off in 2013. Until now, the only routine medical application was treatment of recurrent *Clostridioides difficile* infection. But also, many non-infectious diseases have been extensively studied, such as inflammatory bowel diseases, the microbiota-gut-liver axis, the microbiota-gut-brain axis, and oncologic and hematological diseases [11,12]. The first indication of efficacy for FMT in human metabolic diseases was published in 2012, where in a pilot study in the Netherlands, in nine male adults with metabolic syndrome who received FMT from a lean donor, insulin resistance was improved, whereas no changes were observed in nine controls who received autologous FMT [8]. A potential effect on body weight was first suspected by a "complication" of FMT for *Clostridioides difficile*, where a patient with normal weight received an FMT from an obese donor and gained weight after FMT [13]. While consecutive studies did not report such effects [14], the opposite strategy of increasing the body weight of cachectic cancer patients through FMT from obese donors failed [15]. Nevertheless, body mass index (BMI) in the normal range has been included in the selection criteria for stool donors after the suspicion that obesity could be actually "transferred" by FMT [16].

After these initial proof of concept trials, a number of clinical trials have been performed, where researchers aimed to improve metabolism and reduce weight by FMT in obesity, metabolic syndrome, non-alcoholic fatty liver disease and type 2 diabetes. However, especially in adequately powered, randomized, placebo-controlled trials, mixed results with regards to improvement in metabolic parameters were reported. While most studies demonstrate that FMT is able to change the microbiome composition, clear effects on clinically important endpoints, such as body weight or insulin resistance, are missing. Trials with clinical endpoints are summarized in Table 1. It is notable that the method of FMT may play a role in its effectiveness. Oral capsules, which would be the preferred route of administration for both safety and logistic reasons, have yet to show substantial improvements in metabolic parameters, except when administered together with low fermentable fibers [17–20]. From a mechanistic point of view, it is still not clear what the "effective agent" in FMT is. The notion that bacteria might be the "effective agent" has been challenged, since for the treatment of *Clostridioides difficile* infection, sterile filtrated fecal preparations were similar to or more effective than conventional fecal preparations containing living microorganisms [21]. This indicates that maybe not the living bacteria, but rather bacterial components, metabolites, or bacteriophages may mediate the effects of FMT. From a pilot study, it was suggested that bacteriophages in the recipient that are associated with metabolic syndrome are replaced by new viruses that were not originally present in either the healthy donors or control subjects at detectable abundances and that changes in the bacteriophage population are related to response to FMT [22]. Another study reported an influence of FMT on plasma metabolites related to lipid metabolism and DNA methylation status; however, a clear-cut pathophysiological explanation for a potential mechanism to influence glucose metabolism could not be identified [23]. A recent systematic review did not identify consistent changes in clinically relevant endpoints (such as insulin sensitivity) achieved in the recipient after FMT for metabolic diseases [24].

Table 1. Clinical studies on FMT to treat obesity and metabolic syndrome.

Study	Population	Number of Participants	FMT Mode	FMT Duration and Dose	Primary Endpoint	Result Metabolic	Result Microbiome	Adverse Events
Yu 2020 [20]	obesity and insulin resistance	24 adults	oral capsules	6 weeks, once per week versus placebo	insulin sensitivity	no change in insulin sensitivity	shift towards donor microbiome	no significant difference to placebo, no severe adverse events
Allegretti 2020 [17]	obesity	22 adults	oral capsules versus placebo	8 weeks, 2 doses	safety	no reduction in BMI	shift towards donor microbiome	no significant difference to placebo
Rinott 2021 [25,26]	obesity or dyslipidemia, randomized to healthy dietary guidelines, Mediterranean diet, and green-Mediterranean diet weight-loss groups	90 adults	autologous transplantation of microbiome collected under diet	100 capsules in 8 months versus placebo	weight regain	autologous FMT attenuated weight regain in combination with green Mediterranean diet	green Mediterranean diet caused change in microbiome composition	no treatment attributable adverse events
Mocanu 2021 [19]	obesity and metabolic syndrome	70 adults	oral capsules, + either high fermentable or low fermentable fibres versus placebo	6 weeks, 1 dose	insulin sensitivity	FMT + low fermentable fibers improved insulin sensitivity	FMT + low fermentable fibers increased diversity, shift towards donor microbiome	no treatment attributable adverse events
Leong 2020 [18,27]	obesity	87 adolescents	oral capsules	6 weeks, 1 dose versus placebo	BMI	no effect on BMI, reduction of abdominal obesity, resolution of metabolic syndrome at baseline	greater dissimilarity between baseline and post treatment in FMT group versus placebo, increase in diversity in female participants	no treatment attributable adverse events
Craven 2020 [28]	non-alcoholic fatty liver disease	21 adults	allogenic or autologous FMT in the distal duodenum via endoscopy	6 weeks, 1 dose	insulin resistance	no effect on insulin resistance, reduction of increased intestinal permeability	no changes	not reported

Table 1. *Cont.*

Study	Population	Number of Participants	FMT Mode	FMT Duration and Dose	Primary Endpoint	Result Metabolic	Result Microbiome	Adverse Events
Ng 2022 [29]	type 2 diabetes	61 adult	allogenic FMT via nasogastric tube +, lifestyle intervention	24 weeks, 3 doses	donor microbiome engraftment	FMT + lifestyle intervention reduced total and low-density lipoprotein cholesterol and liver stiffness (secondary endpoints)	FMT + lifestyle intervention significantly better engraftment (primary endpoint)	no differences between groups, several cardiovascular events
Xue 2022 [30]	non-alcoholic fatty liver disease	75 adults	allogenic FMT, 3 doses, 1 via colonoscopy 3 via enema, versus oral probiotics	4 weeks, 3 doses	clinical efficacy and safety (not further specified)	no increase in liver fat content compared to probiotic group	differing responses in lean and obese patients	not reported
Ding 2022 [31]	type 2 diabetes	17 adults	allogenic FMT, transendoscopic jejunal tube, unblinded	12 weeks, 2 doses	insulin resistance	improvement in HbA1c, glucose, uric acid, increase in C-reactive protein	difference between responders and non-responders	no adverse events
Su 2022 [32]	type 2 diabetes	16 adults	allogenic FMT + formula diet versus formula diet alone, oral capsules	12 weeks, 3 doses	health status	Both interventions improved BMI, glucose metabolism and blood pressure	reduction of diversity in both groups, less in FMT group	no adverse events

FMT: fecal microbiome transplantation, BMI: body mass index.

3. Going beyond FMT

Recent studies also explored how the selection of the donor could impact the effect of FMT. A vegan diet is associated with reduced trimethylamine-N-oxide (TMAO) production and therefore lower cardiovascular risk; however, the FMT from vegan donors did not decrease TMAO production in patients with metabolic syndrome, which indicates that despite evidence of compositional changes that resemble the microbiome of the donor, the functional capacity is not easily transferred [33]. FMT from bariatric surgery patients and obese patients with metabolic syndrome, revealed microbiome-driven modulation of brain dopamine and serotonin transporters [34]. And the use of autologous fecal transplants with fecal material obtained at the "weight nadir" of a successful diet was able to delay weight regain after the diet [25,26]. These studies indicate that the concept of FMT most likely needs to be augmented by adequate preparatory measures like diet or additional prebiotic "fertilizers" of the transplanted microbiome. Since studies also show considerable differences between different donors, it is essential to characterize donors and understand the interaction between the dysbiotic recipient microbiome and the donor microbiome [35]. The host microbiome composition determines the efficacy of engraftment of an FMT [36]. "Tuning" the host microbiome, e.g., by special dietary measures such as fiber supplementation, may improve functional engraftment of FMT [19,29].

4. Challenges in FMT for Metabolic Diseases

When transferring living microorganisms to a new host, adverse events have to be considered. It is surprising that some studies do not report safety data (see Table 1). A meta-analysis on the safety of FMT across different disease entities showed no significant differences in the incidence of adverse events between FMT and the control group. Adverse events can be related to the transplanted microbiome or to the route of administration. It seems that administration via oral capsules or endoscopically via the lower gastrointestinal tract is less prone to adverse events. Translocation and infection with transplanted bacteria can occur; the risk seems to be higher in patients with an altered intestinal barrier [37]. In the studies related to metabolic diseases, no bacteremia or sepsis events have been described so far.

From a practical point of view, however, it is unlikely that FMT, which requires highly skilled personnel and is resource intensive, will be applicable to treat the worldwide "obesity pandemic". Although no international figures exist for the workforce requirements to perform large-scale FMT treatments, a shortage in skilled personnel can be extrapolated from studies related to cancer screening colonoscopies, where a severe shortage was noticed [38].

FMT clearly helps to understand the relationship between the gut microbiome and metabolic disorders and facilitates the notion that microbiome modulation can be an effective therapeutic strategy. The low predictability and the resource intensity of the intervention warrant further research towards a better translation or transformation into clinical practice. Further efforts are necessary to improve the timely and personalized diagnosis of the individual dysbiosis in obesity and augment and retain the effect of a diet by influencing the microbiome in a personalized but also "affordable" microbiome modulation strategy. A clear commitment to research from all stakeholders (politics, funding bodies, the health industry, researchers, and society) is necessary to move forward in the direction of developing live microbial agents, next-generation probiotics, and targeted dietary interventions. Table 2 summarizes the pros and cons for FMT in metabolic disorders.

Table 2. Pro and con arguments for/against FMT to treat metabolic diseases.

PRO	CON
Altered microbiomes are likely to contribute to the pathophysiology of metabolic diseases	Our conceptual, mechanistic and ecological understanding of FMTs is poorly developed
Experiments in animal models provide a rationale for FMTs because the clinical phenotype can be transplanted and reverted by FMT	Difficult to rationalize timing and dosing regimens, and it is also unclear whether dysbiosis can be corrected
Known dysbiosis in metabolic diseases provides a rationale for "complete" microbiome restoration through FMT	Safety concerns (both short- and long-term) regarding the exposure to infectious and non-infectious transmissible diseases
FMT has the potential to revert aspects of dysbiosis, engraft health promoting microbes and/or expose the host temporarily to beneficial microbes	Risk of conveyance of unintended characteristics of the donor microbiome that might predispose to chronic diseases
There is already one standard use case for FMT (refractory C. difficile infection)	It is currently unclear which specific measures are necessary to enable FMT to work effectively
Broad acceptance from patients	Application via colonoscopy is costly and bears a small but relevant procedural risk
	Other routes of application, e.g., encapsulation, do not yet provide satisfactory efficacy

Funding: This research received no external funding.

Institutional Review Board Statement: Not applicable.

Informed Consent Statement: Not applicable.

Data Availability Statement: Not applicable.

Acknowledgments: None.

Conflicts of Interest: The author declares no conflict of interest.

References

1. Blüher, M. Obesity: Global epidemiology and pathogenesis. *Nat. Rev. Endocrinol.* **2019**, *15*, 288–298. [CrossRef] [PubMed]
2. Bäckhed, F.; Ding, H.; Wang, T.; Hooper, L.V.; Koh, G.Y.; Nagy, A.; Semenkovich, C.F.; Gordon, J.I. The Gut Microbiota as an Environmental Factor That Regulates Fat Storage. *Proc. Natl. Acad. Sci. USA* **2004**, *101*, 15718–15723. [CrossRef] [PubMed]
3. Maruvada, P.; Leone, V.; Kaplan, L.M.; Chang, E.B. The Human Microbiome and Obesity: Moving beyond Associations. *Cell Host Microbe* **2017**, *22*, 589–599. [CrossRef] [PubMed]
4. Ley, R.E.; Bäckhed, F.; Turnbaugh, P.; Lozupone, C.A.; Knight, R.D.; Gordon, J.I. Obesity alters gut microbial ecology. *Proc. Natl. Acad. Sci. USA* **2005**, *102*, 11070–11075. [CrossRef]
5. Turnbaugh, P.J.; Ley, R.E.; Mahowald, M.A.; Magrini, V.; Mardis, E.R.; Gordon, J.I. An Obesity-Associated Gut Microbiome with Increased Capacity for Energy Harvest. *Nature* **2006**, *444*, 1027–1031. [CrossRef]
6. Cani, P.D.; Amar, J.; Iglesias, M.A.; Poggi, M.; Knauf, C.; Bastelica, D.; Neyrinck, A.M.; Fava, F.; Tuohy, K.M.; Chabo, C.; et al. Metabolic endotoxemia initiates obesity and insulin resistance. *Diabetes* **2007**, *56*, 1761–1772. [CrossRef] [PubMed]
7. Thaiss, C.A.; Levy, M.; Grosheva, I.; Zheng, D.; Soffer, E.; Blacher, E.; Braverman, S.; Tengeler, A.C.; Barak, O.; Elazar, M.; et al. Hyperglycemia drives intestinal barrier dysfunction and risk for enteric infection. *Science* **2018**, *359*, 1376–1383. [CrossRef]
8. Vrieze, A.; Van Nood, E.; Holleman, F.; Salojärvi, J.; Kootte, R.S.; Bartelsman, J.F.; Dallinga-Thie, G.M.; Ackermans, M.T.; Serlie, M.J.; Oozeer, R.; et al. Transfer of Intestinal Microbiota From Lean Donors Increases Insulin Sensitivity in Individuals With Metabolic Syndrome. *Gastroenterology* **2012**, *143*, 913–916.e7. [CrossRef] [PubMed]
9. Petersen, C.; Bell, R.; Klag, K.A.; Lee, S.-H.; Soto, R.; Ghazaryan, A.; Buhrke, K.; Ekiz, H.A.; Ost, K.S.; Boudina, S.; et al. T cell–mediated regulation of the microbiota protects against obesity. *Science* **2019**, *365*, eaat9351. [CrossRef]
10. Ridaura, V.K.; Faith, J.J.; Rey, F.E.; Cheng, J.; Duncan, A.E.; Kau, A.L.; Griffin, N.W.; Lombard, V.; Henrissat, B.; Bain, J.R.; et al. Gut Microbiota from Twins Discordant for Obesity Modulate Metabolism in Mice. *Science* **2013**, *341*, 1241214. [CrossRef]
11. Wang, J.-W.; Kuo, C.-H.; Kuo, F.-C.; Wang, Y.-K.; Hsu, W.-H.; Yu, F.-J.; Hu, H.-M.; Hsu, P.-I.; Wang, J.-Y.; Wu, D.-C. Fecal microbiota transplantation: Review and update. *J. Formos. Med. Assoc.* **2019**, *118* (Suppl. S1), S23–S31. [CrossRef] [PubMed]
12. Wang, Y.; Zhang, S.; Borody, T.J.; Zhang, F. Encyclopedia of fecal microbiota transplantation: A review of effectiveness in the treatment of 85 diseases. *Chin. Med. J.* **2022**, *135*, 1927–1939. [CrossRef] [PubMed]
13. Alang, N.; Kelly, C.R. Weight Gain After Fecal Microbiota Transplantation. *Open Forum Infect. Dis.* **2015**, *2*, ofv004. [CrossRef] [PubMed]
14. Fischer, M.; Kao, D.; Kassam, Z.; Smith, J.; Louie, T.; Sipe, B.; Torbeck, M.; Xu, H.; Ouyang, F.; Mozaffarian, D.; et al. Stool Donor Body Mass Index Does Not Affect Recipient Weight After a Single Fecal Microbiota Transplantation for Clostridium difficile Infection. *Clin. Gastroenterol. Hepatol.* **2017**, *16*, 1351–1353. [CrossRef] [PubMed]
15. de Clercq, N.C.; van den Ende, T.; Prodan, A.; Hemke, R.; Davids, M.; Pedersen, H.K.; Nielsen, H.B.; Groen, A.K.; de Vos, W.M.; van Laarhoven, H.W.M.; et al. Fecal Microbiota Transplantation from Overweight or Obese Donors in Cachectic Patients with Advanced Gastroesophageal Cancer: A Randomized, Double-blind, Placebo-Controlled, Phase II Study. *Clin. Cancer Res.* **2021**, *27*, 3784–3792. [CrossRef] [PubMed]
16. Cammarota, G.; Ianiro, G.; Tilg, H.; Rajilić-Stojanović, M.; Kump, P.; Satokari, R.; Sokol, H.; Arkkila, P.; Pintus, C.; Hart, A.; et al. European consensus conference on faecal microbiota transplantation in clinical practice. *Gut* **2017**, *66*, 569–580. [CrossRef] [PubMed]
17. Allegretti, J.R.; Kassam, Z.; Mullish, B.H.; Chiang, A.; Carrellas, M.; Hurtado, J.; Marchesi, J.R.; McDonald, J.A.K.; Pechlivanis, A.; Barker, G.F.; et al. Effects of Fecal Microbiota Transplantation with Oral Capsules in Obese Patients. *Clin. Gastroenterol. Hepatol.* **2020**, *18*, 855.e2–863.e2. [CrossRef]
18. Leong, K.S.W.; Jayasinghe, T.N.; Wilson, B.C.; Derraik, J.G.B.; Albert, B.B.; Chiavaroli, V.; Svirskis, D.M.; Beck, K.L.; Conlon, C.A.; Jiang, Y.; et al. Effects of Fecal Microbiome Transfer in Adolescents With Obesity: The Gut Bugs Randomized Controlled Trial. *JAMA Netw. Open.* **2020**, *3*, e2030415. [CrossRef] [PubMed]
19. Mocanu, V.; Zhang, Z.; Deehan, E.C.; Kao, D.H.; Hotte, N.; Karmali, S.; Birch, D.W.; Samarasinghe, K.K.; Walter, J.; Madsen, K.L. Fecal microbial transplantation and fiber supplementation in patients with severe obesity and metabolic syndrome: A randomized double-blind, placebo-controlled phase 2 trial. *Nat. Med.* **2021**, *27*, 1272–1279. [CrossRef]
20. Yu, E.W.; Gao, L.; Stastka, P.; Cheney, M.C.; Mahabamunuge, J.; Soto, M.T.; Ford, C.B.; Bryant, J.A.; Henn, M.R.; Hohmann, E.L. Fecal microbiota transplantation for the improvement of metabolism in obesity: The FMT-TRIM double-blind placebo-controlled pilot trial. *PLoS Med.* **2020**, *17*, e1003051. [CrossRef]

21. Ott, S.J.; Waetzig, G.H.; Rehman, A.; Moltzau-Anderson, J.; Bharti, R.; Grasis, J.A.; Cassidy, L.; Tholey, A.; Fickenscher, H.; Seegert, D.; et al. Efficacy of Sterile Fecal Filtrate Transfer for Treating Patients with Clostridium difficile Infection. *Gastroenterology* **2017**, *152*, 799.e7–811.e7. [CrossRef]
22. Manrique, P.; Zhu, Y.; van der Oost, J.; Herrema, H.; Nieuwdorp, M.; de Vos, W.M.; Young, M. Gut bacteriophage dynamics during fecal microbial transplantation in subjects with metabolic syndrome. *Gut Microbes* **2021**, *13*, 1897217. [CrossRef] [PubMed]
23. van der Vossen, E.W.J.; Bastos, D.; Stols-Gonçalves, D.; de Goffau, M.C.; Davids, M.; Pereira, J.P.B.; Yim, A.Y.F.L.; Henneman, P.; Netea, M.G.; de Vos, W.M.; et al. Effects of fecal microbiota transplant on DNA methylation in subjects with metabolic syndrome. *Gut Microbes* **2021**, *13*, 1993513. [CrossRef]
24. Zhang, Z.; Mocanu, V.; Cai, C.; Dang, J.; Slater, L.; Deehan, E.C.; Walter, J.; Madsen, K.L. Impact of Fecal Microbiota Transplantation on Obesity and Metabolic Syndrome—A Systematic Review. *Nutrients* **2019**, *11*, 2291. [CrossRef]
25. Rinott, E.; Youngster, I.; Meir, A.Y.; Tsaban, G.; Kaplan, A.; Zelicha, H.; Rubin, E.; Koren, O.; Shai, I. Autologous fecal microbiota transplantation can retain the metabolic achievements of dietary interventions. *Eur. J. Intern. Med.* **2021**, *92*, 17–23. [CrossRef] [PubMed]
26. Rinott, E.; Youngster, I.; Meir, A.Y.; Tsaban, G.; Zelicha, H.; Kaplan, A.; Knights, D.; Tuohy, K.; Fava, F.; Scholz, M.U.; et al. Effects of Diet-Modulated Autologous Fecal Microbiota Transplantation on Weight Regain. *Gastroenterology* **2021**, *160*, 158–173.e10. [CrossRef]
27. Wilson, B.C.; Vatanen, T.; Jayasinghe, T.N.; Leong, K.S.W.; Derraik, J.G.B.; Albert, B.B.; Chiavaroli, V.; Svirskis, D.M.; Beck, K.L.; Conlon, C.A.; et al. Strain engraftment competition and functional augmentation in a multi-donor fecal microbiota transplantation trial for obesity. *Microbiome* **2021**, *9*, 107. [CrossRef]
28. Craven, L.; Rahman, A.; Nair Parvathy, S.; Beaton, M.; Silverman, J.; Qumosani, K.; Hramiak, I.; Hegele, R.; Joy, T.; Meddings, J.; et al. Allogenic Fecal Microbiota Transplantation in Patients With Nonalcoholic Fatty Liver Disease Improves Abnormal Small Intestinal Permeability: A Randomized Control Trial. *Am. J. Gastroenterol.* **2020**, *115*, 1055–1065. [CrossRef] [PubMed]
29. Ng, S.C.; Xu, Z.; Mak, J.W.Y.; Yang, K.; Liu, Q.; Zuo, T.; Tang, W.; Lau, L.; Lui, R.N.; Wong, S.H.; et al. Microbiota engraftment after faecal microbiota transplantation in obese subjects with type 2 diabetes: A 24-week, double-blind, randomised controlled trial. *Gut* **2021**, *71*, 716–723. [CrossRef]
30. Xue, L.; Deng, Z.; Luo, W.; He, X.; Chen, Y. Effect of Fecal Microbiota Transplantation on Non-Alcoholic Fatty Liver Disease: A Randomized Clinical Trial. *Front. Cell. Infect. Microbiol.* **2022**, *12*, 850. [CrossRef]
31. Ding, D.; Yong, H.; You, N.; Lu, W.; Yang, X.; Ye, X.; Wang, Y.; Cai, T.; Zheng, X.; Chen, H.; et al. Prospective Study Reveals Host Microbial Determinants of Clinical Response to Fecal Microbiota Transplant Therapy in Type 2 Diabetes Patients. *Front. Cell. Infect. Microbiol.* **2022**, *12*, 291. [CrossRef] [PubMed]
32. Su, L.; Hong, Z.; Zhou, T.; Jian, Y.; Xu, M.; Zhang, X.; Zhu, X.; Wang, J. Health improvements of type 2 diabetic patients through diet and diet plus fecal microbiota transplantation. *Sci. Rep.* **2022**, *12*, 1152. [CrossRef] [PubMed]
33. Smits, L.P.; Kootte, R.S.; Levin, E.; Prodan, A.; Fuentes, S.; Zoetendal, E.G.; Wang, Z.; Levison, B.S.; Cleophas, M.C.P.; Kemper, E.M.; et al. Effect of Vegan Fecal Microbiota Transplantation on Carnitine- and Choline-Derived Trimethylamine-N-Oxide Production and Vascular Inflammation in Patients With Metabolic Syndrome. *J. Am. Heart Assoc.* **2018**, *7*, e008342. [CrossRef] [PubMed]
34. Hartstra, A.V.; Schüppel, V.; Imangaliyev, S.; Schrantee, A.; Prodan, A.; Collard, D.; Levin, E.; Dallinga-Thie, G.; Ackermans, M.T.; Winkelmeijer, M.; et al. Infusion of donor feces affects the gut–brain axis in humans with metabolic syndrome. *Mol. Metab.* **2020**, *42*, 101076. [CrossRef] [PubMed]
35. Wilson, B.; Vatanen, T.; Cutfield, W.S.; O'Sullivan, J.M. The Super-Donor Phenomenon in Fecal Microbiota Transplantation. *Front. Cell. Infect. Microbiol.* **2019**, *9*, 2. [CrossRef]
36. Li, S.S.; Zhu, A.; Benes, V.; Costea, P.I.; Hercog, R.; Hildebrand, F.; Huerta-Cepas, J.; Nieuwdorp, M.; Salojärvi, J.; Voigt, A.Y.; et al. Durable coexistence of donor and recipient strains after fecal microbiota transplantation. *Science* **2016**, *352*, 586–589. [CrossRef]
37. Chen, C.; Chen, L.; Sun, D.; Li, C.; Xi, S.; Ding, S.; Luo, R.; Geng, Y.; Bai, Y. Adverse events of intestinal microbiota transplantation in randomized controlled trials: A systematic review and meta-analysis. *Gut Pathog.* **2022**, *14*, 20. [CrossRef]
38. Ravindran, S.; Munday, J.; Veitch, A.M.; Broughton, R.; Thomas-Gibson, S.; Penman, I.D.; McKinlay, A.; Fearnhead, N.S.; Coleman, M.; Logan, R. Bowel cancer screening workforce survey: Developing the endoscopy workforce for 2025 and beyond. *Front. Gastroenterol.* **2021**, *13*, 12–19. [CrossRef]

Journal of *Personalized Medicine*

MDPI

Review

Telemedicine as an Option for Monitoring Metabolic Dysfunction-Associated Fatty Liver Disease (MAFLD) Patients Facing the COVID-19 Pandemic: A Systematic Review and Meta-Analysis

Femmy Nurul Akbar [1,*], Safira Rosiana Choirida [1], Ahmad Zaqi Muttaqin [1], Fika Ekayanti [2], Hoirun Nisa [3] and Hari Hendarto [1]

1 Department of Internal Medicine, Faculty of Medicine, Universitas Islam Negeri Syarif Hidayatullah Jakarta, South Tangerang 15419, Banten, Indonesia; safira.rosiana16@mhs.uinjkt.ac.id (S.R.C.); ahmad.zaqi16@mhs.uinjkt.ac.id (A.Z.M.); hari.hendarto@uinjkt.ac.id (H.H.)

2 Department of Community Medicine, Faculty of Medicine, Universitas Islam Negeri Syarif Hidayatullah Jakarta, South Tangerang 15419, Banten, Indonesia; fikaekayanti@uinjkt.ac.id

3 Department of Public Health, Faculty of Health Sciences, Universitas Islam Negeri Syarif Hidayatullah Jakarta, South Tangerang 15419, Banten, Indonesia; hoirun.nisa@uinjkt.ac.id

* Correspondence: femmy.nurul@uinjkt.ac.id

Abstract: Healthcare visits were reduced during the COVID-19 pandemic, causing disturbances in sustainable MAFLD monitoring. Telemedicine acts to maintain connectivity between patients and healthcare professionals. This review aimed to assess the role of telemedicine in monitoring MAFLD during the pandemic. Databases searched included *l* PubMed Central and ScienceDirect from 2020 to 2023. Assessment with The Cochrane Risk of Bias for randomized controlled trials (RCTs) and the Newcastle-Ottawa scale for non-RCTs systematic reviews. Meta-analyses employing a random-effect model were performed to determine the pooled mean difference (MD) and *p*-value. The results showed three RCT and two non-RCT ($n = 239$) with 56.9% males and a mean age of 51.3 years. The median intervention duration was 5.5 months. The parameters assessed included body weight (BW), body mass index (BMI), waist circumference, liver function (AST/ALT), lipid profile, HbA1c, and others. Meta-analysis revealed that telemedicine had a significant effect on improving outcomes for BW (MD −2.81: 95% CI, −4.11, −1.51, $p < 0.0001$) and BMI (MD −1.01: 95% CI, −1.47, −0.55, $p < 0.0001$) compared to standard care, while the AST/ALT levels were not significantly reduced. Some biochemical markers decreased based on the systematic reviews. In conclusion, telemedicine using mobile-based applications could be an option for monitoring lifestyle modification in MAFLD patients facing the COVID-19 pandemic.

Keywords: telemedicine; metabolic dysfunction-associated fatty liver disease; monitor; lifestyle modification; COVID-19 pandemic

Citation: Akbar, F.N.; Choirida, S.R.; Muttaqin, A.Z.; Ekayanti, F.; Nisa, H.; Hendarto, H. Telemedicine as an Option for Monitoring Metabolic Dysfunction-Associated Fatty Liver Disease (MAFLD) Patients Facing the COVID-19 Pandemic: A Systematic Review and Meta-Analysis. *J. Pers. Med.* **2024**, *14*, 281. https://doi.org/10.3390/jpm14030281

Academic Editors: Konstantinos Tziomalos and Guido Gerken

Received: 26 January 2024
Revised: 21 February 2024
Accepted: 29 February 2024
Published: 2 March 2024

1. Introduction

Telemedicine utilizing mobile-based applications provides integrated health services that connect health workers and patients using virtual communication technology [1]. Before the pandemic era, telemedicine had already been implemented for counseling in the United States. However, both users and providers of telemedicine face numerous obstacles. A huge number of users utilized telemedicine inconsistently, causing a lack of sustainability in medical treatment. Providers encountered challenges, including costly investments required for privacy regulation of patient data and difficulties obtaining approvals from both local and national governments [2]. However, the outbreak of SARS-CoV-2 presented an opportunity for telemedicine to optimize communication between health workers and patients [1,2]. Government policy is necessary to regulate telemedicine so that comprehensive health services can be provided [2].

Approximately 272.5 million people worldwide have been infected with severe acute respiratory syndrome coronavirus 2 (SARS-CoV-2), resulting in approximately 5.3 million deaths due to coronavirus disease 2019 (COVID-19) [3]. This RNA virus belongs to the Coronaviridae family. It interacts with the angiotensin-converting enzyme-2 (ACE-2) binding site on host cells in the heart, lungs, kidneys, and gastrointestinal tract. The mode of transmission of SARS-CoV-2 involves airborne infectious particles and aerosols, which are transmitted from infected individuals to their close contacts [4,5].

Metabolic dysfunction-associated fatty liver disease (MAFLD), previously known as non-alcoholic fatty liver disease (NAFLD), is a chronic liver disease characterized by excessive fat accumulation in the liver, without another obvious cause such as no excessive alcohol consumption, hepatotoxic medications, toxins, viral infections, genetic hepatic diseases [6,7]. MAFLD is defined as the accumulation of hepatic steatosis, as determined using imaging and/or liver biopsy and accompanied by at least one of three conditions, namely type 2 diabetes mellitus (T2DM), obesity, and metabolic dysregulation that can happen in both developed and developing countries [6–8]. The updated clinical guidelines from the American Association for the Research of Liver Diseases (AASLD), the American College of Gastroenterology, and the American Gastroenterological Association stipulate that the diagnosis of MAFLD should be based on three criteria: (1) evidence of increased hepatic fat on imaging, (2) absence of significant alcohol consumption (>30 g/day), and (3) absence of other known causes of chronic liver disease [9,10].

Typically, individuals with MAFLD remain asymptomatic. However, this condition can potentially progress to an end-stage state due to complex liver-cell injury and the accumulation of inflammatory cells, leading to a transformation of normal histology into a more aggressive form called metabolic dysfunction-associated steatohepatitis (MASH). MASH, in turn, can gradually evolve into the formation of fibrous scars characterized by varying degrees of fibrosis and will become liver cirrhosis. This situation will elevate the risk of morbidity and mortality among MAFLD patients due to liver cancer by approximately 1–2% annually in the absence of tailored and personalized treatment for each individual patient [6,8].

The current approach to treating MAFLD starts with identifying and addressing comorbidities according to established guidelines and implementing lifestyle modifications such as dietary changes, weight loss, and increased physical activity. These lifestyle adjustments serve as the cornerstone of MAFLD treatment [11–13]. Specifically, no medication has been approved as the primary treatment option for the disease [12].

Meanwhile, comorbidities such as obesity, diabetes, and fatty liver are associated with an elevated risk of developing severe COVID-19 [14,15]. Recent studies have demonstrated that individuals with MAFLD are at a four to six times higher risk of experiencing a worsening of their respiratory signs and symptoms compared to those who do not have fatty liver disease [15]. A study carried out in China discovered that severe COVID-19 was present in 70 out of 324 patients with fatty liver identified via abdominal computed tomography (CT) scans [16]. Nonetheless, patients with comorbidities were more likely to have compromised immunity, which facilitated robust viral replication. The SARS-CoV-2 virus may directly infiltrate hepatocytes via angiotensin-converting enzyme 2 (ACE2) located on the cytomembrane of hepatocytes. This process induces excessive activation of pro-inflammatory cytokines and exacerbates the cytokine storm associated with COVID-19. The viral and inflammatory processes might infiltrate hepatocytes, leading to hepatocyte damage. This cascade can elevate clinical biomarkers of hepatocyte injury, such as alanine aminotransferase (ALT), aspartate aminotransferase (AST), and bilirubin [17,18].

AST/ALT are essential diagnostic biomarkers for evaluating liver disease and abnormalities [18]. According to a multicenter study conducted in ten different cities in China, 30.7% (n = 103) of patients with COVID-19 and MAFLD exhibited elevated levels of AST and ALT [16]. According to a study conducted at a single site in India, which investigated the outcomes of COVID-19 infection in patients with or without MAFLD, the serum levels of ALT and AST were found to be significantly higher in patients with MAFLD [19].

Due to the COVID-19 pandemic, there were limitations on patients accessing healthcare facilities and restrictions on outdoor movement and exercise [20,21]. However, advancements in telemedicine have significantly progressed to face this pandemic era, offering a solution to monitor patients when social distancing is required. The impact of monitoring patients using telemedicine has enhanced the overall quality of medical services, encompassing both communicable diseases (such as COVID-19) and non-communicable diseases (such as asthma, hypertension, diabetes, etc.) [1,22]. During the pandemic, telemedicine could assist hepatologists in delivering healthcare for lifestyle modification in fatty liver disease, with a majority of patients expressing satisfaction with this approach [23]. This also could serve as a bridge to integrate standardized telemedicine into regular standard care delivery [1,22,23].

Therefore, the objective of this review was to assess how telemedicine could aid in monitoring lifestyle modification as a treatment of MAFLD patients during the COVID-19 pandemic, albeit focusing solely on MAFLD patients without coinfection with COVID-19.

2. Materials and Methods

This article was conducted in accordance with the Preferred Reporting Items for Systematic Reviews and Meta-Analyses (PRISMA) statement.

2.1. Search Strategy

A comprehensive search strategy was employed, utilizing literature databases such as PubMed Central and ScienceDirect, covering the period from June 2020 to July 2023. Manual searches were also conducted in JMIR mHealth and uHealth to enhance the comprehensiveness of the search strategy further. The search strategy is used keywords to retrieve articles, including "telemedicine", "telehealth", "telemessaging", or "digital health intervention", "non-alcoholic fatty liver disease", "NAFLD", MAFLD or "hepatic steatosis", and "weight loss", "lifestyle modification", or "lifestyle changes", along with "liver function test" and "biochemical markers." The following results were exported to Zotero 6.0.26 to identify and manage duplicate records.

2.2. Research Selection and Data Extraction

The inclusion criteria for this literature review were as follows: (1) original research from either randomized controlled trials (RCTs) or non-RCTs; (2) studies involving human subjects; (3) publication between 2020 and 2023; and (4) availability of English language and online access, with full-text articles freely accessible. There were no limitations on patient criteria for inclusion in this review. Two independent reviewers, Safira Rosiana Choirida and Ahmad Zaqi Zaenal Muttaqin, screened the titles and abstracts of the relevant studies and reviewed the full text of the selected studies; a third and fourth reviewer, Femmy Nurul Akbar and Hari Hendarto, reviewed the full text of the selected studies and discussed discrepancy. The collected journal articles were also transferred to Microsoft Excel 2019 for data extraction. This facilitated the organization of specific results, including authorship, year of publication, design, setting, sample size, intervention methods, health worker related, and outcomes.

2.3. Risk of Bias Assessment

Two independent reviewers, Safira Rosiana Choirida and Ahmad Zaqi Zaenal Muttaqin, assessed the risk of bias using the Cochrane risk-of-bias tool for randomized controlled trials (RCTs) and the Newcastle-Ottawa scale (NOS) for non-randomized controlled trials (non-RCTs) [24,25]. Discrepancies in the results were resolved by the third and fourth reviewers (Femmy Nurul Akbar and Hari Hendarto).

2.4. Data Analysis

The outcomes were reported as quantifiable measures to evaluate the effect of the intervention on lifestyle modification concerning body mass index (BMI), body weight (BW), liver function indicators such as alanine aminotransferase (ALT) and aspartate aminotransferase (AST), waist circumference (WC), lipid profile, and hemoglobin A1c (HbA1c). Furthermore, a qualitative analysis was conducted by extracting the data using Microsoft Excel to create descriptive text and tables and analyzing the outcomes of all parameters. For the meta-analysis, studies that did not perform standard care for the comparison with the intervention group were excluded. The analysis was carried out using Review Manager software (version 5.4, Cochrane Collaboration, 2020). The mean, standard deviation (SD), sample size of the intervention, and standard care group were inputted, and mean difference (MD) with 95% confidence interval (CI), while statistical heterogeneity using the I^2 statistic. Finally, the illustrated results of the metanalysis were visually presented in a forest plot (Figures 1–4).

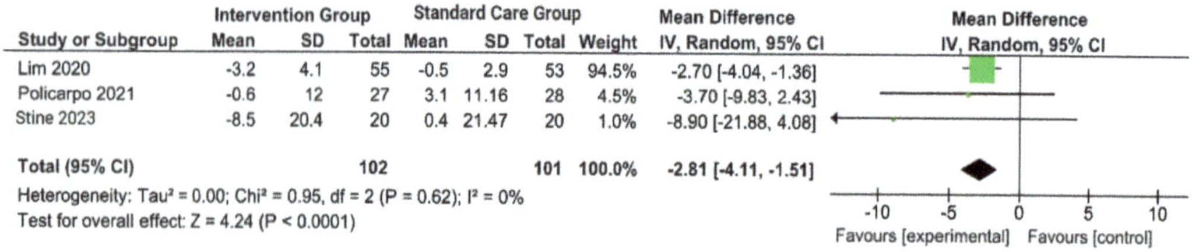

Figure 1. Forest plot of the effects of digital health intervention group compared with standard care group on the body weight (BW) [26–28]. Green square: results of individual studies effect, green dot: weight given in each study, black arrow: represent the 95% confidence interval (CI), black rhombus: overall summary effect.

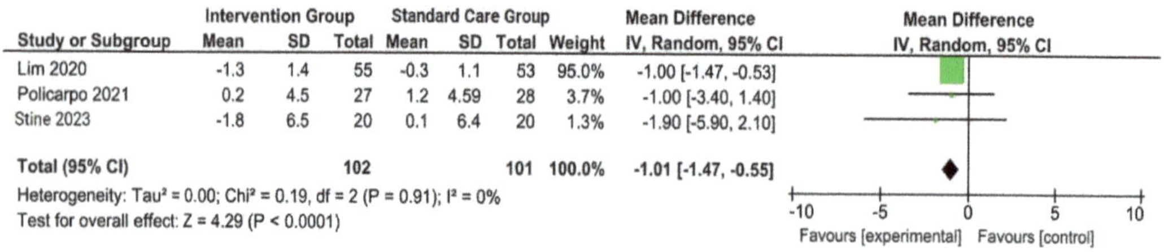

Figure 2. Forest plot of the effects of digital health intervention group compared with standard care group on the body mass index (BMI) [26–28]. Green square: results of individual studies effect, green dot: weight given in each study, black rhombus: overall summary effect.

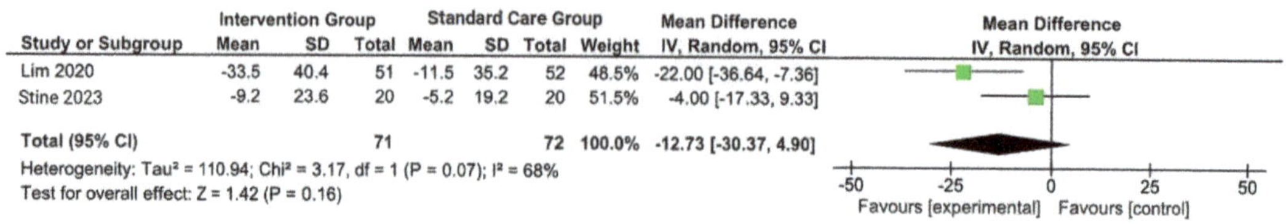

Figure 3. Forest plot of the effects of digital health intervention group compared with standard care group on the alanine aminotransferase (ALT) [26,28]. Green square: results of individual studies effect, black rhombus: overall summary effect.

Study or Subgroup	Intervention Group Mean	Intervention Group SD	Intervention Group Total	Standard Care Group Mean	Standard Care Group SD	Standard Care Group Total	Weight	Mean Difference IV, Random, 95% CI
Lim 2020	-17.4	27.5	52	-7.4	17.6	52	56.5%	-10.00 [-18.87, -1.13]
Stine 2023	-4.3	17.88	20	-3.9	18.2	20	43.5%	-0.40 [-11.58, 10.78]
Total (95% CI)			**72**			**72**	**100.0%**	**-5.83 [-15.15, 3.50]**

Heterogeneity: Tau² = 19.56; Chi² = 1.74, df = 1 (P = 0.19); I² = 42%
Test for overall effect: Z = 1.22 (P = 0.22)

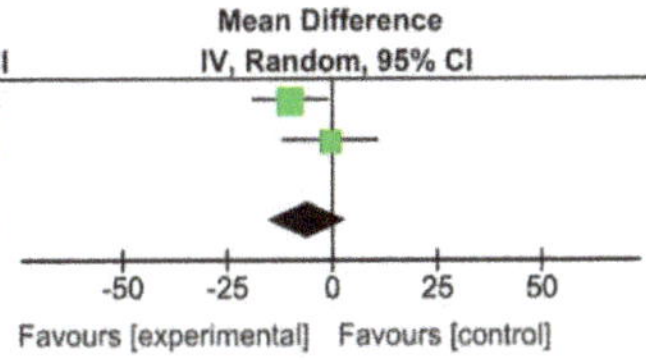

Figure 4. Forest plot of the effects of digital health intervention group compared with standard care group on the aspartate aminotransferase (AST) [26,28]. Green square: results of individual studies effect, black rhombus: overall summary effect.

3. Results

3.1. Research Selection

Out of the initial 90 studies sourced from databases, 88 studies were retrieved from PubMed Central and ScienceDirect. Additionally, two studies were manually added from JMIR, Mhealth, and Uhealth to maximize the tracing. After the removal of duplicates and application of filters, followed by screening for full-text eligibility screening, nine articles were prepared for manual screening. After excluding four articles that did not use telemedicine-based intervention, we included five studies as systematic reviews and meta-analyses and added 29 articles to explain this idea of the research.

PRISMA (Preferred Reporting Items for Systematic Reviews and Meta-Analyses) Flow Chart is depicted in Figure 5. Four studies were identified that reported data on body weight (BW), body mass index (BMI), alanine aminotransferase (ALT), and aspartate aminotransferase (AST).

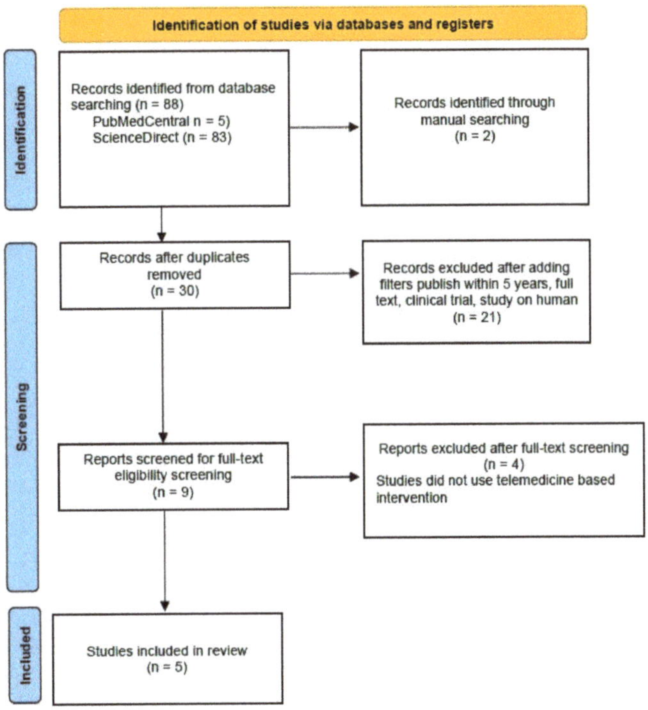

Figure 5. PRISMA Flow Chart of Literature Search.

3.2. Research Characteristics

The characteristics of the included studies were presented in Table 1, supplemented with additional information. Predominantly, the studies were conducted in the USA [26,29,30], followed by Portugal [27] and Singapore [31]. Three out of the five studies, namely Lim et al., Policarpo et al., and Stine et al., were randomized controlled trials (RCTs) conducted at a single center within outpatient hospital and clinic settings. The remaining non-RCT studies were conducted by Motz et al. and Tincopa et al. The total sample size consisted of 239 patients, divided into intervention and standard care groups, with the exception of Motz et al. and Tincopa et al., who did not provide information on standard care. Overall, 136 (56.9%) males and 103 (43.1%) females were recruited; Motz et al. exclusively recruited female participants due to COVID-19 restrictions. The majority of participants were obese patients of Caucasian descent, followed by Chinese individuals. The mean age of the participants was 51.3 years. The median duration of the intervention was 5.5 months. Most studies indicated their adherence to inclusion criteria in line with recent guidelines of MAFLD. Additionally, all studies shared the same objective of assessing the potential benefits of telemedicine for facilitating lifestyle changes among patients with MAFLD.

3.3. Assessment of Bias

Only RCT studies were considered for inclusion in the Cochrane Risk of Bias (RoB) tool assessment [25]. Among these three studies, Lim et al. reported the highest risk of detection bias, as they mentioned that the assessors were not blinded to the study-allocated groups. Policarpo et al. and Stine et al. did not clearly specify whether participants and personnel were blinded (performance bias) or how detection bias was addressed.

The bias assessment outcomes for non-RCTs indicated that Tincopa et al. achieved the most favorable outcomes. The risk-of-bias assessments are depicted in Figure 6.

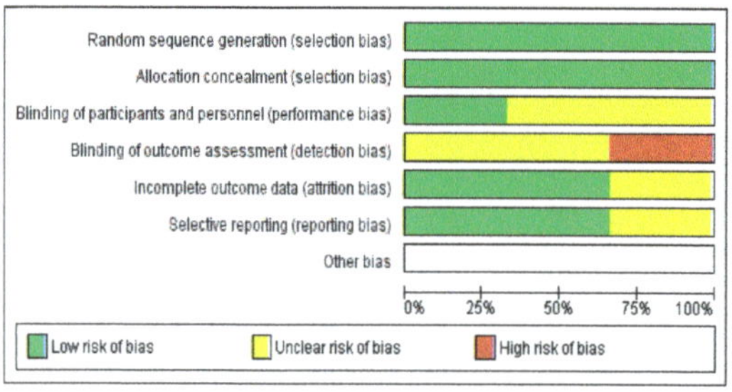

(a)

	Poins	Tincopa, 2021	Motz, 2021
SELECTION (maximum 4 stars)	Representativeness of the exposed cohort	*	*
	Selection of the non-exposed cohort	*	
	Ascertainment of exposure	*	*
	Outcome of interest was not present at start of study	*	*
COMPARABILITY (maximum 2 stars)	Comparability cases and controls on the basis of the design or analysis	**	*
OUTCOME (maximum 3 stars)	Assessment of outcome	*	*
	Was follow -up long enough for outcomes to occur	*	*
	Adequacy of follow -up of cohorts	*	*
	Total Score =	9	7

(b)

Figure 6. The risk of bias graph displays the review authors' assessments of each risk of bias item presented as percentages across all included studies. This includes (**a**) Cochrane risk-of-bias assessment for randomized controlled trials (RCTs) and (**b**) New-castle Ottawa risk of bias assessment for non-RCT studies [29,30]. * and ** : indicate the rating for these categories.

3.4. Meta-Analysis: The Pooled Effects of Telemedicine

Out of the five included studies, three were included in the meta-analysis, while two studies were conducted with single-arm designs. Three studies (Lim et al., Policarpo et al., Stine et al.) reported outcomes regarding BW and BMI [26,31]. However, one study (Policarpo et al.) did not provide the standard deviation outcomes for alanine aminotransferase (ALT) and aspartate aminotransferase (AST) [27].

Table 1. Summary of the included studies.

Author, Year, Country.	Setting	Sample Size	Dur. (Mon.)	Group		Types of Telemedicine	The Health Coworker Related	Outcome	Outcomes
				Intervention	Control				
Lim SL, 2020, [28] Singapore RCT, single center	National University Hospital Singapore	108 adults 55 IG, 53 SC	6	Patients underwent private consultations offline with a dietitian, followed by online consultations via a mobile app (nBuddy). They were instructed on how to utilize the app to log and track their food intake, physical activity, and behavioral changes.	Standard dietary and physical activity recom-mendations based on AHA guidelines	Mobile telehealth (nBuddy)	Single dietician, trained nurse, assessors.	BMI, BW, AST, ALT, WC, SBP, DBP	BMI *, BW *, AST *, ALT *, WC *, SBP *, DBP *
Motz V, 2021 [29] Non-RCT, single center	RCT patients	3 adults of IG	5	Moderate-intensity aerobic exercise for a duration of 30 min, conducted on 5 days per week.	NA	Direct supervision using audio-visual telehealth platform	Exercise physiologist	BMI, BW, ALT, AST, WC, HbA1c	BMI #, BW #, ALT # AST #, WC #, HbA1C, Homeostasis model assessment for insulin resistance (HOMA-IR)3
Tincopa, 2021, USA [30] Non—RCT, single-center	General hepatology outpatient clinic	40 adults of IG	6	Participants are required to engage in physical activity via walking or step counting, with a target range of a minimum of 800 steps per week to 10,000 steps per week. Additionally, personalized feedback on physical activity is provided.	NA	FitBit mobile application, phone call	The researchers	BMI, BW, ALT, AST, WC, SBP, Lipid Profile, HbA1c	ALT *, WC *, lipid profile *, HbA1c * BW ^, ALT ^ Fibroscan ^

Table 1. *Cont.*

Author, Year, Country.	Setting	Sample Size	Dur. (Mon.)	Group		Types of Telemedicine	The Health Coworker Related	Outcome	Outcomes
				Intervention	Control				
Policarpo S, 2021, Portugal [27] RCT, single-cener.	Outpatients Infectious Disease Clinic	55 adults 27 IG, 28 SC	6	The nutritional plan is founded on the Mediterranean diet and is provided in the format of written healthy eating guidelines and tips. It entails a reduction of 500 kilocalories in daily calorie intake, abstaining from alcohol consumption, and limiting fruit consumption to less than 5 units per day.	General dietary recommendations	Video and/or phone	Single dietician	BMI, BW, Lipid profile, Blood glucose Liver stiffness (Fibroscan)	BMI *, BW *, Lipid profile Blood glucose, FIB-4 index Fibroscan $^{\$}$
Stine JG, 2022, USA [26] RCT, single-center	Penn State Milton S. Hershey Medical Center and the Penn State College of Medicine	40 adults 20 IG, 20 SC	4	• Human coaching from an academic hepatologist regarding the Mediterranean diet and moderate-intensity physical activity, with a minimum of 150 min per week. • NoomWeight's curriculum	General dietary recommendations	Mobile application	Hepatologist	BMI, BW, Albumin, ALT, AST, fasting blood glucose Platelet, FIB-4, HbA1c	BMI *, BW *, Platelet * Albumin, ALT, AST, ALP, FIB-4 index, NFS, fasting glucose. HbA1c

IG: intervention group, SC: standard care, MAFLD: metabolic dysfunction-associated fatty liver disease, BMI: body mass index, BW: body weight, AST: aspartate aminotransferase, ALT: alanine aminotransferase, WC: weight circumference, SDP: systolic blood pressure, DBP: diastolic blood pressure, FIB: fibrosis-4 index, HbA1c: hemoglobin A1c, RCT: randomized controlled trial, AHA: American Heart Association, ALP: alkali phosphatase, NFS; NAFLD fibrosis score. *: the outcomes of the studies were significant, #: significant without *p*-value, ^: reduced but not significant, $^{\$}$: cannot be done after intervention.

A concise summary of the included studies was compiled into a forest plot illustrating body weight, BMI, ALT, and AST. Outcomes from the meta-analyses (Figure 1) suggested no heterogeneity in body weight and BMI ($I^2 = 0\%$). The pooled MD for the effects of digital health intervention on body weight (MD −2.81: 95% CI, −4.11 to −1.51, $p = 0.0001$) and BMI (MD −1.01: 95% CI, −1.47 to −0.55, $p = 0.0001$) were statistically significant. However, no statistically significant relationship between ALT (MD −12.73: 95% CI, −30.37 to −4.90, $p = 0.16$) and AST (MD −5.83: 95% CI, −15.15 to −3.50, $p = 0.22$), with a moderate level of heterogeneity ($I^2 = 68\%$ for ALT and $I^2 = 42\%$ for AST).

3.5. Result of Individual Studies

All the included studies focused on lifestyle modifications, including weight loss, physical activity, and stress management. We identified four studies concerning BW, BMI, AST, and ALT. The mobile application provided access focusing on body weight loss and behavioral changes, which technically involved monitoring and providing personalized feedback to patients. Additionally, three studies addressed WC, while two studies investigated lipid profiles and HbA1c.

However, each study had its main topic, including lifestyle change programs. Stine et al. study focused on a lifestyle change program encompassing nutrition, physical activity, and sustainable behavioral changes facilitated using a mobile-based health application or telemedicine called NoomWeight (NW), while Lim et al. was composed of dietary and lifestyle guidance via telemedicine known as nBuddy [26,31]. Moreover, Tincopa et al., which employed the utilization of FitBit, a mobile digital health technology, had the objective of monitoring daily steps and providing nutritional assessments. Patients could easily log their daily meals, daily step counts, and body weight and access a wealth of health and stress management information [30]. In contrast, studies by Motz et al. and Policarpo et al. utilized video and/or phone communication as their primary digital health intervention for monitoring the intervention group (IG) [27,29].

In most of the included studies, participants were instructed to engage in physical activities, at least with moderate-intensity aerobic training. Regarding dietary intervention, three studies considered providing recommendations for the Mediterranean diet [26,27,29,30], one study represented nutritional counseling according to the original research protocol [29], and one study represented dietary intervention via general advice from a dietitian [31].

Stine et al. included 40 participants in their study, equally divided between an intervention group and a standard care group. The trial lasted for four months, during which both groups received education from a hepatologist regarding the current guidelines for MASH in clinical practice, focusing on the Mediterranean diet and 150 min of moderate-intensity physical activity per week. The IG received a dietary and physical activity program via the NoomWeight application. At the end of the trials, the IG exhibited a significant decrease in body weight ($p = 0.0008$) and BMI ($p = 0.037$) compared to the standard care (SC) group. Moreover, 45% of the patients successfully reduced 5% of their body weight. The IG also received a significant reduction in platelet count compared to SC (−28 vs. −5.7 × 109, $p = 0.038$). Other laboratory parameters were also evaluated, such as liver function test, albumin level, blood sugar level, and Fibrosis-4 index (FIB-4), but the results were not statistically significant, respectively ($p > 0.05$). Furthermore, 70% and 75% of IG met the criteria for feasibility and acceptability concerning the use of NoomWeight [26]. Stine et al. stated 43% of patients had stage F0/F1, and 40% had stage F2.

In a study conducted by Lim et al., participants underwent a 6-month RCT where the IG received guidance on dietary choices and physical activity from a dietitian using nBuddy mobile application to record their dietary intake, physical activity, and behavioral modifications while the SC group received guidance from a trained nurse at the clinic. All groups received dietary programs based on the guidelines provided by the American Heart Association. At the end of the trial period, the IG demonstrated notable reductions in body weight (mean 3.2, SD 4.1 kg vs. mean 0.5, SD 2.9 kg; $p < 0.001$), waist circumference (mean 2.9, SD 5.0 cm vs. mean −0.7, SD 4.4 cm; $p < 0.001$), systolic blood pressure (mean 12.4, SD 14.8 mmHg vs. mean 2.4, SD 12.4 mmHg; $p = 0.003$), diastolic blood pressure (mean 6.8, SD 8.9 mmHg vs. mean −0.9, SD 10.0 mmHg; $p = 0.001$), ALT (mean 33.5, SD 40.4 IU/L vs. mean 11.5, SD 35.2 IU/L; $p = 0.004$), and AST (mean 17.4, SD 27.5 U/L vs. mean 7.4, SD 17.6 IU/L, $p = 0.03$) compared to SC group [31].

Policarpo et al. conducted a study using NAFLD-HIV patients from an outpatient infectious disease clinic assigned to the IG or the SC group. Both groups received uniform guidance pertaining to a structured dietary intervention, emphasizing caloric restriction and weight loss strategies rooted in the Mediterranean diet. During the 3rd and 4th months of the research, patients underwent a review of dietary advice, eating habits, and lifestyle modifications during the pandemic via video and/or phone calls, which also included the completion of four stress questionnaires. Before the implementation of the lockdown, the IG exhibited a decrease in body weight, with a median loss of 1.5 kg compared to a median loss of 0.65 kg in the Standard Care (SC) group ($p < 0.001$). Following 3 months of lockdown, both groups experienced weight gain, with the SC group showing a higher weight gain of around 3 kg compared to less than 1 kg in the IG ($p < 0.001$) [27]. Policarpo et al. also presented that 63.6% of patients had no evidence of fibrosis, 27.3% of patients had a moderate degree of fibrosis (F2–F3), and 9.1% had severe fibrosis (F4) using Liver transient elastography (Fibroscan©, Echosense, France) that underwent before intervention [26]. However, the researcher could not perform a Fibroscan after intervention due to local and government issues.

Some studies used the Fibrosis-4 (FIB-4) index or NAFLD fibrosis score (NFS), which is also used to predict the severity of fibrosis in MALFD patients. FIB-4 evaluated the risk of liver fibrosis by calculating on the basis of age, AST/ALT levels, and platelet count. Moreover, NFS predicted the severity of fibrosis based on six variables: age (years), BMI (kg/m^2), fasting blood glucose, AST/ALT ratio, platelet count, and albumin levels. As the results, Policarpo et al. presented a low probability for liver fibrosis FIB-4 in 87.3% of patients [27], while Stine et al. revealed no significant changes in clinical scoring such as NFS and FIB-4 ($p > 0.05$) after intervention [26].

Motz et al. conducted a study involving 20 weeks of moderate-intensity aerobic training delivered via telehealth, which was administered to patients with non-alcoholic fatty liver disease (NAFLD) under direct supervision using an audio-visual telehealth platform. Body weight, body fat percentage, and waist circumference improved with exercise. Mean reduction was also shown in HbA1c levels, AST/ALT levels, and homeostatic model assessment for insulin resistance (HOMA-IR) with respective values. Subsequently, all participants met the priori definition of feasibility [29].

Furthermore, Tincopa et al. also illustrated the implementation of physical activity utilizing the FitBit mobile application and conducted nutritional assessments to evaluate the feasibility and acceptability of lifestyle modifications among NAFLD patients using mobile technology. They intervened with the participants to assess secondary outcomes, encompassing body weight, physical fitness, liver transient elastography, laboratory testing, and quality of life. They had statistically significant improvements in waist circumference lipid profile parameters, including HDL, LDL, and triglycerides, along with reductions in hemoglobin A1c levels, with a p-value of < 0.05. Approximately 50% of the participants exhibited reductions in BW and ALT and a 42.4% reduction in liver stiffness or fibrosis by liver transient elastography (Fibroscan). Roughly 59% of the participants reported that the mobile application was easy to use, and 66% of the patients expressed motivation to enhance their physical activity while utilizing the daily step count tracker. These outcomes suggest that these tools are not only feasible but also acceptable and hold promise for future interventions [30].

4. Discussion

The COVID-19 pandemic has imposed limitations on patients accessing healthcare facilities, leading to a decrease in treatment sustainability. One breakthrough is the utilization of telemedicine to enhance connectivity between patients and healthcare professionals.

Lifestyle modification remains the cornerstone of treatment for MAFLD due to insufficient evidence supporting pharmacological interventions [6,12].

This systematic review summarizes the usage of telemedicine for monitoring lifestyle modification interventions in adults with MAFLD. Most participants were males, similar to the incidence of fatty liver disease globally that males are higher than females, although one included study exclusively recruited females [29]. The mean age was 51.3 years, similar to the most prevalent worldwide, which was 51–60 years. All included studies were published between 2020 and 2023.

Multiple guidelines have indicated that achieving a reduction in body weight of 3–5% could lead to improvement in hepatic steatosis, while a reduction of 7% in body weight could induce changes in the histological features of metabolic-associated steatohepatitis (MASH) [16]. For MAFLD, dietary recommendations include reducing daily calorie intake by 500–1000 kcal per day (hypocaloric diet), adherence to the Mediterranean diet, adoption of a low-carbohydrate diet, and adherence to a low-fructose diet. Additionally, moderate-intensity aerobic physical activity with a minimum duration of 150–200 min per week or 30 min per day and a frequency of more than twice per week (3–5 days a week) is recommended [12].

Based on our meta-analysis, the intervention group employing telemedicine was significantly more effective in achieving reductions in body weight and BMI compared to standard care in MAFLD patients. It is also observed from all included studies that reducing body weight is associated with a subsequent decrease in biochemical markers such as AST and ALT. However, this meta-analysis revealed that telemedicine did not significantly decrease AST and ALT levels. Waist circumference was reported in three studies, indicating improvement in the intervention groups. One study showed improvements in other markers of MAFLD risk factors, including a panel of lipids, liver stiffness, and HbA1c, but they were not statistically significant [30]. Other studies showed improvement in blood pressure systolic as well as diastolic pressure [31]. Motz's study also revealed improvement in' BW, BMI, waist circumference by exercise, and a mean reduction in AST/ALT, HbA1c, and HOMA-IR with respective values.

The degree of liver fibrosis was assessed in three studies using liver transient elastography (Fibroscan) [26,27,30,31]. It was concluded that most of the degrees of fibrosis in the included studies had no evidence of mild fibrosis (F0/F1), and there is no study that showed statistically significant improvement in liver stiffness or fibrosis by Fibroscan [27,30]. FIB-4 index and NFS also showed no clinically significant changes after intervention [26].

Based on outcomes, the duration and frequency of digital interventions varied widely, ranging from 4 to 6 months, but the majority of articles implemented interventions lasting 5–6 months. At present, there are no established standard regulations regarding the duration of digital interventions in MAFLD patients and their potential impact on the clinical and biochemical outcomes of MAFLD. However, the included studies showed that reducing body weight by 5% was found to be more effective when the intervention was implemented for more than 4 or 6 months [29–31]. The changes in biochemical markers were also notably reduced over a duration of 5–6 months [28,29]. Hence, the intervention for 5–6 months was found to be more sustainable.

Telemedicine serves as a bridge that facilitates health workers, including medical students, in monitoring patients indirectly and virtually. Additionally, there were advantages for patients in terms of cost savings, as they could avoid visiting onsite healthcare facilities [28,32]. This demonstrates that telemedicine is both feasible and acceptable for monitoring patients with MAFLD, primarily via the use of mobile applications [33–35].

The role of telemedicine may commence with primary healthcare practitioners for early detection using risk factors, progressing to further evaluation and categorizing patients into risk groups for referral to higher-level healthcare facilities. If the patient is deemed to have a low risk of disease progression, they will be referred back to their primary healthcare for lifestyle modification and to receive ongoing laboratory monitoring. Meanwhile, patients in the intermediate and high-risk stages of disease progression will be referred to tertiary healthcare for monitoring and therapy. Additionally, they will receive lifestyle modification guidance from primary healthcare providers using telemedicine [23,33–35].

The limitation of our study resided in the relatively small number of included studies and participants, as well as the fact that the majority of studies were conducted in Western countries (3 in the USA, 1 in Portugal, and 1 in Singapore). Hence, the interventions in those studies remain applicable in real-world settings. Another limitation of the included studies was the variability in the competencies of the workers involved in making recommendations or implementing interventions.

Our study strengths in its focus solely on current studies conducted during the pandemic, ensuring its up-to-dateness. Therefore, despite the challenges in real-world settings associated with its implementation, telemedicine could be utilized for monitoring body weight and BMI, given the presence of adequate facilities, infrastructure, and government support.

5. Conclusions

This systematic review and meta-analysis revealed that telemedicine is beneficial in reducing body weight and BMI in patients with metabolic dysfunction-associated fatty liver disease (MAFLD). However, our meta-analysis indicated that telemedicine did not significantly lower liver function markers such as AST and ALT levels in MAFLD patients. Additionally, waist circumference was found to be lower in the intervention groups across three studies. Moreover, other markers associated with MAFLD risk factors, such as a panel of lipids and HbA1c levels. Telemedicine has the potential to facilitate healthcare professionals in remotely monitoring patients, including tracking their body weight, dietary intake, daily physical activities, and biochemical markers. Patients could save money by avoiding visits to healthcare facilities. This demonstrates the feasibility and acceptability of telemedicine for monitoring patients with MAFLD, particularly using mobile applications. Telemedicine enables patients to maintain connectivity to healthcare professionals while mitigating the risks of disease exposure, including the transmission of COVID-19 infection in traditional offline healthcare settings.

In conclusion, telemedicine could serve as a viable option for monitoring lifestyle modification, including body weight and BMI in MAFLD patients amidst the COVID-19 pandemic, and offer the patients convenience to remain connected with their healthcare providers while preventing the spread of COVID-19 infection. Future studies should be conducted on larger populations to assess the generalizability of telemedicine utilization in clinical settings.

Author Contributions: Conceptualization, F.N.A., S.R.C., A.Z.M. and H.H.; methodology, F.N.A., S.R.C., A.Z.M. and H.N.; software, S.R.C., A.Z.M. and H.N.; validation, F.N.A., S.R.C., A.Z.M., H.N. and H.H.; formal analysis F.N.A., S.R.C., A.Z.M., F.E., H.N. and H.H.; investigation F.N.A., S.R.C. and A.Z.M.; data curation, F.N.A., S.R.C. and A.Z.M.; resolved discrepancies, F.N.A., H.H. and F.E.; writing—original draft preparation F.N.A., S.R.C., A.Z.M. and F.E.; writing—reviewing and editing, F.N.A., S.R.C., A.Z.M., F.E. and H.H.; visualization, S.R.C.; supervision, H.H.; project administration, F.N.A. All authors have read and agreed to the published version of the manuscript.

Funding: This research did not receive any external funding.

Institutional Review Board Statement: Not applicable.

Informed Consent Statement: Not applicable.

Data Availability Statement: The data depicted in this research are obtainable upon request from the corresponding author.

Conflicts of Interest: The authors affirm that there are no conflicts of interest, whether in the interpretation of data, the drafting of the manuscript, or the decision to publish the outcomes.

References

1. Giacalone, A.; Marin, L.; Febbi, M.; Franchi, T.; Tovani-Palone, M.R. eHealth, telehealth, and telemedicine in the management of the COVID-19 pandemic and beyond: Lessons learned and future perspectives. *World J. Clin. Cases* **2022**, *8*, 2363–2368. [CrossRef] [PubMed]
2. Shaver, J. The state of telehealth before and after the COVID-19 pandemic. *Prim. Care Clin. Office Pract.* **2022**, *49*, 517–530. [CrossRef] [PubMed]
3. Chang, D.; Chang, X.; He, Y.; Tan, K.J.K. The determinants of COVID-19 morbidity and mortality across countries. *Sci. Rep.* **2022**, *12*, 5888. [CrossRef] [PubMed]
4. Buwono, P.W.; Akbar, F.N.; Hendarto, H.; Parlindungan, F.; Khomeini, A.; Prasetyo, A.H.T.; Nufus, H.; Niswara, C.; Suseno, D.; Syahlan, R.H.; et al. Association between serum d-dimer levels and COVID-19 severity among vaccinated and non-vaccinated individuals. *Mal. J. Med. Health Sci.* **2022**, *18*, 67–71.
5. Erfani, Z.; Alizadeh, N.; Faraji, N.; Teymouri, A. Vitamin B 12 deficiency and increased susceptibility and severity of COVID-19: A review. *Int. J. Creat. Res. Thoughts* **2023**, *10*, 623–651.
6. Shen, K.; Singh, A.D.; Modaresi, E.J.; Wakim, F.J. Therapies for non-alcoholic fatty liver disease: A 2022 update. *World J. Hepatol.* **2022**, *14*, 1718–1729. [CrossRef]
7. Ge, X.; Zheng, L.; Wang, M.; Du, Y.; Jiang, J. Prevalence trends in non-alcoholic fatty liver disease at the global, regional and national levels, 1990–2017: A population-based observational study. *BMJ Open* **2020**, *10*, e036663. [CrossRef]
8. Boccatonda, A.; Andreetto, L.; D'Ardes, D.; Cocco, G.; Rossi, I.; Vicari, S.; Schiavone, C.; Cipollone, F.; Guagnano, M.T. From NAFLD to MAFLD: Definition, pathophysiological basis and cardiovascular implications. *Biomedicines* **2023**, *11*, 883. [CrossRef]
9. Cusi, K.; Isaacs, S.; Barb, D.; Basu, R.; Caprio, S.; Garvey, W.T. American association of clinical endocrinology clinical practice guideline for the diagnosis and management of nonalcoholic fatty liver disease in primary care and endocrinology clinical settings. *Endocr. Pract.* **2022**, *28*, 528–562. [CrossRef]
10. Gallacher, J.; McPherson, S. Practical diagnosis and staging of nonalcoholic fatty liver disease: A narrative review. *EMJ* **2018**, *6*, 108–118. [CrossRef]
11. Ahmed, A.; Wong, R.J.; Harrison, S.A. Nonalcoholic fatty liver disease review: Diagnosis, treatment, and outcomes. *Clin. Gastroenterol. Hepatol.* **2015**, *13*, 2062–2070. [CrossRef]
12. Paternostro, R.; Trauner, M. Current treatment of non-alcoholic fatty liver disease. *J. Intern. Med.* **2022**, *2*, 190–204. [CrossRef]
13. Dufour, J.F.; Anstree, Q.M.; Bugianesi, E.; Harrison, S.; Loomba, R.; Paradis, V.; Tilg, H.; Wong, V.W.; Zelber-Sagi, S. Curent therapies and new developments in NASH. *Gut* **2022**, *71*, 2123–2134. [CrossRef]
14. Nowroozi, A.; Momtazmanesh, S.; Rezaei, N. COVID-19 and MAFLD/NAFLD: An updated review. *Front. Med.* **2023**, *24*, 1126491. [CrossRef]
15. Jagirdhar, G.S.K.; Qasba, R.K.; Pattnaik, H.; Rama, K.; Banga, A.; Reddy, S.T.; Flumignan, B.A.C.; Kashyap, R.; Elmati, P.R.; Bansal, V.; et al. Association of non-alcoholic fatty liver and metabolic-associated fatty liver with COVID-19 outcomes: Systematic review and meta-analysis. *World J. Gastroenterol.* **2023**, *21*, 3363–3378. [CrossRef]
16. Huang, R.; Zhu, L.; Wang, J.; Xue, L.; Liu, L.; Yan, X.; Huang, S.; Li, Y.; Yan, X.; Zhang, B.; et al. Clinical features of COVID-19 patients with non-alcoholic fatty liver disease. *Hepatol. Commun.* **2020**, *4*, 1758–1768. [CrossRef]
17. Imam, M.T.; Almalki, Z.S.; Alzahrani, A.R.; Al-Ghamdi, S.; Falemban, A.H.; Alanazi, I.M.; Shahzad, N.; Alrooqi, M.M.; Jabeen, Q.; Shahid, I. COVID-19 and severity of liver diseases: Possible crosstalk and clinical implications. *Int. Immunopharmacol.* **2023**, *121*, 110439. [CrossRef] [PubMed]
18. Nagarajan, R.; Krishnamoorthy, Y.; Rajaa, S.; Hariharan, V.S. COVID-19 severity and mortality among chronic liver disease patients: A systematic review and meta-analysis. *Prev. Chronic Dis.* **2022**, *19*, 5–6. [CrossRef] [PubMed]
19. Nath, P.; Kumar, R.; Mallick, B.; Das, S.; Anand, A.; Panigrahi, S.C.; Duseja, A.; Acharya, S.K.; Chawla, Y.K.; Praharaj, D.L.; et al. Effect of nonalcoholic fatty liver disease (NAFLD) on COVID-19: A single-center study of 3983 patients with review of literature. *Cureus* **2022**, *14*, e26683. [CrossRef] [PubMed]
20. Esteban, J.R.; Nuñez, R.M.; Broquetas, T.; Serra-Matamala, I.; Basegoda, O.; Soriano-Varela, A.; Espin, G.; Castillo, J.; Banares, J.; Carrion, J.A.; et al. Impact of the COVID-19 pandemic on the care and outcome of people with NAFLD-related cirrhosis. *JHEP Rep.* **2022**, *4*, 100574. [CrossRef]
21. Aditya, D.M.N.; Kalanjati, V.P.; Pamungkas, D.B.B. The use of telemedicine in COVID-19 pandemic era: A systematic review. *Bali Med. J.* **2022**, *11*, 1987–1995.
22. Guarino, M.; Cossiga, V.; Cutolo, F.M.; Attanasio, M.R.; Lieto, R.; Morisco, F. COVID-19 and fatty liver disorders. *J. Clin. Med.* **2023**, *12*, 4316. [CrossRef] [PubMed]
23. Kim, R.; Medina, S.P.; Magee, C.; Khalili, M. Fatty liver and the coronavirus disease 2019 pandemic: Health behaviors, social factors, and telemedicine satisfaction in vulnerable populations. *Hepatol. Commun.* **2022**, *6*, 1045–1055. [CrossRef] [PubMed]
24. Lo, C.K.L.; Mertz, D.; Loeb, M. Newcastle-Ottawa scale: Comparing reviewers to authors assessments. *BMC Med. Res. Methodol.* **2014**, *14*, 45. [CrossRef] [PubMed]
25. Higgins, J.P.T.; Altman, D.G.; Gotzsche, P.C.; Juni, P.; Moher, D.; Oxman, A.D.; Savovic, J.; Schulz, K.F.; Weeks, L.; Sterne, J.A.; et al. The Cochrane collaboration's tool for assessing risk of bias in randomised trials. *BMJ* **2011**, *18*, 343. [CrossRef] [PubMed]

26. Stine, J.G.; Rivas, G.; Hummer, B.; Duarte-Rojo, A.; May, C.N.; Geyer, N.; Chincilli, V.M.; Conroy, D.E.; Mitchell, E.S.; McCallum, M.; et al. Mobile health lifestyle intervention program leads to clinically significant loss of body weight in patients with NASH. *Hepatol. Commun.* **2023**, *7*, e0052. [CrossRef] [PubMed]
27. Policarpo, S.; Machado, M.V.; Cortez, P.H. Telemedicine as a tool for dietary intervention in NAFLD-HIV patients during the COVID-19 lockdown: A randomized controlled trial. *Clin. Nutr. ESPEN* **2021**, *43*, 329–334. [CrossRef]
28. Lim, S.L.; Johal, J.; Ong, K.W.; Han, C.Y.; Chan, Y.H.; Lee, Y.M.; Loo, W.M. Lifestyle Intervention Enabled by Mobile Technology on Weight Loss in Patients with nonalcoholic fatty liver disease: Randomized Controlled Trial. *JMIR mHealth uHealth* **2020**, *8*, e14802. [CrossRef]
29. Motz, V.; Faust, A.; Dahmus, J.; Stern, B.; Soriano, C.; Stine, J.G. Utilization of a directly supervised telehealth-based exercise training program in patients with nonalcoholic steatohepatitis: Feasibility study. *JMIR Form. Res.* **2021**, *5*, e30239. [CrossRef]
30. Tincopa, M.A.; Lyden, A.; Wong, J.; Jackson, E.A.; Richardson, C.; Lok, A.S. Impact of a pilot structured mobile technology Based lifestyle ontervention for patients with nonalcoholic fatty liver disease. *Dig. Dis. Sci.* **2022**, *67*, 481–491. [CrossRef]
31. Akbar, F.N.; Tedjasaputra, T.R.; Makmun, D.; Akbar, N. Correlation between the degree of esophageal varices and liver stiffness in liver cirrhosis patients. *Asian J. Microbiol. Biotechnol. Environ. Sci.* **2015**, *17*, 401–404.
32. Snoswell, C.L.; Taylor, M.L.; Comans, T.A.; Smith, C.; Gray, L.C.; Caffery, L.J. Determining if telehealth can reduce health system costs: Scoping Review. *J. Med. Internet Res.* **2020**, *10*, e17298. [CrossRef] [PubMed]
33. Moein, H.R.; Arteaga, E.K.; Sarkar, S. Using telehealth for diagnosis and management of non-alcoholic fatty liver disease. *Clin. Liver Dis.* **2022**, *19*, 171–174. [CrossRef] [PubMed]
34. Zhou, R.; Gu, Y.; Zhang, B.; Kong, T.; Zhang, W.; Li, J.; Shi, J. Digital therapeutics: Emerging new therapy for nonalcoholic fatty liver disease. *Clin. Transl. Gastroenterol.* **2023**, *14*, e00575. [CrossRef]
35. Faust, A.; Stine, J.G. Time to step it up: Mobile health intervention for lifestyle modification in patients with nonalcoholic fatty liver disease. *Digest Dis. Sci.* **2022**, *67*, 403–405. [CrossRef]

Journal of Personalized Medicine

Review

MASLD-Related HCC—Update on Pathogenesis and Current Treatment Options

Catherine Leyh [1], Jason D. Coombes [2], Hartmut H. Schmidt [3], Ali Canbay [4], Paul P. Manka [4] and Jan Best [3,*]

[1] Department of Gastroenterology, Hepatology and Infectious Diseases, University Hospital Düsseldorf, Medical Faculty of Heinrich Heine University Düsseldorf, 40225 Düsseldorf, Germany
[2] Internal Medicine, Division of Gastroenterology and Hepatology, Saint Louis University School of Medicine, Saint Louis, MO 63104, USA; jason.coombes@health.slu.edu (J.D.C.)
[3] Department of Gastroenterology, Hepatology and Transplant Medicine, Medical Faculty, University of Duisburg-Essen, 45147 Essen, Germany
[4] Department of Internal Medicine, University Hospital Knappschaftskrankenhaus Bochum, Ruhr University Bochum, 44801 Bochum, Germany
* Correspondence: jan.best@uk-essen.de; Tel.: +49-201-723-3611; Fax: +49-201-723-5971

Abstract: Hepatocellular carcinoma (HCC) is a common complication of chronic liver diseases and remains a relevant cause of cancer-related mortality worldwide. The global prevalence of metabolic dysfunction-associated steatotic liver disease (MASLD) as a risk factor for hepatocarcinogenesis is on the rise. Early detection of HCC has been crucial in improving the survival outcomes of patients with metabolic dysfunction-associated steatohepatitis (MASH), even in the absence of cirrhosis. Understanding how hepatocarcinogenesis develops in MASH is increasingly becoming a current research focus. Additive risk factors such as type 2 diabetes mellitus (T2DM), genetic polymorphisms, and intestinal microbiota may have specific impacts. Pathophysiological and epidemiological associations between MASH and HCC will be discussed in this review. We will additionally review the available tumor therapies concerning their efficacy in MASH-associated HCC treatment.

Keywords: hepatocellular carcinoma; HCC; MASLD; MASH; NAFLD; NASH; biomarker; intestinal microbiota

Citation: Leyh, C.; Coombes, J.D.; Schmidt, H.H.; Canbay, A.; Manka, P.P.; Best, J. MASLD-Related HCC—Update on Pathogenesis and Current Treatment Options. *J. Pers. Med.* **2024**, *14*, 370. https://doi.org/10.3390/jpm14040370

Academic Editor: Manuel Rodríguez-Perálvarez

Received: 28 February 2024
Revised: 25 March 2024
Accepted: 26 March 2024
Published: 30 March 2024

1. Introduction

Hepatocellular carcinoma (HCC) is one of the leading causes of cancer-related deaths worldwide [1]. Hepatocarcinogenesis is predisposed by liver cirrhosis of any etiology. However, chronic viral hepatitis B and metabolic dysfunction-associated steatohepatitis (MASH), formerly known as non-alcoholic steatohepatitis (NASH) [2], already significantly increase the risk of developing HCC, even in the absence of cirrhosis [3–5].

The increasing prevalence of metabolic dysfunction-associated steatotic liver disease (MASLD), previously known as non-alcoholic fatty liver disease (NAFLD) [2], and its progression to MASH raise the incidence of progressive liver fibrosis, cirrhosis, and HCC. Current data support the notion that MASH is the leading cause of the predicted increase in HCC incidence in the next decades [6].

Causal therapeutic approaches to HCC-predisposing MASH are taking place in the context of clinical trials, which is why weight normalization, exercise, and optimization of concomitant diseases such as diabetes mellitus are still the current focus. Understanding how hepatocarcinogenesis develops in MASH is increasingly becoming a key research focus. Here, the roles of additive risks such as type 2 diabetes mellitus (T2DM), genetic polymorphisms, and the intestinal microbiome are of particular interest.

The importance of HCC surveillance in the MASH population is controversial. On the one hand, there is a lack of international consensus regarding the risk population to be defined. On the other hand, it has not been clarified whether standard screening methods using sonography in patients with MASH have sufficient sensitivity in early HCC detection [7,8].

In recent years, there has finally been a breakthrough in the systemic therapy of HCC, so several therapeutic options are now available in addition to current locoregional procedures. Since MASH is associated with a multitude of comorbidities, it remains challenging to extract the individual impact of MASH on the efficacy of available treatment options. This is particularly important when comparing MASH-related HCC with the HCC of other etiologies.

2. Epidemiology of HCC in MASH

Primary liver carcinomas represent the sixth most common cancer worldwide and the third leading cause of cancer-related death [1]. Approximately 75–90% of these tumors are classified as primary HCC, and only 10–15% of cases are classified as cholangiocarcinoma (CCA) [9]. The annual HCC incidence is almost identical to its annual mortality, illustrating the high mortality of this disease [10,11].

In the Western world, chronic viral hepatitis C is currently shown to be responsible for the majority of all new HCC cases. However, a gradual decline is expected in the coming years due to the success of direct antiviral therapies (DAAs) [12]. Alcohol-induced liver disease also accounts for 10–20% of relevant HCC predisposing factors, and chronic viral hepatitis B accounts for 10–15% of cases. In older epidemiological analyses, a large proportion of patients with MASH were not recorded separately and were, thus, incorrectly assigned to the cohort of "cryptogenic" cirrhosis. This can partly explain divergent data on the frequency of cryptogenic liver disease (15–50%) as an HCC risk factor [13].

According to the most recent epidemiological studies, hepatic steatosis is the leading cause of the increasing incidence of HCC in Western industrialized nations. An analysis of the US Liver Transplant Registry between 2002 and 2017 found that the prevalence of patients with MASLD awaiting liver transplantation increased by 16% over the past 15 years, from 6% in 2002 to 22% in 2017, with MASH emerging as the fastest growing cause. The proportion of MASH as a cause of HCC increased 8.5-fold during the study period, from 2% in 2002 to 18% in 2017 [14]. It should be noted here that a high number of HCC cases arise against a background of MASH in the absence of cirrhosis [3]. This population has not been subject to international consensus screening for early HCC detection.

Another analysis of patients transplanted for end-stage liver disease between January 2002 and December 2016 using the European Liver Transplant Registry database showed a greater proportion of patients transplanted for MASH-associated HCC (39.1%) than non-MASH patients (28.9%, $p < 0.001$) [15].

A US cohort study of over 500,000 participants, published in 2018, showed a cumulative 5- (or 10)-year HCC risk of 0.8 (or 1.7) per 1000 patients in the MASLD cohort. Compared with healthy controls, patients with MASLD had an 8.6-fold increase in HCC risk. Patients with MASLD-associated cirrhosis showed a further marked increase in HCC risk, with an annual incidence of 0.8–2.3% per year. Approximately 20% of all hepatic steatosis-associated HCC cases occurred without predisposing cirrhosis [11].

A large German monocentric study with 1119 HCC patients, in 2015, demonstrated epidemiological differences between patients with MASH-related HCC and those with other HCC-predisposing liver diseases. For instance, patients with MASH-related HCC were older at initial diagnosis than others (68 vs. 65 years). The MASH-HCC cohort showed a higher prevalence of obesity (31.1% vs. 14.7%) and T2DM (66.7% vs. 37.85%), with higher body mass index (BMI) correlating with worse overall survival. In MASH patients, there was a trend toward multifocality (80% vs. 70%) with larger lesions overall (6.0 cm vs. 4.8 cm). In addition, there was a tendency toward an increased extrahepatic metastasis rate at the initial diagnosis. Interestingly, liver function was preserved to a greater extent than in other etiologies of HCC patients. Therefore, it can be postulated that the potential diagnosis of MASH-associated cirrhosis and its complications occurred later. As a result, HCC could only be detected at more advanced stages [16]. Consequently, optimization of current HCC screening algorithms and a more precise definition of the population to be screened seem all the more urgent.

Projections using dynamic Markov modeling for accounting for obesity and diabetes trends claim that the annual incidence of MASLD-related HCC in the USA would increase by 137% from 5160 cases in 2015 to 12,240 cases in 2030 [17].

3. Pathogenesis of HCC in MASH

Chronic hepatic inflammation in the setting of MASH represents a potential trigger of hepatocarcinogenesis even in the absence of predisposing cirrhosis. It is a complex, multifactorial process involving various risk factors (genomic instability, obesity, diabetes mellitus, and others) (Figure 1). Metabolic alterations (lipid and glucose metabolism) contribute to hepatic steatosis. However, additional factors such as genetic variants, oxidative/endoplasmic reticulum stress, mitochondrial dysfunction, altered immune response, and microbiome conditions fuel disease progression (hepatic inflammation, fibrogenesis, and carcinogenesis) [18]. For conciseness, we focus on the influence of genetics and polymorphisms. In addition, the microbiome's impact on the progression from MASH to HCC will be considered separately.

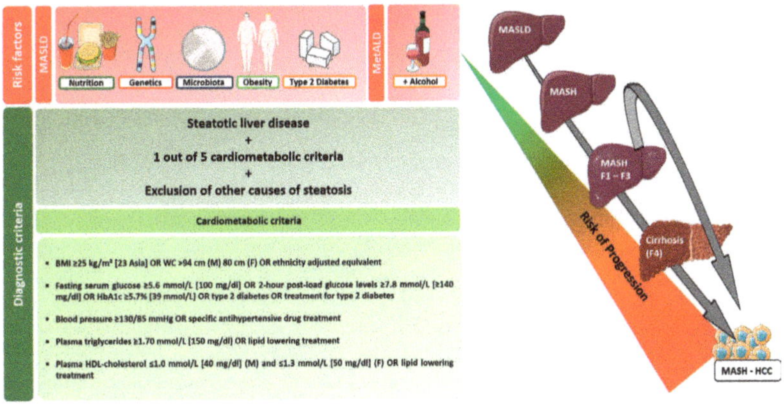

Figure 1. Diagnostic criteria and risk factors for MASLD and progression of MASLD to MASH-HCC [19]. In the progression from MASLD to MASH, MASH fibrosis, and finally cirrhosis, a variety of factors (type 2 diabetes mellitus, obesity, microbiome, genetic and epigenetic factors, lifestyle) have a similar effect on hepatocarcinogenesis. HCC risk appears to correlate with the extent of fibrosis, but MASH-associated HCC may develop even without cirrhosis. The time to which the above factors, particularly type 2 diabetes, have direct carcinogenic potential has not been conclusively determined. Abbreviations: MetALD: metabolic dysfunction and alcohol-associated steatotic liver disease, BMI: body mass index; WC: waist circumference.

3.1. Genetic Factors

Several gene polymorphisms have demonstrated an association between the prevalence of MASLD per se and the risk of progression to advanced MASH fibrosis [20]. The best known is the polymorphism of the PNPLA3 (patatin-like phospholipase domain containing 3) gene (variant rs738409 c.444 C>G, p.I148M) on chromosome 22. This PNPLA3 variant leads to impaired triglyceride mobilization of hepatic lipid droplets with increased hepatic lipid accumulation, but also the alteration of retinol storage in the liver, and altered retinol serum levels, especially in obese patients. Of note, the PNPLA3 polymorphism is associated with a 3-fold increased risk of HCC in its carriers, independent of other risk factors such as BMI, diabetes, and advanced fibrosis [21].

The TM6SF2 (transmembrane 6 superfamily member 2) gene polymorphism (variant rs58542926, c.449 C>T, p.E167K) on chromosome 19 manifests as a transport disorder of pre-VLDL particles. It correlates with the extent of steatosis and progression of fibrosis in

MASH, independent of obesity, diabetes, and the PNPLA3 genotype. However, the direct role of this TM6SF2 variant in hepatocarcinogenesis is controversial; the profibrogenic effect might indirectly promote progression to HCC [22,23].

Recent data also suggest that a loss-of-function variant in the 17-beta hydroxysteroid dehydrogenase 13 gene (HSD17B13) is associated with a reduced risk of chronic liver disease and of progression from steatosis to steatohepatitis and, thus, may represent another factor in HCC development [24–26].

In addition, individual studies of different mutations have demonstrated an unfavorable influence on the course of MASH. Mutations of the hereditary hemochromatosis protein-encoding gene (HFE) on chromosome 6 are associated with a complicated course of MASH, potentially favoring the development of HCC [27].

In conclusion, it is important to note that both hepatocarcinogenic single mutations and the increased genetic instability in patients with MASH, compared to those with MASLD, favor the development of HCC. Other mechanisms promoting the progression of MASH to HCC include epigenetic alterations causing aberrant DNA methylation and the expression of diverse microRNAs (miR-21, miR-23, miR-29, miR-93, miR-106, miR-155, miR-221, miR-222, and miR-519). Mechanisms and pathways involved are the major tumor-associated signaling cascades (transforming growth factor (TGF-), Wingless and INT-1 (Wnt)/catenin, mitogen-activated protein kinase (MAPK), Hedgehog, NF-κB, phosphatidylinositol 3-kinase (PI3K)/AKT/Mechanical (mammalian) target of rapamycin (mTOR)) [28], and CD44 [29].

3.2. Intestinal Microbiome

The gut microbiome is considered a key modulator of metabolism. It not only facilitates the extraction of nutrients and energy from food but is also essential for producing numerous metabolites, including bile acids, regulating various metabolic pathways. The gut microbiome outnumbers the human genome many times over. It plays a vital role in metabolism, immune system formation, health maintenance, tolerance development, and the prevention of colonization by pathogens. Its alteration, called dysbiosis, has been described for different intestinal metabolic and inflammatory diseases, including MASLD/MASH, alcoholic liver disease, cirrhosis, and complications [30–33].

Shifts of specific bacterial strains affect the production of bacterial-derived metabolic active components, including bile acids, ethanol, cytokines, short-chain fatty acids, or other inflammatory metabolites. Those may affect the host and possibly promote cancer-related risk factors or diseases [34]. In rodent studies, fecal microbiota transplantation increased the abundance of beneficial bacterial groups and alleviated the progression of MASH development [35].

Modulating the gut microbiota, e.g., with antibiotic treatment, may reduce the risk of hepatic carcinogenesis [36,37]. After antibiotic treatment, mice fed a high-fat diet showed reduced toxic secondary bile acids [38]. In MASLD and especially MASH, nutrition, metabolic disturbances, and related comorbidities such as diabetes may influence gut microbiota composition. Changes in the abundances of different bacterial groups have been described within other patient groups. The metabolism of specific bacterial groups affects the mucosal barrier, hepatic inflammation, fibrogenesis, and tumorigenesis [39]. The gut microbiota impacts energy balance, altering the uptake of calories derived from food and alcohol [40]. Emerging data indicate that specific characteristic changes in the gut microbiome are associated with MASLD and even cirrhosis, which is the primary driver of HCC development [31,41,42]. In MASLD-related HCC and viral hepatitis-related HCC (hepatitis B), specific modification of gut microbiota may represent a potential therapeutic option for HCC treatment [43]. A study comparing MASH and MASH-HCC patients with or without cirrhosis showed that changes in bacterial groups regulating bile acid metabolism affected hepatic fibrogenesis and liver injury. The changes in the bile acid pool were associated with an increased abundance of several bacterial strains, particularly Lactobacilli and Bacteroides, which were related to altered liver injury and liver stiffness [44].

In patients with MASH, the abundance of bile salt hydrolase-expressing bacteria is shifted, resulting in increased bile acid levels and the altered composition of the bile acid pool, which tends to increase the amount of secondary conjugated bile acids [45]. Alterations in bile acid composition may be associated with advanced fibrosis in MASH-HCC, suggesting an essential role in fibrosis-related tumorigenesis [46,47].

4. HCC Surveillance in MASLD/MASH

It is well established that patients participating in surveillance programs are diagnosed with HCC at less advanced stages, resulting in a survival advantage. Successful screening, however, requires reliable screening methods, on the one hand, and a definition of the population at risk, on the other hand.

In the context of a steady worldwide increase in HCC incidence on the background of MASH, there is a compelling need for a clear recommendation regarding HCC surveillance for this patient population. While the risk of HCC in MASH cirrhosis is sufficiently high to warrant surveillance, there is still no clear consensus among international guidelines for patients with MASH without cirrhosis. The German HCC S3 guideline provides sonographic screening with or without additional AFP determination every six months in MASH patients, even in non-cirrhotic livers. The current AASLD (American Association for the Study of Liver Diseases) HCC guidelines do not recommend surveillance in MASLD alone; only in MASH cirrhosis should sonography (±AFP determination) be offered every six months [48]. The 2018 EASL (European Association for the Study of the Liver) HCC guidelines also do not recommend general HCC surveillance in MASH; however, surveillance may be considered across etiologies in patients with advanced fibrosis (F3) without cirrhosis based on the individual risk profile. The EASL guidelines only recommend sonographic controls every six months without additional determination of AFP (European Association for the Study of the Liver).

As early as 2011, Ertle et al., in a retrospective epidemiological HCC monocenter study, described that 42% of patients in the MASLD/MASH-HCC group did not have cirrhosis at the time of HCC diagnosis [3]. Later publications showed a slightly lower proportion of non-cirrhotic MASH-HCC patients ranging from 23% to 37% [49,50].

The sensitivity of ultrasound-based surveillance in MASH patients is limited by investigator dependence, frequent concomitant obesity, and by MASH itself. A recent US meta-analysis examined the sensitivity of sonography in HCC detection, with or without concurrent AFP determination, in a cohort of 13,367 high-risk HCC patients of any etiology. Ultrasound alone showed a sensitivity of only 45%, which significantly increased to 63% with the addition of AFP. Thus, whether sonography alone can reduce HCC-associated mortality remains inconclusive and urgently requires data from prospective studies [51].

To address the unsatisfactory performance of sonography-based HCC surveillance, the GALAD score was developed, which includes age, sex, and the biomarkers AFP, AFP isoform L3 (AFP-L3), and Des-γ-Carboxy-Prothrombin (DCP), also known as protein induced by vitamin K absence or antagonist-II (PIVKA-II). In a recent Japanese–German multicenter study, it was shown in MASH patient collectives that the GALAD score can detect HCC with high sensitivity while comprising reasonable specificity in these patients in the presence and absence of cirrhosis. Even early-stage HCC could be detected with high reliability [52]. However, a limitation of this multicenter study is the lack of a direct comparison with an ultrasound-based surveillance strategy in the populations studied. A future prospective study should, therefore, investigate the performance of ultrasound alone and in combination with the GALAD score in NASH patients. A retrospective American multicenter study shows that this approach may be promising, with the combination of GALAD and ultrasound achieving an AUC of 0.98 (US alone AUC 0.82 vs. GALAD alone AUC 0.95) [53].

Recent data show that hepatocarcinogenesis is accompanied by epigenetic changes and mutations in various genes. These molecular changes, which have so far been described using tissue biopsies and cell models, can also be detected in circulating fragmented DNA (cfDNA) in the blood. As with other tumor entities, the molecular cfDNA can be used to analyze the molecular changes [54]. In this context, it would be all the more obvious to establish the use of liquid biopsy (LB) in HCC on a larger scale in order to be able to do justice to the still insufficient early diagnosis. The use of LB appears to be particularly useful in HCC, as the arterial hyperperfusion of tumor tissue means that there is a particularly high probability of circulating tumor cells (CTCs) and cfDNA being distributed in the peripheral blood.

5. Therapeutical Challenges in Patients with MASH-Related HCC

In unifocal HCC (Barcelona Clinic Liver Cancer (BCLC) stage 0/A), therapy of smaller lesions with curative intent is possible by radiofrequency ablation (RFA) or microwave ablation (MWA). For larger lesions, liver resection is indicated, assuming there is preserved liver synthesis function in the absence of portal hypertension. A large US monocenter study demonstrated that patients with HCC and MASH had better liver function at the time of resection and showed a lower rate of cirrhosis than patients with hepatitis C or alcohol-related liver disease. As a result, a more significant proportion of patients could be resected in MASH-associated HCC compared with the other etiologies. In addition, improved overall survival (OS) has been detected compared with the HCV/alcohol cohort [55].

For functionally irresectable HCC with a single lesion <5 cm or a maximum of 3 lesions <3 cm (Milan criteria), evaluation for liver transplantation (LT) is recommended. It should be emphasized at this point that, as mentioned at the outset, MASH is primarily responsible for the sharp increase in HCC-related transplant listings in the United States.

The resulting increase in demand for donor organs requires more optimized allocation of organs. However, current data suggest that the Milan criteria, which are still considered the standard, could deny many potential organ recipients suffering from MASLD-associated HCC from receiving a donor organ due to the more advanced tumor stages at initial diagnosis. In recent years, several different approaches have been taken to extend these current criteria. Among the best known are the UCSF criteria, defined as a single tumor that is less than or equal to 6.5 cm, or up to 3 lesions with the largest lesion less than or equal to 4.5 cm, with a total tumor diameter no greater than 8 cm. Here, patients transplanted within these criteria had a 5-year survival rate of 75.2% [56]. In contrast, the Up-to-Seven criteria propose that for HCC, with 7 as the total of the size of the largest lesion in cm and the number of lesions, and without vascular invasion, could have survival outcomes as good as those within the Milan criteria. In the patient cohort without microvascular invasion but within the Up-to-Seven criteria, 5-year overall survival was 71.2% [57]. Later, Mazzaferro et al. developed the Metroticket 2.0 model, incorporating the size of the largest HCC lesion, the total number of nodules, and the AFP level in 2018. According to this model, the overall 5-year survival rate reached 79.7%, and tumor recurrence could be more accurately estimated than with the Milan, UCSF, or Up-to-7 criteria [58].

The additional use of biomarkers in decision algorithms for LT is a matter of debate. However, there is a correlation between microvascular invasion (MVI) [59], histological differentiation, and the risk of post-LT HCC recurrence, as reported in several publications. AFP and DCP levels are reliable surrogate markers of the biological behavior of HCC due to their strong correlation with MVI and degree of differentiation [60] and micro-intrahepatic metastasis. The inclusion of these markers in the development of criteria has become widely accepted in the last decade.

Even at the listing stage, MASH presents unique features compared to other etiologies. It is well known that metabolic syndrome (especially T2DM and obesity), related coronary artery disease, and chronic renal failure are predictors of worse postoperative and long-term outcomes. On the LT waiting list, patients with MASH frequently exhibit numerous metabolically related risk factors simultaneously. This results in several potential pitfalls for LT-listed patients with MASH: an overall older patient age at the time of listing for liver transplantation, comorbidities, and a lower model for end-stage liver disease (MELD) score than other etiologies. In the context of obesity and transplantation, the results of different meta-analyses are wildly divergent. Some studies report worse post-LT survival in patients with high BMI, whereas other papers report similar long-term outcomes as in normal-weight patients. The postoperative complication rate is higher in obese patients. Whether combining bariatric surgery with LT may improve overall survival needs to be investigated using more extensive randomized trials. However, T2DM should be emphasized as a predictor of poor graft and long-term patient survival. In the future, the integrated weighting of MASH-associated comorbidities in selecting MASH-HCC patients for potential LT listing will be a significant challenge [61].

5.1. Treatment of Intermediate-Stage HCC

For patients presenting non-diffuse, multifocal tumor nodules without extrahepatic tumor manifestations or macrovascular invasion and compensated liver function (BCLC stage B), a transarterial chemoembolization (TACE) treatment is recommended, according to the EASL guidelines [62]. A survival benefit from TACE was first demonstrated in 2002 in a randomized controlled trial comparing TACE to symptomatic treatment [63]. Since then, the technique has undergone several improvements, e.g., super-selective chemoembolization of the target volume can facilitate optimal sparing of healthy liver tissue surrounding the tumor and, thus, help preserve liver function. As a different concept for locoregional therapy, transarterial radioembolization (TARE) is an alternative treatment option for patients with locally advanced HCC who are not eligible for TACE [62]. To the best of our knowledge, little evidence exists regarding MASLD's influence on the efficacy and safety of locoregional therapies.

In a retrospective study comparing the outcome of TACE in cirrhotic patients with MASH-related HCC to patients with HCV- or alcohol-related HCC, no significant differences in OS, response rate, or time to progression (TTP) could be identified. Apart from adverse events, no significant differences between the groups were also observable [64]. Concerning MASLD-related comorbidities, another retrospective study could identify obesity (BMI $\geq$ 25 kg/m^2) as a risk factor for worse tumor response after TACE [65]. Furthermore, T2DM could be identified as an independent predictor for worse OS in patients with non-viral HCC undergoing TACE [66]. The negative influence of T2DM could be demonstrated in a further retrospective study. Patients suffering from liver cirrhosis and concomitant T2DM had a worse long-term prognosis after TACE [67]. Interestingly, the drug Metformin, frequently prescribed for T2DM, seems to have beneficial effects on the outcomes after TACE in patients with T2DM and early-stage HCC [68]. Regarding TARE, a retrospective study conducted by Schotten and colleagues found MASLD-related comorbidities not to influence the critical outcomes of TARE [69].

5.2. The Treatment Landscape for Advanced-Stage HCC

Over the past few years, the landscape of systemic therapy options for patients with advanced-stage HCC and compensated liver function has significantly expanded. In 2008, for the first time, two independently conducted studies (SHARP trial and Asia-Pacific trial) demonstrated a survival benefit for HCC patients treated with systemic therapy, leading to the approval of the tyrosine kinase inhibitor (TKI) sorafenib [70,71]. Based on the results of the REFLECT trial, lenvatinib was approved in 2018 as a further first-line treatment option. Although lenvatinib did not achieve superior OS (OS lenvatinib: 13.6 months vs. OS sorafenib: 12.3 months, hazard ratio (HR) 0.92, 95% confidence interval (CI) 0.79–1.06), progression-free survival (PFS), time to progression (TTP), as well as the objective response rate (ORR) significantly improved in the lenvatinib group compared to sorafenib [72]. Further, two more TKIs, regorafenib and cabozantinib, have been approved as second (or third) line therapies after pretreatment with sorafenib [73,74]. With the introduction of ramucirumab, therapy options in second-line treatment further expanded to include a monoclonal antibody directed against vascular endothelial growth factor receptor type 2 (VEGFR2). Ramucirumab, the only drug that has to be guided by a biomarker, has been approved for patients with tumor progression on sorafenib and an alpha fetoprotein (AFP) value of at least 400 ng/mL [75]. Developing the aforementioned new TKIs and monoclonal antibodies has significantly improved the OS of HCC patients eligible for systemic therapy. However, the improved OS is not consistently matched by an improved ORR. Furthermore, in HCC patients with MASH, TKI treatment approaches are potentially limited by extrahepatic manifestations of the metabolic syndrome such as obesity, diabetes, hyperlipidemia, and arterial hypertension, frequently accompanied by chronic kidney and cardiovascular disease. Here, TKIs may aggravate arterial hypertension and increase the risk of myocardial infarction, potentially necessitating treatment de-escalation or discontinuation. Therefore, in this patient cohort, alternative treatment approaches with more favorable safety profiles are required.

Regarding the critical endpoint of ORR, a clear improvement was achieved with the combination of the programmed death-ligand 1 (PD-L1) inhibitor atezolizumab and the VEGF inhibitor bevacizumab tested against sorafenib as first-line therapy (IMbrave 150 trial) [76]. In addition to a survival benefit of nearly six months (OS atezolizumab plus bevacizumab: 19.2 months vs. OS sorafenib 13.4 months), this new combination therapy induced an ORR in 30% of all participants, and 25 patients (8%) even showed a complete response. As a result, the combination of atezolizumab and bevacizumab became the new standard of care for first-line systemic treatment in 2020. Supported by the results of the above-mentioned IMbrave150 study, an immune checkpoint inhibitor (ICI)-based combination therapy is seen as having great potential in treating liver cancer. Based on the positive results of the phase III HIMALAYA trial, the dual immune checkpoint blockade consisting of the cytotoxic T-lymphocyte-associated Protein 4 (CTLA-4) antibody tremelimumab and the PD-1 inhibitor durvalumab was recently approved as an additional first-line therapy [77]. The so-called STRIDE regimen (single tremelimumab regular interval durvalumab) showed both an OS benefit (OS STRIDE 16.4 months vs. OS sorafenib 13.8 months, HR 0.78, 96% CI 0.65–0.92, $p = 0.0035$) and an improved treatment response rate (ORR STRIDE 20.1% vs. sorafenib 5.1%). In addition to combination therapy, durvalumab was also tested as a monotherapy and was found to be non-inferior to sorafenib [77].

Apart from the dual ICI blockade, the combination of an ICI with TKIs has also been tested. Recently, the results of LEAP-002, a phase III study evaluating the combination of pembrolizumab plus lenvatinib versus lenvatinib as monotherapy, were reported. Although the combination therapy did not reach the pre-specified statistical significance for OS, it achieved the longest OS among systemic treatments for advanced HCC to date (OS lenvatinib plus pembrolizumab: 21.2 months vs. OS lenvatinib: 19 months, HR 0.84, 95% CI 0.708–0.997, $p = 0.0227$) [78]. The combination of atezolizumab and cabozantinib investigated in the COSMIC 312 trial only improved PFS, but not OS [79]. In addition to their use in the first-line setting, ICIs have also been investigated in the second-line setting. For example, the CheckMate 040 phase I/II trial found that the combination of ipilimumab and nivolumab induced a durable response in a subset of patients with HCC who had previously received sorafenib treatment [34].

5.3. Immunotherapy and Targeted Therapies in Non-Viral HCC

Regarding the expanding therapeutic armamentarium, a new challenge arises—namely, to identify the most suitable treatment sequence for each patient as systemic treatment has been conducted in a "one-size-fits-all" fashion for many years. To what extent the etiology of the underlying chronic liver disease contributes to therapy success or failure is currently gaining increased attention. In light of the rapidly increasing incidence, patients with MASLD-associated HCC are of particular interest. A closer look into the patient cohorts of the above-mentioned phase III trials reveals that only small subgroups had non-viral HCC. Furthermore, only a few studies reported a precise number of MASLD-related etiology (Table 1) [70–76,80]. Interestingly, to our knowledge, only the CELESTIAL trial considered MASLD as a stratification parameter. The current knowledge regarding therapy implications for non-viral HCC is primarily based on findings from retrospective cohorts, potentially biased subgroup analyses, and meta-analyses, but no specific RCTs have been conducted assessing this issue.

Table 1. Role of HCC etiologies in selected clinical trials on systemic therapy with subsequent approval of the therapy regimen by the EMA.

Trial And Treatment Arms	Etiology *	Stratification Criteria	Primary Endpoints	Secondary Endpoints
SHARP—sorafenib vs. placebo [70]	HCV 29% HBV 19% Alcohol 26% Unknown 16% Other 9%	Geographical region ECOG PS (0 vs. 1–2) Macrovascular invasion or extrahepatic spread (presence vs. absence)	OS 10.7 vs. 7.9 (HR 0.69, 95%-CI 0.55–0.87, $p < 0.0019$) TTSP 4.1 vs. 4.9 (HR 1.08, 95% CI 0.88–1.31, $p = 0.77$)	TTRP 5.5 vs. 2.8 (HR 0.58, 95% CI 0.45–0.74, $p < 0.001$) DCR 43% vs. 32%; $p = 0.002$
Asia-Pacific—sorafenib vs. placebo [71]	HCV 70.7% HBV 10.7%	Geographical region Macrovascular invasion and/or extrahepatic spread (presence vs. absence) ECOG PS (0–2)	OS 6.5 vs. 4.2 (HR 0.68 95% CI 0.50–0.93. $p = 0.014$)	TTP 2.8 vs. 1.4 (HR 0.57, 95% CI 0.42–0.79, $p = 0.0005$) TTSP 3.5 vs. 3.4 (HR 0.90, 95% CI 0.67–1.22, $p = 0.50$) DCR 35.3% vs. 15.8% ($p = 0.0019$)
IMbrave150—atezolizumab + bevacizumab vs. sorafenib [76,80]	HCV 21% HBV 49% Non-viral 30% #	Geographical region (Asia excluding Japan vs. rest of the world) Macrovascular invasion or extrahepatic spread (presence vs. absence) Baseline AFP < 400 ng/mL vs. ≥400 ng/mL ECOG PS (0 vs. 1)	OS 19.2 vs. 13.4 (HR 0.66, 95% CI 0.52–0.85, $p < 0.001$) PFS 6.9 vs. 4.3 (HR 0.65, 95% CI 0.53–0.81, $p < 0.001$)	ORR 30% vs. 11% ($p < 0.001$) DoR 18.1 (95% CI 14.6-NE) vs. 14.9 (95% CI 4.9–17.0)
HIMALAYA—durvalumab vs. sorafenib and durvalumab + tremelimumab vs. sorafenib [77]	HBV 31% HCV 28% Nonviral 41%	Asia (excluding Japan) 39.7% and rest of world 60.3%. ECOG PS (0 vs. 1), AFP ≥ 400 (yes vs. no)), macrovascular invasion (yes vs. no), extrahepatic disease (yes vs. no), PD-L1 status pos. vs. neg.)	OS STRIDE 16.4 vs. sorafenib 13.8 (HR 0.78, 96% CI 0.65–0.92, $p = 0.0035$)	ORR STRIDE 20.1% vs. sorafenib 5.1% TTP 5.4 (95% CI, 3.8 to 5.6) in STRIDE arm, 3.8 (95% CI, 3.7 to 5.4) in durvalumab arm, and 5.6 (95% CI, 5.1 to 5.8) in Sorafenib arm
REFLECT—lenvatinib vs. sorafenib [72]	HCV 19% HBV 52.5% Alcohol 7.5% Other 7.9% Unknown 13%	Geographical region (Asia-Pacific or Western) ECOG PS (0 vs. 1) Presence or absence of macroscopic portal vein invasion and/or extrahepatic spread Body weight (<60 kg or ≥60 kg)	OS 13.6 vs. 12.3 (HR 0.92, 95% CI 0.79–1.06	PFS 7.4 vs. 3.7 (HR 0.66, 95% CI 0.57–0.77, $p < 0.0001$) TTP 8.9 vs. 3.7 (HR 0.63, 95% CI 0.53–0.73, $p < 0.0001$) ORR 24.1% vs. 9.2% (OR 3.13, 95% CI 2.15–4.56, $p < 0.0001$)

Table 1. *Cont.*

Trial And Treatment Arms	Etiology *	Stratification Criteria	Primary Endpoints	Secondary Endpoints
RESORCE—regorafenib vs. placebo [73]	HCV 21% HBV 38% Alcohol 24% Unknown 17% MASH 7% Other 7%	Geographical region (Asia vs. rest of world) Macrovascular invasion (yes vs. no) Extrahepatic spread (yes vs. no) Baseline AFP < 400 ng/mL vs. ≥400 ng/mL ECOG PS (0 vs. 1)	OS 10.6 vs. 7.8 (HR 0.68, 95% CI 0.50–0.79, $p < 0.0001$)	PFS 3.1 vs. 1.5 (HR 0.46, 95% CI 0.37–0.56, $p < 0.0001$) TTP 3.2 vs. 1.5 (HR 0.44, 95% CI 0.36–0.55, $p < 0.0001$) ORR 11% vs. 4% ($p = 0.0047$) DCR 65% vs. 36% ($p < 0.0001$)
CELESTIAL—cabozantinib vs. placebo [74]	HCV 24% HBV 38% HBV + HCV 2% Alcohol 24% MASH 9% Other 5% Unknown 16%	Etiology (HBV with or without HCV vs. HCV without HBV, or other) Geographical region (Asia or other) Extrahepatic spread and/or macrovascular invasion (yes vs. no)	OS 10.2 vs. 8 (HR 0.76; 95% CI 0.63–0.92, $p = 0.005$)	PFS 5.2 vs. 1.9 (HR 0.44, 95% CI 0.36–0.52, $p < 0.001$) ORR 4% vs. <1% ($p = 0.009$)
REACH-2—ramucirumab vs. placebo [75]	HCV 24% HBV 36% Alcohol 24% MASH 10% Cryptogenic 6% Other 9%	Geographical region (America, Europe, Australia, Israel vs. Asia, excluding Japan vs. Japan) Macrovascular invasion (yes vs. no) ECOG PS (0 vs. 1)	OS 8.5 vs. 7.3 (HR 0.710. 95% CI 0.53–0.95, $p = 0.0199$)	PFS 2.8 vs. 1.6 (HR 0.452, 95% CI 0.34–0.60, $p < 0.0001$) ORR 5% vs.1%, $p = 0.1697$) TTRP 3 vs. 1.6 (HR 0.427, 95% CI 0.31–0.58, $p < 0.0001$)

* Distribution is given for the treatment arm. # Non-viral included alcohol, other, unknown non-hepatitis B or C. SHARP: Sorafenib Hepatocellular Carcinoma Assessment Randomized Protocol, OS: overall survival, TTSP: time to symptomatic progression, TTRP: time to radiologic progression, DCR: disease control rate, HCV: hepatitis C virus, HBV: hepatitis B virus, ECOG: Eastern Cooperative Oncology Group Performance Score, PFS: progression-free survival, ORR: objective response rate, DoR: duration of response, NE: not evaluable, AFP: alpha fetoprotein, MASH non-alcoholic steatohepatitis.

It was hypothesized that patients with non-viral HCC may benefit less from ICI-based immunotherapy. Subgroup analyses of the IMbrave150 trial indicate a clear survival benefit for viral HCC, but the HR for non-viral HCC was not statistically significant (HR 1.05, 95% CI 0.68–1.63) [76]. These findings are further supported by two meta-analyses that merged data from the three large ICI phase III trials (IMbrave 150, Checkmate 459, and Keynote 240). The analyses showed that ICI-based therapies seemed to be less effective in patients with non-viral HCC [81,82]. The possible underlying mechanisms of resistance against immunotherapy in MASLD-associated HCC are not fully understood. It seems that CD8 + PD1 + -T-cells contribute to the pathophysiology of MASLD, and interestingly, anti-PD1 therapy promotes adverse tumor necrosis factor-alpha (TNFα)-secretion by these cells in the MASLD setting [81]. However, in contrast to the aforementioned studies, newer trials reported more promising results, questioning the assumption that ICI-based therapy is not the optimal choice for patients with non-viral HCC. The recently published HIMALAYA trial demonstrated a significant survival benefit for patients with non-viral HCC receiving the STRIDE regime (HR 0.74, 95% CI 0.57–0.95) [77]. Another meta-analysis did not demonstrate a notable difference in treatment efficacy associated with the underlying HCC etiology [83].

Regarding TKI-based therapy, the role of the etiology of the underlying chronic liver disease is also not fully understood. A recent meta-analysis as well as a real-world study, which included patients treated with lenvatinib from German tertiary cancer centers, did not identify any differences in HCC treatment success according to the etiology of chronic liver disease [82,84]. In the SHARP trial, the survival benefit of sorafenib over the placebo was consistent for the different etiologies HBV, HCV, and alcohol. However, the effect seemed more pronounced in patients suffering from HCC due to HCV infection [85]. In a multicentric retrospective study conducted in a Japanese cohort investigating the impact of lenvatinib in accordance with the HCC etiology (MASLD/MASH group vs. Viral/Alcohol group), OS and PFS tended to be even better in the MASLD/MASH group [86]. Another retrospective study featuring a European cohort of HCC patients receiving lenvatinib treatment, conducted by Sacco and colleagues, also found that patients with non-viral HCC had longer OS compared to patients with viral-related HCC [87]. In further retrospective studies, two other working groups investigated the effect of lenvatinib compared to the combination therapy of atezolizumab and bevacizumab in patients with non-viral HCC and were able to identify a favorable effect for lenvatinib in this patient group [88,89]. In summary, etiology cannot currently serve as a selection criterion for the preferred therapy due to insufficient high-quality data from prospective studies using MASLD as a stratification criterion or even considering MASLD as a distinct subgroup. It is important to note that MASLD has also not been considered as a separate patient group in phase III studies to date and it can also be challenging to identify MASH as the sole risk factor for the development of HCC. Therefore, the combination of atezolizumab and bevacizumab remains the recommended first-line therapy for all patients.

6. Future Directions

The relationship between obesity and type II diabetes mellitus in the progression of MASLD to fibrosis, cirrhosis, and HCC is well-established. Adipokines and insulin resistance are among the factors that orchestrate this progression. By effectively treating the additive risk factors of obesity and diabetes, the risk of progression to HCC can be mitigated.

Drug therapy for morbid obesity remains a clinical challenge, which is why bariatric surgery continues to play an important role here. According to a recent large meta-analysis by Ramai and colleagues (9 studies, 18,423,546 controls vs. 1,091,204 bariatric patients), surgical treatment of obesity can significantly reduce the risk of concomitant progression from MASLD to HCC. The pooled rate per 1000 person-years was 0.05 (95% CI: 0.02–0.07) in bariatric surgery patients versus 0.34 (95% CI: 0.20–0.49) in the control group with an incidence rate ratio of 0.28 (95% CI: 0.18–0.42). Bariatric weight reduction reduces the risk of HCC in obese patients, as indicated by the data [90].

The efficacy of glucagon-like peptide-1 (GLP-1) receptor agonists such as liraglutide in the treatment of MASLD/MASH remains unclear. In an animal study, mice with streptozotocin- and high-fat diet-induced diabetes with MASH were treated with liraglutide or saline for 14 days in the control arm. The two groups were compared in terms of glycemic control: liver histology and hepato-carcinogenesis. While fasting plasma glucose was significantly lower in the liraglutide group than in the control group, fasting insulin levels were significantly higher in the test group than in the control group. Impressively, in contrast to the control group, liraglutide completely suppressed the development of HCC and also significantly attenuated steatosis and inflammation [91].

Another treatment option for the effective management of T2DM, one of the major risk factors for MASLD, is sodium-glucose co-transporter (SGLT2) inhibitors. Experimental studies in animal models have suggested that SGLT2 inhibitors may have beneficial modulatory effects on MASLD, and several studies in patients have demonstrated their beneficial effects on liver enzymes, BMI, hyperlipidemia, hyperglycemia, and insulin resistance in MASLD patients, potentially inhibiting the progression of liver damage to HCC in these patients [92]. In the future, effective control of predisposing factors for HCC will undoubtedly be an essential preventive component in reducing the increasing incidence of HCC worldwide.

7. Conclusions

MASH is considered to be a significant cause of the increasing incidence of HCC worldwide. Chronic hepatic inflammation in MASH triggers hepatocarcinogenesis even without predisposing cirrhosis. It is a complex, multifactorial process involving various risk factors (genomic instability, obesity, diabetes mellitus, etc.). The heterogeneity of HCCs resulting from this multitude of MASH-associated risks makes it difficult to define a risk population to be screened clearly. In consequence, MASLD is associated with lower HCC surveillance receipt, lower early-stage cancer detection, and modestly worse OS. Furthermore, previous approaches to ultrasound-based HCC surveillance show clear limitations.

Therefore, the evaluation of alternative, e.g., biomarker-based, screening methods is mandatory. The selection of appropriate HCC treatment options for MASH patients must also consider MASH-associated and potentially therapy-limiting comorbidities, as non-cancer mortality significantly impacts OS even in curatively treated patients. Here future systemic therapy trials urgently need to address MASLD as a separate subgroup to avoid underrepresentation of this worldwide epidemic disease.

Author Contributions: Conceptualization, J.B., P.P.M. and C.L.; investigation, J.B., P.P.M. and C.L.; performed the literature review, J.B., P.P.M. and C.L. writing—original draft preparation, J.B., P.P.M. and C.L.; writing—review and editing, J.B., P.P.M., C.L. and J.D.C.; designed the project, J.B., P.P.M. and C.L. All authors provided intellectual input. All authors have read and agreed to the published version of the manuscript.

Funding: This research received no external funding.

Institutional Review Board Statement: Not applicable.

Informed Consent Statement: Not applicable.

Data Availability Statement: Not applicable.

Acknowledgments: Figures were created with "Biological illustration" (http://smart.servier.com, accessed on 10 February 2024) by Servier, used under Creative Commons Attribution 3.0 Unported License/modified by Paul Manka.

Conflicts of Interest: C.L. and J.D.C. certify that they have NO affiliations with or involvement in any organization or entity with any financial interest (such as honoraria; educational grants; participation in speakers' bureaus; membership, employment, consultancies, stock ownership, or other equity interest; and expert testimony or patent-licensing arrangements), or nonfinancial interest (such as personal or professional relationships, affiliations, knowledge, or beliefs) in the subject matter or materials discussed in this manuscript. P.P.M. reports personal fees and non-financial support from Intercept Pharmaceuticals. J.B. reports personal fees and non-financial support from Roche, BMS, Novartis, EISAI, Falk, Ipsen, MSD, Boston Scientific. No other potential conflict of interest relevant to this article was reported.

References

1. Ferlay, J.; Colombet, M.; Soerjomataram, I.; Mathers, C.; Parkin, D.M.; Piñeros, M.; Znaor, A.; Bray, F. Estimating the global cancer incidence and mortality in 2018: GLOBOCAN sources and methods. *Int. J. Cancer* **2019**, *144*, 1941–1953. [CrossRef] [PubMed]
2. Rinella, M.E. Examining the Nomenclature Change from NAFLD and NASH to MASLD and MASH. *Gastroenterol. Hepatol.* **2023**, *19*, 697–699.
3. Ertle, J.; Dechêne, A.; Sowa, J.-P.; Penndorf, V.; Herzer, K.; Kaiser, G.; Schlaak, J.F.; Gerken, G.; Syn, W.-K.; Canbay, A. Non-alcoholic fatty liver disease progresses to hepatocellular carcinoma in the absence of apparent cirrhosis. *Int. J. Cancer* **2011**, *128*, 2436–2443. [CrossRef] [PubMed]
4. Stine, J.G.; Wentworth, B.J.; Zimmet, A.; Rinella, M.E.; Loomba, R.; Caldwell, S.H.; Argo, C.K. Systematic review with meta-analysis: Risk of hepatocellular carcinoma in non-alcoholic steatohepatitis without cirrhosis compared to other liver diseases. *Aliment. Pharmacol. Ther.* **2018**, *48*, 696–703. [CrossRef] [PubMed]
5. Desai, A.; Sandhu, S.; Lai, J.-P.; Sandhu, D.S. Hepatocellular carcinoma in non-cirrhotic liver: A comprehensive review. *World J. Hepatol.* **2019**, *11*, 1–18. [CrossRef] [PubMed]
6. Balakrishnan, M.; El-Serag, H.B. Editorial: NAFLD-related hepatocellular carcinoma—Increasing or not? With or without cirrhosis? *Aliment. Pharmacol. Ther.* **2018**, *47*, 437–438. [CrossRef] [PubMed]

7. Loomba, R.; Lim, J.K.; Patton, H.; El-Serag, H.B. AGA Clinical Practice Update on Screening and Surveillance for Hepatocellular Carcinoma in Patients with Nonalcoholic Fatty Liver Disease: Expert Review. *Gastroenterology* **2020**, *158*, 1822–1830. [CrossRef] [PubMed]
8. Singal, A.G.; El-Serag, H.B. Rational HCC screening approaches for patients with NAFLD. *J. Hepatol.* **2022**, *76*, 195–201. [CrossRef]
9. Center, M.M.; Jemal, A. International trends in liver cancer incidence rates. *Cancer Epidemiol. Biomark. Prev.* **2011**, *20*, 2362–2368. [CrossRef]
10. Ascha, M.S.; Hanouneh, I.A.; Lopez, R.; Tamimi, T.A.; Feldstein, A.F.; Zein, N.N. The incidence and risk factors of hepatocellular carcinoma in patients with nonalcoholic steatohepatitis. *Hepatology* **2010**, *51*, 1972–1978. [CrossRef]
11. Kanwal, F.; Kramer, J.R.; Mapakshi, S.; Natarajan, Y.; Chayanupatkul, M.; Richardson, P.A.; Li, L.; Desiderio, R.; Thrift, A.P.; Asch, S.M.; et al. Risk of Hepatocellular Cancer in Patients with Non-Alcoholic Fatty Liver Disease. *Gastroenterology* **2018**, *155*, 1828–1837. [CrossRef]
12. Sayiner, M.; Golabi, P.; Younossi, Z.M. Disease Burden of Hepatocellular Carcinoma: A Global Perspective. *Dig. Dis. Sci.* **2019**, *64*, 910–917. [CrossRef]
13. White, D.L.; Kanwal, F.; El–Serag, H.B. Association Between Nonalcoholic Fatty Liver Disease and Risk for Hepatocellular Cancer, Based on Systematic Review. *Clin. Gastroenterol. Hepatol.* **2012**, *10*, 1342–1359.e2. [CrossRef] [PubMed]
14. Younossi, Z.; Stepanova, M.; Ong, J.P.; Jacobson, I.M.; Bugianesi, E.; Duseja, A.; Eguchi, Y.; Wong, V.W.; Negro, F.; Yilmaz, Y.; et al. Nonalcoholic Steatohepatitis Is the Fastest Growing Cause of Hepatocellular Carcinoma in Liver Transplant Candidates. *Clin. Gastroenterol. Hepatol.* **2019**, *17*, 748–755. [CrossRef] [PubMed]
15. Haldar, D.; Kern, B.; Hodson, J.; Armstrong, M.J.; Adam, R.; Berlakovich, G.; Fritz, J.; Feurstein, B.; Popp, W.; Karam, V.; et al. Outcomes of liver transplantation for non-alcoholic steatohepatitis: A European Liver Transplant Registry study. *J. Hepatol.* **2019**, *71*, 313–322. [CrossRef] [PubMed]
16. Weinmann, A.; Koch, S.; Niederle, I.M.; Schulze-Bergkamen, H.; König, J.; Hoppe-Lotichius, M.; Hansen, T.; Pitton, M.B.; Düber, C.; Otto, G.; et al. Trends in epidemiology, treatment, and survival of hepatocellular carcinoma patients between 1998 and 2009: An analysis of 1066 cases of a German HCC Registry. *J. Clin. Gastroenterol.* **2014**, *48*, 279–289. [CrossRef] [PubMed]
17. Estes, C.; Anstee, Q.M.; Arias-Loste, M.T.; Bantel, H.; Bellentani, S.; Caballeria, J.; Colombo, M.; Craxi, A.; Crespo, J.; Day, C.P.; et al. Modeling NAFLD disease burden in China, France, Germany, Italy, Japan, Spain, United Kingdom, and United States for the period 2016–2030. *J. Hepatol.* **2018**, *69*, 896–904. [CrossRef] [PubMed]
18. Anstee, Q.M.; Reeves, H.L.; Kotsiliti, E.; Govaere, O.; Heikenwalder, M. From NASH to HCC: Current concepts and future challenges. *Nat. Rev. Gastroenterol. Hepatol.* **2019**, *16*, 411–428. [CrossRef] [PubMed]
19. Rinella, M.E.; Lazarus, J.V.; Ratziu, V.; Francque, S.M.; Sanyal, A.J.; Kanwal, F.; Romero, D.; Abdelmalek, M.F.; Anstee, Q.M.; Arab, J.P.; et al. A multisociety Delphi consensus statement on new fatty liver disease nomenclature. *Hepatology* **2023**, *78*, 1966–1986. [CrossRef]
20. Ioannou, G.N. Epidemiology and risk-stratification of NAFLD-associated HCC. *J. Hepatol.* **2021**, *75*, 1476–1484. [CrossRef]
21. Liu, Y.L.; Patman, G.L.; Leathart, J.B.; Piguet, A.C.; Burt, A.D.; Dufour, J.F.; Day, C.P.; Daly, A.K.; Reeves, H.L.; Anstee, Q.M. Carriage of the PNPLA3 rs738409 C > G polymorphism confers an increased risk of non-alcoholic fatty liver disease associated hepatocellular carcinoma. *J. Hepatol.* **2014**, *61*, 75–81. [CrossRef] [PubMed]
22. Stickel, F.; Buch, S.; Nischalke, H.D.; Weiss, K.H.; Gotthardt, D.; Fischer, J.; Rosendahl, J.; Marot, A.; Elamly, M.; Casper, M.; et al. Correction: Genetic Variants in PNPLA3 and TM6SF2 Predispose to the Development of Hepatocellular Carcinoma in Individuals with Alcohol-Related Cirrhosis. *Am. J. Gastroenterol.* **2018**, *113*, 1099. [CrossRef] [PubMed]
23. Yang, J.; Trépo, E.; Nahon, P.; Cao, Q.; Moreno, C.; Letouzé, E.; Imbeaud, S.; Gustot, T.; Deviere, J.; Debette, S.; et al. PNPLA3 and TM6SF2 variants as risk factors of hepatocellular carcinoma across various etiologies and severity of underlying liver diseases. *Int. J. Cancer* **2019**, *144*, 533–544. [CrossRef] [PubMed]
24. Gellert-Kristensen, H.; Nordestgaard, B.G.; Tybjaerg-Hansen, A.; Stender, S. High Risk of Fatty Liver Disease Amplifies the Alanine Transaminase-Lowering Effect of a HSD17B13 Variant. *Hepatology* **2020**, *71*, 56–66. [CrossRef] [PubMed]
25. Abul-Husn, N.S.; Cheng, X.; Li, A.H.; Xin, Y.; Schurmann, C.; Stevis, P.; Liu, Y.; Kozlitina, J.; Stender, S.; Wood, G.C.; et al. A Protein-Truncating HSD17B13 Variant and Protection from Chronic Liver Disease. *N. Engl. J. Med.* **2018**, *378*, 1096–1106. [CrossRef] [PubMed]
26. Ma, Y.; Belyaeva, O.V.; Brown, P.M.; Fujita, K.; Valles, K.; Karki, S.; de Boer, Y.S.; Koh, C.; Chen, Y.; Du, X.; et al. 17-Beta Hydroxysteroid Dehydrogenase 13 Is a Hepatic Retinol Dehydrogenase Associated with Histological Features of Nonalcoholic Fatty Liver Disease. *Hepatology* **2019**, *69*, 1504–1519. [CrossRef] [PubMed]
27. Pietrangelo, A. Iron in NASH, chronic liver diseases and HCC: How much iron is too much? *J. Hepatol.* **2009**, *50*, 249–251. [CrossRef] [PubMed]
28. Kutlu, O.; Kaleli, H.N.; Ozer, E. Molecular Pathogenesis of Nonalcoholic Steatohepatitis- (NASH-) Related Hepatocellular Carcinoma. *Can. J. Gastroenterol. Hepatol.* **2018**, *2018*, 8543763. [CrossRef] [PubMed]
29. Dhar, D.; Antonucci, L.; Nakagawa, H.; Kim, J.Y.; Glitzner, E.; Caruso, S.; Shalapour, S.; Yang, L.; Valasek, M.A.; Lee, S.; et al. Liver Cancer Initiation Requires p53 Inhibition by CD44-Enhanced Growth Factor Signaling. *Cancer Cell* **2018**, *33*, 1061–1077.e1066. [CrossRef]
30. Brenner, D.A.; Paik, Y.-H.; Schnabl, B. Role of Gut Microbiota in Liver Disease. *J. Clin. Gastroenterol.* **2015**, *49*, S25–S27. [CrossRef]

31. Etienne-Mesmin, L.; Chassaing, B.; Gewirtz, A.T. Tryptophan: A gut microbiota-derived metabolites regulating inflammation. *World J. Gastrointest. Pharmacol. Ther.* **2017**, *8*, 7–9. [CrossRef] [PubMed]
32. Konturek, P.C.; Harsch, I.A.; Konturek, K.; Schink, M.; Konturek, T.; Neurath, M.F.; Zopf, Y. Gut–Liver Axis: How Do Gut Bacteria Influence the Liver? *Med. Sci.* **2018**, *6*, 79. [CrossRef] [PubMed]
33. Kostic, A.D.; Xavier, R.J.; Gevers, D. The microbiome in inflammatory bowel disease: Current status and the future ahead. *Gastroenterology* **2014**, *146*, 1489–1499. [CrossRef]
34. Kudo, M.; Matilla, A.; Santoro, A.; Melero, I.; Gracian, A.C.; Acosta-Rivera, M.; Choo, S.P.; El-Khoueiry, A.B.; Kuromatsu, R.; El-Rayes, B.; et al. CheckMate 040 cohort 5: A phase I/II study of nivolumab in patients with advanced hepatocellular carcinoma and Child-Pugh B cirrhosis. *J. Hepatol.* **2021**, *75*, 600–609. [CrossRef]
35. Zhou, D.; Pan, Q.; Shen, F.; Cao, H.X.; Ding, W.J.; Chen, Y.W.; Fan, J.G. Total fecal microbiota transplantation alleviates high-fat diet-induced steatohepatitis in mice via beneficial regulation of gut microbiota. *Sci. Rep.* **2017**, *7*, 1529. [CrossRef]
36. Dapito, D.H.; Mencin, A.; Gwak, G.Y.; Pradere, J.P.; Jang, M.K.; Mederacke, I.; Caviglia, J.M.; Khiabanian, H.; Adeyemi, A.; Bataller, R.; et al. Promotion of hepatocellular carcinoma by the intestinal microbiota and TLR4. *Cancer Cell* **2012**, *21*, 504–516. [CrossRef]
37. Nowak, A.; Paliwoda, A.; Błasiak, J. Anti-proliferative, pro-apoptotic and anti-oxidative activity of Lactobacillus and Bifidobacterium strains: A review of mechanisms and therapeutic perspectives. *Crit. Rev. Food Sci. Nutr.* **2019**, *59*, 3456–3467. [CrossRef]
38. Yoshimoto, S.; Loo, T.M.; Atarashi, K.; Kanda, H.; Sato, S.; Oyadomari, S.; Iwakura, Y.; Oshima, K.; Morita, H.; Hattori, M.; et al. Obesity-induced gut microbial metabolite promotes liver cancer through senescence secretome. *Nature* **2013**, *499*, 97–101. [CrossRef] [PubMed]
39. Ponziani, F.R.; Bhoori, S.; Castelli, C.; Putignani, L.; Rivoltini, L.; Del Chierico, F.; Sanguinetti, M.; Morelli, D.; Paroni Sterbini, F.; Petito, V.; et al. Hepatocellular Carcinoma Is Associated with Gut Microbiota Profile and Inflammation in Nonalcoholic Fatty Liver Disease. *Hepatology* **2019**, *69*, 107–120. [CrossRef]
40. Schonewille, M.; de Boer, J.F.; Groen, A.K. Bile salts in control of lipid metabolism. *Curr. Opin. Lipidol.* **2016**, *27*, 295–301. [CrossRef]
41. Mouzaki, M.; Wang, A.Y.; Bandsma, R.; Comelli, E.M.; Arendt, B.M.; Zhang, L.; Fung, S.; Fischer, S.E.; McGilvray, I.G.; Allard, J.P. Bile Acids and Dysbiosis in Non-Alcoholic Fatty Liver Disease. *PLoS ONE* **2016**, *11*, e0151829. [CrossRef] [PubMed]
42. Neuschwander-Tetri, B.A.; Van Natta, M.L.; Tonascia, J.; Brunt, E.M.; Kleiner, D.E. Trials of obeticholic acid for non-alcoholic steatohepatitis—Authors' reply. *Lancet* **2015**, *386*, 28–29. [CrossRef] [PubMed]
43. Liu, Q.; Li, F.; Zhuang, Y.; Xu, J.; Wang, J.; Mao, X.; Zhang, Y.; Liu, X. Alteration in gut microbiota associated with hepatitis B and non-hepatitis virus related hepatocellular carcinoma. *Gut Pathog.* **2019**, *11*, 1. [CrossRef] [PubMed]
44. Sydor, S.; Best, J.; Messerschmidt, I.; Manka, P.; Vilchez-Vargas, R.; Brodesser, S.; Lucas, C.; Wegehaupt, A.; Wenning, C.; Aßmuth, S.; et al. Altered Microbiota Diversity and Bile Acid Signaling in Cirrhotic and Noncirrhotic NASH-HCC. *Clin. Transl. Gastroenterol.* **2020**, *11*, e00131. [CrossRef] [PubMed]
45. Jiao, N.; Baker, S.S.; Chapa-Rodriguez, A.; Liu, W.; Nugent, C.A.; Tsompana, M.; Mastrandrea, L.; Buck, M.J.; Baker, R.D.; Genco, R.J.; et al. Suppressed hepatic bile acid signalling despite elevated production of primary and secondary bile acids in NAFLD. *Gut* **2018**, *67*, 1881–1891. [CrossRef] [PubMed]
46. Bechmann, L.P.; Kocabayoglu, P.; Sowa, J.P.; Sydor, S.; Best, J.; Schlattjan, M.; Beilfuss, A.; Schmitt, J.; Hannivoort, R.A.; Kilicarslan, A.; et al. Free fatty acids repress small heterodimer partner (SHP) activation and adiponectin counteracts bile acid-induced liver injury in superobese patients with nonalcoholic steatohepatitis. *Hepatology* **2013**, *57*, 1394–1406. [CrossRef] [PubMed]
47. Caussy, C.; Tripathi, A.; Humphrey, G.; Bassirian, S.; Singh, S.; Faulkner, C.; Bettencourt, R.; Rizo, E.; Richards, L.; Xu, Z.Z.; et al. A gut microbiome signature for cirrhosis due to nonalcoholic fatty liver disease. *Nat. Commun.* **2019**, *10*, 1406. [CrossRef] [PubMed]
48. Heimbach, J.K.; Kulik, L.M.; Finn, R.S.; Sirlin, C.B.; Abecassis, M.M.; Roberts, L.R.; Zhu, A.X.; Murad, M.H.; Marrero, J.A. AASLD guidelines for the treatment of hepatocellular carcinoma. *Hepatology* **2018**, *67*, 358–380. [CrossRef] [PubMed]
49. Dyson, J.; Jaques, B.; Chattopadyhay, D.; Lochan, R.; Graham, J.; Das, D.; Aslam, T.; Patanwala, I.; Gaggar, S.; Cole, M.; et al. Hepatocellular cancer: The impact of obesity, type 2 diabetes and a multidisciplinary team. *J. Hepatol.* **2014**, *60*, 110–117. [CrossRef]
50. van Meer, S.; van Erpecum, K.J.; Sprengers, D.; Coenraad, M.J.; Klümpen, H.J.; Jansen, P.L.; Ijzermans, J.N.; Verheij, J.; van Nieuwkerk, C.M.; Siersema, P.D.; et al. Hepatocellular carcinoma in cirrhotic versus noncirrhotic livers: Results from a large cohort in the Netherlands. *Eur. J. Gastroenterol. Hepatol.* **2016**, *28*, 352–359. [CrossRef]
51. Tzartzeva, K.; Obi, J.; Rich, N.E.; Parikh, N.D.; Marrero, J.A.; Yopp, A.; Waljee, A.K.; Singal, A.G. Surveillance Imaging and Alpha Fetoprotein for Early Detection of Hepatocellular Carcinoma in Patients with Cirrhosis: A Meta-analysis. *Gastroenterology* **2018**, *154*, 1706–1718.e1. [CrossRef] [PubMed]
52. Best, J.; Bechmann, L.P.; Sowa, J.P.; Sydor, S.; Dechêne, A.; Pflanz, K.; Bedreli, S.; Schotten, C.; Geier, A.; Berg, T.; et al. GALAD Score Detects Early Hepatocellular Carcinoma in an International Cohort of Patients with Nonalcoholic Steatohepatitis. *Clin. Gastroenterol. Hepatol.* **2020**, *18*, 728–735.e4. [CrossRef] [PubMed]
53. Yang, J.D.; Addissie, B.D.; Mara, K.C.; Harmsen, W.S.; Dai, J.; Zhang, N.; Wongjarupong, N.; Ali, H.M.; Ali, H.A.; Hassan, F.A.; et al. GALAD Score for Hepatocellular Carcinoma Detection in Comparison with Liver Ultrasound and Proposal of GALADUS Score. *Cancer Epidemiol. Biomark. Prev.* **2019**, *28*, 531–538. [CrossRef] [PubMed]

54. Okajima, W.; Komatsu, S.; Ichikawa, D.; Miyamae, M.; Ohashi, T.; Imamura, T.; Kiuchi, J.; Nishibeppu, K.; Arita, T.; Konishi, H.; et al. Liquid biopsy in patients with hepatocellular carcinoma: Circulating tumor cells and cell-free nucleic acids. *World J. Gastroenterol.* **2017**, *23*, 5650–5668. [CrossRef] [PubMed]
55. Reddy, S.K.; Steel, J.L.; Chen, H.W.; DeMateo, D.J.; Cardinal, J.; Behari, J.; Humar, A.; Marsh, J.W.; Geller, D.A.; Tsung, A. Outcomes of curative treatment for hepatocellular cancer in nonalcoholic steatohepatitis versus hepatitis C and alcoholic liver disease. *Hepatology* **2012**, *55*, 1809–1819. [CrossRef] [PubMed]
56. Yao, F.Y.; Ferrell, L.; Bass, N.M.; Watson, J.J.; Bacchetti, P.; Venook, A.; Ascher, N.L.; Roberts, J.P. Liver transplantation for hepatocellular carcinoma: Expansion of the tumor size limits does not adversely impact survival. *Hepatology* **2001**, *33*, 1394–1403. [CrossRef] [PubMed]
57. Mazzaferro, V.; Llovet, J.M.; Miceli, R.; Bhoori, S.; Schiavo, M.; Mariani, L.; Camerini, T.; Roayaie, S.; Schwartz, M.E.; Grazi, G.L.; et al. Predicting survival after liver transplantation in patients with hepatocellular carcinoma beyond the Milan criteria: A retrospective, exploratory analysis. *Lancet Oncol.* **2009**, *10*, 35–43. [CrossRef] [PubMed]
58. Mazzaferro, V.; Sposito, C.; Zhou, J.; Pinna, A.D.; De Carlis, L.; Fan, J.; Cescon, M.; Di Sandro, S.; Yi-Feng, H.; Lauterio, A.; et al. Metroticket 2.0 Model for Analysis of Competing Risks of Death after Liver Transplantation for Hepatocellular Carcinoma. *Gastroenterology* **2018**, *154*, 128–139. [CrossRef]
59. Shirabe, K.; Itoh, S.; Yoshizumi, T.; Soejima, Y.; Taketomi, A.; Aishima, S.; Maehara, Y. The predictors of microvascular invasion in candidates for liver transplantation with hepatocellular carcinoma-with special reference to the serum levels of des-gamma-carboxy prothrombin. *J. Surg. Oncol.* **2007**, *95*, 235–240. [CrossRef]
60. Duvoux, C.; Roudot-Thoraval, F.; Decaens, T.; Pessione, F.; Badran, H.; Piardi, T.; Francoz, C.; Compagnon, P.; Vanlemmens, C.; Dumortier, J.; et al. Liver transplantation for hepatocellular carcinoma: A model including alpha-fetoprotein improves the performance of Milan criteria. *Gastroenterology* **2012**, *143*, 986–994. [CrossRef]
61. Mikolasevic, I.; Filipec-Kanizaj, T.; Mijic, M.; Jakopcic, I.; Milic, S.; Hrstic, I.; Sobocan, N.; Stimac, D.; Burra, P. Nonalcoholic fatty liver disease and liver transplantation—Where do we stand? *World J. Gastroenterol.* **2018**, *24*, 1491–1506. [CrossRef] [PubMed]
62. Reig, M.; Forner, A.; Rimola, J.; Ferrer-Fàbrega, J.; Burrel, M.; Garcia-Criado, Á.; Kelley, R.K.; Galle, P.R.; Mazzaferro, V.; Salem, R.; et al. BCLC strategy for prognosis prediction and treatment recommendation: The 2022 update. *J. Hepatol.* **2022**, *76*, 681–693. [CrossRef]
63. Llovet, J.M.; Real, M.I.; Montaña, X.; Planas, R.; Coll, S.; Aponte, J.; Ayuso, C.; Sala, M.; Muchart, J.; Solà, R.; et al. Arterial embolisation or chemoembolisation versus symptomatic treatment in patients with unresectable hepatocellular carcinoma: A randomised controlled trial. *Lancet* **2002**, *359*, 1734–1739. [CrossRef] [PubMed]
64. Young, S.; Sanghvi, T.; Rubin, N.; Hall, D.; Roller, L.; Charaf, Y.; Golzarian, J. Transarterial Chemoembolization of Hepatocellular Carcinoma: Propensity Score Matching Study Comparing Survival and Complications in Patients with Nonalcoholic Steatohepatitis Versus Other Causes Cirrhosis. *Cardiovasc. Interv. Radiol.* **2020**, *43*, 65–75. [CrossRef] [PubMed]
65. Wu, S.E.; Charles, H.W.; Park, J.S.; Goldenberg, A.S.; Deipolyi, A.R. Obesity conveys poor outcome in patients with hepatocellular carcinoma treated by transarterial chemoembolization. *Diagn. Interv. Imaging* **2017**, *98*, 37–42. [CrossRef] [PubMed]
66. Zhang, J.; Gong, F.; Li, L.; Zhao, M.; Song, J. Diabetes mellitus and the neutrophil to lymphocyte ratio predict overall survival in non-viral hepatocellular carcinoma treated with transarterial chemoembolization. *Oncol. Lett.* **2014**, *7*, 1704–1710. [CrossRef] [PubMed]
67. Liu, G.; Xia, F.; Fan, G.; Yu, J.; Bao, L.; Zhang, C.; Chi, R.; Zhang, T.; Wang, L.; Shen, F.; et al. Type 2 diabetes mellitus worsens the prognosis of intermediate-stage hepatocellular carcinoma after transarterial chemoembolization. *Diabetes Res. Clin. Pract.* **2020**, *169*, 108375. [CrossRef]
68. Jung, W.J.; Jang, S.; Choi, W.J.; Park, J.; Choi, G.H.; Jang, E.S.; Jeong, S.H.; Choi, W.S.; Lee, J.H.; Yoon, C.J.; et al. Metformin administration is associated with enhanced response to transarterial chemoembolization for hepatocellular carcinoma in type 2 diabetes patients. *Sci. Rep.* **2022**, *12*, 14482. [CrossRef] [PubMed]
69. Schotten, C.; Bechmann, L.P.; Manka, P.; Theysohn, J.; Dechêne, A.; El Fouly, A.; Barbato, F.; Neumann, U.; Radünz, S.; Sydor, S.; et al. NAFLD-Associated Comorbidities in Advanced Stage HCC Do Not Alter the Safety and Efficacy of Yttrium-90 Radioembolization. *Liver Cancer* **2019**, *8*, 491–504. [CrossRef]
70. Llovet, J.M.; Ricci, S.; Mazzaferro, V.; Hilgard, P.; Gane, E.; Blanc, J.F.; de Oliveira, A.C.; Santoro, A.; Raoul, J.L.; Forner, A.; et al. Sorafenib in advanced hepatocellular carcinoma. *N. Engl. J. Med.* **2008**, *359*, 378–390. [CrossRef]
71. Cheng, A.L.; Kang, Y.K.; Chen, Z.; Tsao, C.J.; Qin, S.; Kim, J.S.; Luo, R.; Feng, J.; Ye, S.; Yang, T.S.; et al. Efficacy and safety of sorafenib in patients in the Asia-Pacific region with advanced hepatocellular carcinoma: A phase III randomised, double-blind, placebo-controlled trial. *Lancet Oncol.* **2009**, *10*, 25–34. [CrossRef] [PubMed]
72. Kudo, M.; Finn, R.S.; Qin, S.; Han, K.H.; Ikeda, K.; Piscaglia, F.; Baron, A.; Park, J.W.; Han, G.; Jassem, J.; et al. Lenvatinib versus sorafenib in first-line treatment of patients with unresectable hepatocellular carcinoma: A randomised phase 3 non-inferiority trial. *Lancet* **2018**, *391*, 1163–1173. [CrossRef] [PubMed]
73. Bruix, J.; Qin, S.; Merle, P.; Granito, A.; Huang, Y.H.; Bodoky, G.; Pracht, M.; Yokosuka, O.; Rosmorduc, O.; Breder, V.; et al. Regorafenib for patients with hepatocellular carcinoma who progressed on sorafenib treatment (RESORCE): A randomised, double-blind, placebo-controlled, phase 3 trial. *Lancet* **2017**, *389*, 56–66. [CrossRef] [PubMed]

74. Abou-Alfa, G.K.; Meyer, T.; Cheng, A.L.; El-Khoueiry, A.B.; Rimassa, L.; Ryoo, B.Y.; Cicin, I.; Merle, P.; Chen, Y.; Park, J.W.; et al. Cabozantinib in Patients with Advanced and Progressing Hepatocellular Carcinoma. *N. Engl. J. Med.* **2018**, *379*, 54–63. [CrossRef] [PubMed]
75. Zhu, A.X.; Kang, Y.K.; Yen, C.J.; Finn, R.S.; Galle, P.R.; Llovet, J.M.; Assenat, E.; Brandi, G.; Pracht, M.; Lim, H.Y.; et al. Ramucirumab after sorafenib in patients with advanced hepatocellular carcinoma and increased α-fetoprotein concentrations (REACH-2): A randomised, double-blind, placebo-controlled, phase 3 trial. *Lancet Oncol.* **2019**, *20*, 282–296. [CrossRef] [PubMed]
76. Cheng, A.L.; Qin, S.; Ikeda, M.; Galle, P.R.; Ducreux, M.; Kim, T.Y.; Lim, H.Y.; Kudo, M.; Breder, V.; Merle, P.; et al. Updated efficacy and safety data from IMbrave150: Atezolizumab plus bevacizumab vs. sorafenib for unresectable hepatocellular carcinoma. *J. Hepatol.* **2022**, *76*, 862–873. [CrossRef] [PubMed]
77. Abou-Alfa, G.K.; Lau, G.; Kudo, M.; Chan, S.L.; Kelley, R.K.; Furuse, J.; Sukeepaisarnjaroen, W.; Kang, Y.-K.; Dao, T.V.; Toni, E.N.D.; et al. Tremelimumab plus Durvalumab in Unresectable Hepatocellular Carcinoma. *NEJM Evid.* **2022**, *1*, EVIDoa2100070. [CrossRef] [PubMed]
78. Finn, R.S.; Kudo, M.; Merle, P.; Meyer, T.; Qin, S.; Ikeda, M.; Xu, R.; Edeline, J.; Ryoo, B.Y.; Ren, Z.; et al. LBA34 Primary results from the phase III LEAP-002 study: Lenvatinib plus pembrolizumab versus lenvatinib as first-line (1L) therapy for advanced hepatocellular carcinoma (aHCC). *Ann. Oncol.* **2022**, *33*, S1401. [CrossRef]
79. Kelley, R.K.; Rimassa, L.; Cheng, A.L.; Kaseb, A.; Qin, S.; Zhu, A.X.; Chan, S.L.; Melkadze, T.; Sukeepaisarnjaroen, W.; Breder, V.; et al. Cabozantinib plus atezolizumab versus sorafenib for advanced hepatocellular carcinoma (COSMIC-312): A multicentre, open-label, randomised, phase 3 trial. *Lancet Oncol.* **2022**, *23*, 995–1008. [CrossRef]
80. Finn, R.S.; Qin, S.; Ikeda, M.; Galle, P.R.; Ducreux, M.; Kim, T.Y.; Kudo, M.; Breder, V.; Merle, P.; Kaseb, A.O.; et al. Atezolizumab plus Bevacizumab in Unresectable Hepatocellular Carcinoma. *N. Engl. J. Med.* **2020**, *382*, 1894–1905. [CrossRef]
81. Pfister, D.; Núñez, N.G.; Pinyol, R.; Govaere, O.; Pinter, M.; Szydlowska, M.; Gupta, R.; Qiu, M.; Deczkowska, A.; Weiner, A.; et al. NASH limits anti-tumour surveillance in immunotherapy-treated HCC. *Nature* **2021**, *592*, 450–456. [CrossRef] [PubMed]
82. Haber, P.K.; Puigvehí, M.; Castet, F.; Lourdusamy, V.; Montal, R.; Tabrizian, P.; Buckstein, M.; Kim, E.; Villanueva, A.; Schwartz, M.; et al. Evidence-Based Management of Hepatocellular Carcinoma: Systematic Review and Meta-analysis of Randomized Controlled Trials (2002–2020). *Gastroenterology* **2021**, *161*, 879–898. [CrossRef] [PubMed]
83. Meyer, T.; Galani, S.; Lopes, A.; Vogel, A. Aetiology of liver disease and response to immune checkpoint inhibitors: An updated meta-analysis confirms benefit in those with non-viral liver disease. *J. Hepatol.* **2023**, *79*, e73–e76. [CrossRef] [PubMed]
84. Welland, S.; Leyh, C.; Finkelmeier, F.; Jefremow, A.; Shmanko, K.; Gonzalez-Carmona, M.A.; Kandulski, A.; Jeliazkova, P.; Best, J.; Fründt, T.W.; et al. Real-World Data for Lenvatinib in Hepatocellular Carcinoma (ELEVATOR): A Retrospective Multicenter Study. *Liver Cancer* **2022**, *11*, 219–232. [CrossRef] [PubMed]
85. Bruix, J.; Raoul, J.L.; Sherman, M.; Mazzaferro, V.; Bolondi, L.; Craxi, A.; Galle, P.R.; Santoro, A.; Beaugrand, M.; Sangiovanni, A.; et al. Efficacy and safety of sorafenib in patients with advanced hepatocellular carcinoma: Subanalyses of a phase III trial. *J. Hepatol.* **2012**, *57*, 821–829. [CrossRef] [PubMed]
86. Hiraoka, A.; Kumada, T.; Tada, T.; Tani, J.; Kariyama, K.; Fukunishi, S.; Atsukawa, M.; Hirooka, M.; Tsuji, K.; Ishikawa, T.; et al. Efficacy of lenvatinib for unresectable hepatocellular carcinoma based on background liver disease etiology: Multi-center retrospective study. *Sci. Rep.* **2021**, *11*, 16663. [CrossRef] [PubMed]
87. Sacco, R.; Ramai, D.; Tortora, R.; di Costanzo, G.G.; Burlone, M.E.; Pirisi, M.; Federico, P.; Daniele, B.; Silletta, M.; Gallo, P.; et al. Role of Etiology in Hepatocellular Carcinoma Patients Treated with Lenvatinib: A Counterfactual Event-Based Mediation Analysis. *Cancers* **2023**, *15*, 381. [CrossRef] [PubMed]
88. Casadei-Gardini, A.; Rimini, M.; Tada, T.; Suda, G.; Shimose, S.; Kudo, M.; Cheon, J.; Finkelmeier, F.; Lim, H.Y.; Rimassa, L.; et al. Atezolizumab plus bevacizumab versus lenvatinib for unresectable hepatocellular carcinoma: A large real-life worldwide population. *Eur. J. Cancer* **2023**, *180*, 9–20. [CrossRef]
89. Rimini, M.; Rimassa, L.; Ueshima, K.; Burgio, V.; Shigeo, S.; Tada, T.; Suda, G.; Yoo, C.; Cheon, J.; Pinato, D.J.; et al. Atezolizumab plus bevacizumab versus lenvatinib or sorafenib in non-viral unresectable hepatocellular carcinoma: An international propensity score matching analysis. *ESMO Open* **2022**, *7*, 100591. [CrossRef]
90. Ramai, D.; Singh, J.; Lester, J.; Khan, S.R.; Chandan, S.; Tartaglia, N.; Ambrosi, A.; Serviddio, G.; Facciorusso, A. Systematic review with meta-analysis: Bariatric surgery reduces the incidence of hepatocellular carcinoma. *Aliment. Pharmacol. Ther.* **2021**, *53*, 977–984. [CrossRef]
91. Kojima, M.; Takahashi, H.; Kuwashiro, T.; Tanaka, K.; Mori, H.; Ozaki, I.; Kitajima, Y.; Matsuda, Y.; Ashida, K.; Eguchi, Y.; et al. Glucagon-Like Peptide-1 Receptor Agonist Prevented the Progression of Hepatocellular Carcinoma in a Mouse Model of Nonalcoholic Steatohepatitis. *Int. J. Mol. Sci.* **2020**, *21*, 5722. [CrossRef] [PubMed]
92. Mirarchi, L.; Amodeo, S.; Citarrella, R.; Licata, A.; Soresi, M.; Giannitrapani, L. SGLT2 Inhibitors as the Most Promising Influencers on the Outcome of Non-Alcoholic Fatty Liver Disease. *Int. J. Mol. Sci.* **2022**, *23*, 3668. [CrossRef] [PubMed]

Journal of *Personalized Medicine*

MDPI

Systematic Review

Long-Term Outcomes of Liver Transplantation for the Management of Neuroendocrine Neoplasms: A Systematic Review

Varun Palaniappan [1,†], Chun Hei Li [2,†], Andrea Frilling [3] and Ashley Kieran Clift [3,4,*]

1 Cardiff and Vale University Health Board, Cardiff CF14 4XW, UK
2 St George's University Hospitals NHS Foundation Trust, London SW17 0QT, UK
3 Department of Surgery & Cancer, Imperial College London, London SW7 2BX, UK
4 Cancer Research UK Oxford Centre, University of Oxford, Oxford OX1 2JD, UK
* Correspondence: ashley.clift@phc.ox.ac.uk
† These authors contributed equally to this work.

Abstract: Liver transplantation is an uncommonly used, controversially debated therapeutic approach for highly selected individuals with neuroendocrine liver metastases. Synthesising evidence regarding outcomes from this approach is crucial to understand its position within the broad neuroendocrine liver metastases armamentarium. In this narrative systematic review of studies published in PubMed, Scopus and OVID until 1 July 2021, we summarise and critically appraise the existing literature regarding this modality, with a special focus on long-term outcomes data where possible. Fourteen studies were identified that reported outcomes from the use of liver transplantation for metastatic neuroendocrine neoplasms. No randomised trials were identified. Generally, indications and selection criteria were poorly articulated, with the notable exception of studies using the Milan criteria. The median 5-year overall survival was 65% (ranging from 36% to 97.2%, 11 studies), and the median 10-year overall survival was 50% (ranging from 46.1% to 88.8%, 3 studies). One additional study focussed on treatments and outcomes following post-transplant recurrence. No studies reported outcomes past 10 years. Further follow-up of the largest series with explicit selection criteria will deepen our understanding of the role that transplantation has to play in this setting.

Keywords: neuroendocrine; transplantation; outcomes; liver metastases; systematic review

Citation: Palaniappan, V.; Li, C.H.; Frilling, A.; Clift, A.K. Long-Term Outcomes of Liver Transplantation for the Management of Neuroendocrine Neoplasms: A Systematic Review. *J. Pers. Med.* **2023**, *13*, 1428. https://doi.org/10.3390/jpm13101428

Academic Editor: Guido Gerken

Received: 18 August 2023
Revised: 18 September 2023
Accepted: 18 September 2023
Published: 23 September 2023

1. Introduction

Neuroendocrine neoplasms (NEN) are an increasingly prevalent class of tumour that arise from multiple organs, but most commonly in the gastroenteropancreatic tract, or bronchopulmonary system [1]. Clinical challenges include the symptomatic burdens of hormonally active ("functional") tumours, their propensity to metastasise despite their generally accepted relative indolent growth, and high incidence of nodal and/or distant metastases at initial presentation [2]. Specialist centre experience documents that up to 90% of small intestinal NEN display evidence of nodal metastases at diagnosis [3], with up to 91% of small intestinal NEN patients and up to 77% of pancreatic NEN patients developing hepatic metastases [4–6]. Gold-standard imaging also significantly understages metastatic disease in the liver [7], mandating careful treatment strategy and according follow-up. Whilst patients harbouring neuroendocrine liver metastases (NELM) may have protracted survival, particularly when compared to expected prognosis of stage IV gastrointestinal adenocarcinomas, the presence of NELM exerts a major, negative prognostic effect, and multimodal treatment is often required to attain disease control [2,8].

Recent advances, driven by increased centralisation of care into expert centres/networks and inter-centre collaboration, have expanded the therapeutic armoury, such as evidence from randomised controlled trials supporting the use of somatostatin analogues (SSAs) [9], peptide receptor radionuclide therapy (PRRT) [10], molecularly targeted agents [11,12], and interventional "trans-arterial" procedures such as selective internal radioembolisation for hepatic disease [13]. However, whilst initial reports were promising regarding the anti-proliferative effects of SSAs and PRRT in terms of progression-free survival, there is no evidence that they significantly affect overall survival [14].

Radical surgical intervention is therefore the only modality that possesses the opportunity to attain cure—this is possible in the setting of NELM, with the resection of locoregional tumour burden and liver metastases if oncologically and technically appropriate [15]. A minority of patients with NELM are candidates for resection with curative intent, and even if R0 margins are attained, disease recurrence is a significant hindrance [16], leading some to posit that for many cases, "curative resection" is a palliative endeavour albeit with a sustained duration of disease control [15]. The other aspect of the surgical armamentarium for NELM [17] or advanced primary hepatic NEN [18,19] is liver transplantation, either in the classic orthotopic fashion, or as part of a multivisceral graft [20–23]. Indications for OLT generally comprise control of medically intractable symptoms from functional tumours, amelioration of effects of hepatic tumour bulk, or for oncological control [24]. Initial outcomes with OLT for NELM were very poor [17,24]; however, technical progress and implementation of stringent selection criteria have been shown to be associated with improved prognosis [25–27]. Other centres have demonstrated excellent results exceeding those seen for liver resection based on meticulous selection criteria [28], and some reports have demonstrated the possibility of excellent long-term survival. However, given the rarity of the indication (<1% of all liver transplant activity) [22], summarising the available evidence into a cohesive platform is necessary to understand the divergences in practice and outcomes that generate debate around this contested modality. Specifically, there is lack of robust data on long-term survival outcomes, e.g., 10 years or longer.

Here, we undertake a systematic review of the literature regarding the use of liver transplantation for the treatment of NEN, seeking to update the findings of a previous review [17], extend prior work by critically appraising the evidence limitations and outline avenues for further study. This is done with a specific focus on long-term outcomes of this approach.

2. Materials and Methods

2.1. Protocol Registration and Study Conduct

The protocol for this systematic review was registered on the PROSPERO database (reference CRD42021267963; https://www.crd.york.ac.uk/prospero/display_record.php?ID=CRD42021267963, accessed on 1 August 2021) prior to the commencement of the study. The systematic review was undertaken cognisant of and reported according to the PRISMA guidelines [29] (checklist in Supplementary File S1).

We formulated our review considering the "PICO" framework:

Population: patients with advanced neuroendocrine neoplasms (liver metastases);
Intervention: liver transplantation, either alone (orthotopic) or part of a multivisceral graft;
Comparison: other, non-transplant treatment strategies (data availability permitting);
Outcome: overall and disease-free survival at 1-, 3-, 5-, 10-, and 10+ years.

2.2. Search Strategy, Data Sources and Inclusion

We undertook a systematic review of three databases: PubMed (MEDLINE), Scopus Web of Science and EMBASE. Search terms focussed on "neuroendocrine" and "liver and "transplant"—the full search strings are available via the link on the registered protocol webpage on PROSPERO (https://www.crd.york.ac.uk/PROSPEROFILES/267963_STRATEGY_20210714.pdf, accessed on 1 August 2021) and in Supplementary File S2. There were no amendments to the protocol.

Papers published until 1 July 2021 written in English or German were considered for inclusion. We considered publications reporting on liver transplantation for NEN or NELM, whether or not they were purely NEN cohorts (provided that NEN-specific data were reported), or reports discussing several primary tumour types. Papers were excluded if they were animal studies, review articles, another systematic review with or without meta-analysis, or editorials or other non-research studies not reporting outcomes data. Case reports were not included for data extraction. If data from the same centre or registry were presented in multiple reports, we selected the most recently published paper.

Three authors each performed the search string on one source (VP EMBASE, DL on PubMed, AKC Scopus). The number of articles was noted, and all records uploaded onto the Rayyan platform for duplicate identification and screening. Three independent reviewers (VP, DL and AKC) screened all articles (titles and abstracts), with each record assessed twice. Conflicts were resolved in group teleconference. AKC also reviewed reference lists of the finally selected papers for additional potential references.

2.3. Data Extraction and Summary

All selected articles passing title and abstract screening were assessed by two independent reviewers for potential data extraction: VP and DL reviewed 50% each of all articles, with AKC performing the second review. Disagreements were discussed in group teleconference. A pre-developed data extraction template was trialled using two large, previously known studies to assess appropriateness and ease of use.

Thereafter, data were extracted regarding first author, year of publication, country, study design, number of patients included with NEN, median age of participants (or other summary statistics if presented), male:female numbers, primary tumour types, tumour histology, median follow-up (or other summary statistic), overall survival at 1, 3, 5, 10, and 10+ years, recurrence-free survival at 1, 3, 5, 10, and 10+ years, type of immunosuppression, pre-transplant treatment information, and "miscellaneous" data felt to be of relevance by individual reviewers.

Per-study rows were permitted to be split if the study authors reported outcomes for distinct temporal periods, or if they reported outcomes for specific forms of transplantation (e.g., orthotopic liver transplantation, or multivisceral transplantation). Outcome data were extracted "as is"—in cases where no Kaplan–Meier/other actuarial survival metrics were reported, we did not compute by hand using raw data (if provided).

2.4. Synthesis and Meta-Analysis

The findings of this systematic review were synthesised narratively. We had considered pooling of identified Kaplan–Meier estimates using random effects meta-analysis. However, due to the high heterogeneity in centre selection criteria and follow-up duration, and notable trends in outcomes over time, we felt that this was inappropriate and would lead to imprecise estimates. As the included estimates were overwhelmingly case series, we used the Institute for Health Economics (Edmonton, AB, Canada) quality appraisal checklist: (https://www.ihe.ca/download/development_of_a_quality_appraisal_tool_for_case_series_studies_using_a_modified_delphi_technique.pdf, accessed on 1 August 2021).

3. Results

The reference selection process is illustrated in the PRISMA flowchart in Figure 1. From the initially extracted 982 records, we included 15 studies in the final evidence synthesis: 14 studies reporting survival outcomes after LT for NEN, and 1 reporting on outcomes after post-OLT recurrence. These studies are presented in Table 1. Four papers that reported relevant data were deemed to be older, preceding reports of the same/highly overlapping patient pools [25,30–32] and therefore not included in the final summary. As all records were retrospective case series, this review remained "narrative" and without meta-analysis. A relative quality appraisal is summarised in Supplementary File S3.

Table 1. Summary of included studies regarding liver transplantation for the management of neuroendocrine neoplasms. OS = overall survival, DFS = disease-free survival, OLT = orthotopic liver transplantation, MVT = multivisceral transplantation, NR = not reported.

First Author	Year of Publication	Study Period	Country/Ies	Study Design	Sample Size	Median Age	Gender (M:F)	Median Follow-Up	1-Year OS	3-Year OS	5-Year OS	10-Year OS	1-Year DFS	3-Year DFS	5-Year DFS	10-Year DFS
Routley [33]	1995	1983–1997	United Kingdom	Multicentre, retrospective case series	11	NR	6:5	NR	82%		57%					
Rosenau [34]	2002	1982–1997	Germany	Single centre, retrospective case series	19	Median 47 years (range 18–61)	9:10	Mean 59 months (0.5–146)	89%		80%	50%	56%		21%	21%
Florman [35]	2004	1992–2002	United States	Single centre, retrospective case series	11	Mean 51.2 +/− 6.3 yrs	4:7	Mean 34 +/− 40 months	73%		36%					
van Vilsteren [36]	2006	1998–2002	United States	Single centre, retrospective case series	17	Median 47 years (range 22–64)	15:4	Mean 22 months (range 0–84)	87%				77%			
Marin [37]	2007	1996–2006	Spain	Single centre, retrospective case series	10	Mean 42 years (range 30–62)	5:5	Mean 3 years, range 1 month–6 years	86%	57%						
Olausson [23]	2007	1997–2001	Sweden	Single centre, retrospective case series	15 (10 OLT, 5 MVT)	Median 51.5 years (range 39–64) OLT. Median 43 years (range 38–57) for MVT	11:4	Mean 53.8 months (+/−9.5)			90% OLT		Approx. 70% for all patients		20%	
Dhupar [38]	2009	1991–2006	United States	Single centre, retrospective case series	5	Median 44 years (range 17–53)	2:3	NR	100%							
Frilling [39]	2009	NR	Germany	Single centre, retrospective case series	17	NR	NR	NR			67%				48%	
Bonaccorsi-Riani [40]	2010	NR	Belgium	Single centre, retrospective case series	9	Median 54 years (range 26.6–61)	7:2	NR	88%	77%	33%		67%	33%	11%	
Le Treut [24]	2013	1982–2009	Multiple in Europe	Multicentre, retrospective case series	213	Mean 46 years +/− 11. Median 48 years (range 16–71)	114:99	Mean 56 +/− 49 months (range 0–283)	81%	65%	52%		65%	40%	30%	
Sher [20]	2015	1988–2012	United States, Canada, Europe	Multicentre, retrospective case series	85	Median 48 years (range 16–75)	51:34	Median 2.7 years (range 0.05–21.4)	83%	60%	52%					
Mazzaferro [28]	2016	1995 onwards	Italy	Single centre, retrospective case series	42	Median 40.5 (range 13–62)	26:16	NR			97.2%	88.8%			86.9%	86.9%

Table 1. *Cont.*

First Author	Year of Publication	Study Period	Country/Ies	Study Design	Sample Size	Median Age	Gender (M:F)	Median Follow-Up	1-Year OS	3-Year OS	5-Year OS	10-Year OS	1-Year DFS	3-Year DFS	5-Year DFS	10-Year DFS
Korda [41]	2019	1995–2018	Hungary	Single centre, retrospective case series	10	Median 49.5 years (range 38–62)	4:6	Median 33 months (range 9–104)	89%		71%		80%		43%	
Valvi [42]	2021	1988–2018	United States	Multicentre, retrospective case series	206	Mean 48.2 years (SD 11.7, range 19–75)	117:89	NR	89%	75.3%	65%	46.1%	74.9%	55.7%	43.9%	
Post-OLT recurrence																
Sposito [43]	2021	2004–2018	Italy	Single centre, retrospective case series	53 had LT; 32 recurred	At recurrence, median 55 (range 48.5–60.3)	16:15	Median 73.7 months after recurrence			76.3%	45.5%				

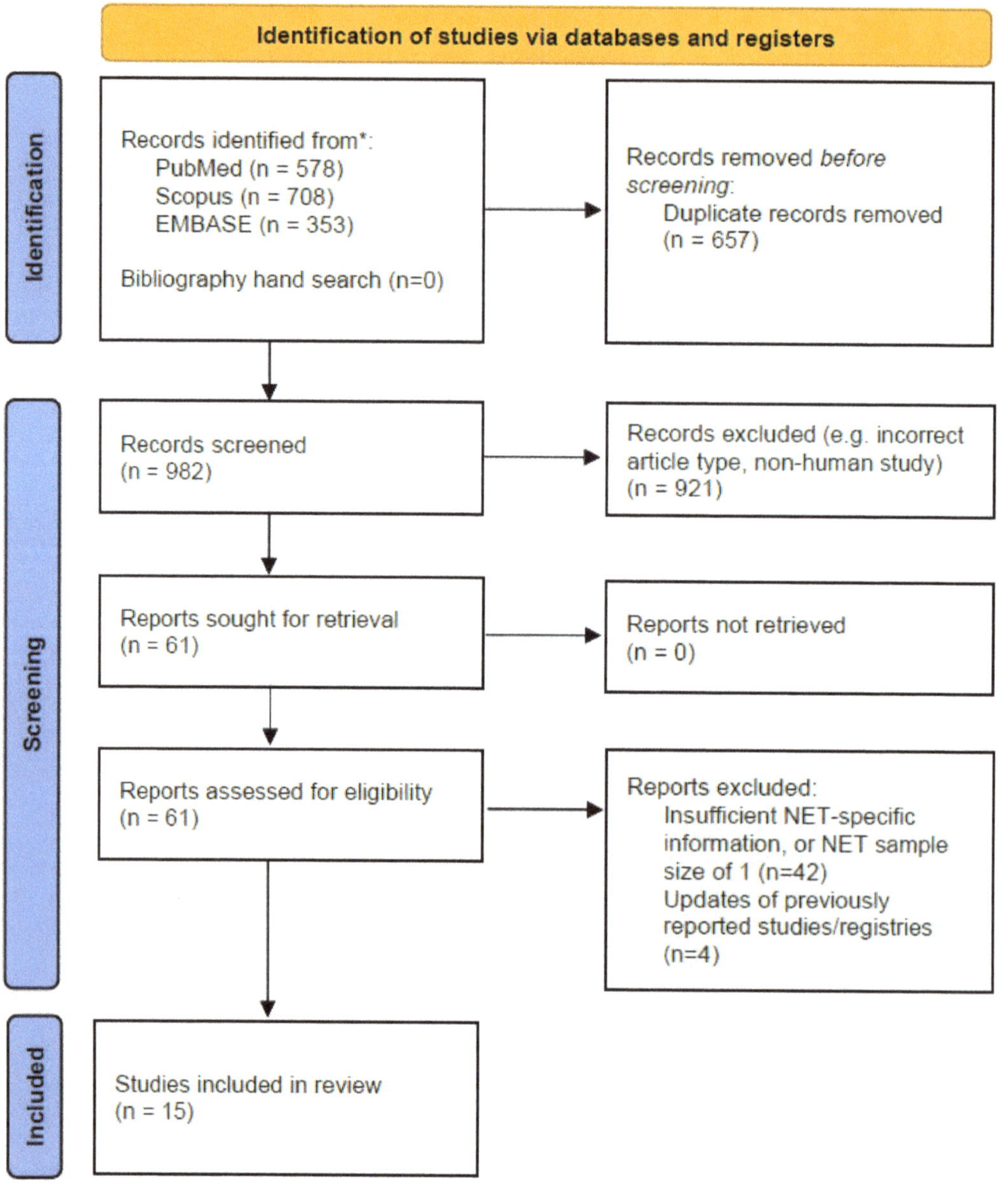

Figure 1. PRISMA flowchart describing the steps in the systematic review from literature search to included studies. * = distinct record databases.

The publication date of the included studies ranged from 1995 to 2021, covering study periods starting in 1983 onwards. Sample sizes ranged between 5 [38] and 213 [24]. There was non-uniform reporting of median age of the cohorts, summary characteristics of follow-up, sex of participants, the survival outcomes reported (intervals and whether this was overall survival or disease-free), prior treatment strategies and selection criteria. Whilst most studies reported solely on outcomes following orthotopic liver transplantation for NELM [24], some included cases of primary hepatic NEN, and some included patients undergoing multivisceral transplantation for advanced tumours including hepatic metastases [20,23,40]. Interestingly, the recent case series of Sposito et al. reported outcomes following recurrence after OLT [43].

Generally, selection criteria for OLT (or MVT) were poorly described, although some studies provided a breakdown of indications, e.g., for intractable symptoms [24]. The exception was the experience of the Mazzaferro et al. with the modified Milan NET criteria [28], which are discussed below.

Across studies, median overall survival at 1 year was 87% (range 73%–100%, 11 studies), and at 5 years this was 65% (range 36%–97.2%, 11 studies). Only four studies [28,34,42,43] reported overall survival at 10 years—two of these had overlapping patient cohorts [28,43], so after retaining the source/main study [28] (the other focussed on post-OLT-recurrence outcomes, rather than outcomes post OLT in general), the median 10 year OS was 50%, ranging from 46.1% to 88.8% (three studies). None reported overall survival outcomes past 10 years. Overall, the overall survival data appeared more optimistic than those reported for the accompanying disease-free/recurrence-free survival: at 1 year, this was 70% (range 56%–80%, seven studies), and at 5 years, this was 36.5% (range 11%–86.9%, eight studies). Similarly, only two studies reported 10-year disease-free survival which was wide ranging: 21% [34] and 86.9% [28]. None reported outcomes after 10 years. Due to the profound heterogeneity in selection criteria, time periods, temporal trends in outcomes over time [42] (e.g., with implementation of scoring systems/organ allocation methods), a random effects meta-analysis was not performed due to the low likelihood of providing meaningful pooled estimate that translates into current practice. Furthermore, if such a meta-analysis was feasible, a meta-regression to examine factors associated with the expected high degree of heterogeneity would be infeasible due to poor recording of potentially attributable factors.

The largest study identified was the multicentric report of the European Liver Transplant Registry, which evaluated outcomes in 213 patients with NEN metastatic to the liver between 1982 and 2009 [24]. The treatment of disease prior to LT included resection of the primary tumour and/or hepatic deposits in over 80% of cases, and a high use of prior chemotherapy (71%, including systemic or intra-arterial liver-directed modalities). Indications for LT were for oncological control (54%), to treat the effects of tumoral bulk (24%, presumably predominantly hepatic burden leading to pain), and control of hormonal excess/functional syndrome (17%). Alongside reporting long-term outcomes (e.g., 5-year OS 52%, 5-year DFS 30%), the authors noted improved longer-term overall survival in the most recent temporal period (5-year OS 59% after 2000, n = 106, compared to 46% prior to 2000). Le Treut and colleagues undertook an evaluation of prognostic factors using data for the 106 patients treated after 2000, and generated a simple 4-point score based thereon. This score incorporated three baseline/pre-treatment information factors: hepatomegaly, age over 45 years, and concomitant additional resection, with prognostic scores ranging between 0 and 3. Furthermore, the authors demonstrated significantly divergent overall survival curves when separated into two groups (0–1 factor versus 2–3 factors present, 79% OS versus 38%, respectively, and 5-year DFS 38% and 19%, respectively). However, whilst this prognostic score is attractive in its simplicity and apparent ability to stratify, it presents several methodological issues. These include the dichotomization of age (which risks information waste and step artefacts [44]), univariate screening of predictors [44], and that the tool is only evaluated in terms of separated survival curves, rather than being assessed in terms of calibration discrimination and clinical utility. Whilst the prognostic score was developed, and the indications for OLT reported, clear centre-specific selection criteria were not available [24].

Evidence from multi-centre series in the United States supports the evidence from European centres regarding improvement in long-term outcomes with LT for NELM over time. For example, in their analysis of the United States Organ Sharing/Organ Procurement and Transplantation Network (UNOS/OPTN) covering recorded transplants between 187 and 2011 (184 for NELM), Nguyen and colleagues [27] reported that survival significantly improved after introduction of the MELD/PELD score in 2002. Indeed, initially observed differences in long-term outcomes between transplants for NELM and transplants for hepatocellular carcinoma were negated to non-significance after this time point, with 5-year OS for NELM patients at 57.8%, versus 64.4% for HCC ($p = 0.109$) [27]. Iteratively updated reports using the UNOS/OPTN database have been published, such as those by Gedaly et al. [25], and most recently, Valvi et al. [42]. The latter study of 206 patients undergoing "isolated" liver transplantation for metastatic NEN (of a total 160,360 total liver transplants between 1988 and 2018) reported a 5-year OS of 64.9% and a 10-year OS of 46.1%, and explored the role of potential prognostic factors on risk of death or recurrence [42]. Propensity score matching (on MELD score and gender) was used to match NEN to HCC and cholangiocarcinoma patients at a ratio of 1:3 to perform comparisons of outcomes in these groups. The NELM group was observed to have a higher incidence of recurrence (34%) versus HCC (8%) or cholangiocarcinoma patients (19.6%), however there were no significant differences in overall survival between these groups. For example, 5-year survival for NELM was 75.4%, 79.9% for HCC, and 70.4% for the cholangiocarcinoma group. Furthermore, an effect of duration on transplant waiting list was observed for recurrence in NEN patients—in those that recurred after liver transplantation, 74.3% waited for under 6 months, whereas 25.7% were on the waiting list for longer than 6 months. Limitations of propensity score matching include reduced sample size, counterintuitive risk of imbalance, dependence on the matching model, and the fact that it does not eliminate confounding.

The only study to meticulously define and adhere to a single set of selection criteria is that of Mazzaferro et al. who documented their prospective experience with patient selection according to their modified Milan criteria for NELM [28]. These criteria were set in 1995 and comprise confirmed histology of G1 or G2 NEN, primary tumour drained by the hepatic portal system and removed as well as extra-hepatic deposits in a separate curative resection prior to consideration for OLT, <50% total liver involvement, at least 6 months of stable disease/disease response prior to consideration of OLT and age < 60 years (relative criterion). Eighty-eight patients with NELM eligible for OLT were included, with two sub-groups analysed: those that underwent transplantation (n = 42), and those that did not (n = 46; 22 refusals/non-compliant, 24 due to transplant list unavailability). Regarding follow-up imaging strategy, 3–4 monthly CT or MRI imaging was used, with somatostatin receptor-based imaging (OctreoScan or 68-Gallium PET/CT) only used in cases of suspicion of recurrence on CT or MRI. Patients undergoing OLT had significantly smaller tumours, were younger (median age 40.5 years versus 55.5 years), and underwent more locoregional therapy (40.5% versus 21.7%) than those that did not receive a transplant. Statistical analyses incorporated adjustment for propensity scores (i.e., propensity to receive a transplant), the models for which incorporated a suite of clinicopathological characteristics and underwent robust derivation, assessment, and implementation. Excellent long-term outcomes were observed with OLT, with 97.2% and 88.8% OS at 5 and 10 years, respectively, compared to 50.9% and 22.4%, respectively in the non-transplant group. Post-transplant disease recurrence was very low, at 13.1% at 10 years. Furthermore, the survival benefits of OLT increased over time, with the adjusted benefit at 5-year follow-up estimated at 6.82 months (95% CI: 1.10 months to 12.54 months) and 38.43 months (95% CI: 21.41 months to 55.45 months). It has been posited that many of the patients undergoing OLT in this study may have been suitable candidates for liver resection, and naturally, these patients are highly selected. Nevertheless, the 86.9% DFS at 10 years with the Milan approach appears exemplary and is worthy of further robust consideration.

The same group reported their experience managing patients that were selected for OLT, but developed post-LT recurrence (study period 2004 to 2018) [43]. This retrospective case series comprised 32 patients treated at the same centre as the previous discussed study [28], and thus, in this review, we focussed separately on their post-recurrence outcomes [43]. Follow-up imaging included OctreoScan or 68-Gallium PET/CT. Recurrence was most commonly at a single site (81.2% of cases), particularly in the distant lymph nodes (40.6%) or locoregional lymph nodes (18.8%), but this also manifested as peritoneal or pulmonary lesions. The inadequacy of chromogranin A for post-transplant surveillance was suggested, as only 12 patients (37.5%) had elevated levels at the time recurrence was ascertained. Fourteen patients (43.8%) underwent treatment with radical intent, with 13/14 having no evidence of disease at follow-up radiology at 3 months. Other individuals that were not candidates for aggressive recurrence treatment due to non-resectability received chemotherapy, PRRT or SSAs. Within a median follow-up from recurrence of 73.7 months, 5- and 10-year post-recurrence overall survival was estimated to be 76.3% and 45.5%, respectively, suggesting that, even in cases of post-LT recurrence, favourable long-term outcome is attainable. However, this must be considered in terms of the relatively small sample size at a single institution, as must their analyses of prognostic factors (not discussed here).

4. Discussion

Liver transplantation may be associated with favourable long-term outcomes in NET patients with advanced disease—for example, the median 10 year OS was 50%, range 46.1%–88.8%. Recurrence-free survival was inconsistently reported, with two studies (varying in veracity of patient selection) reporting 10-year DFSs of 21% and 86.9%.

Liver transplantation offers a seemingly attractive yet uncommonly used approach for the aggressive management of metastatic neuroendocrine liver metastases. Whilst hepatectomy with curative intent is associated with favourable outcomes in NELM, very high recurrence rates are a major limitation of this approach [15,16], likely driven by understaging of hepatic disease, which is observed even with gold-standard imaging with ensuing residual micro-disease burden [7]. Thus, aggressive treatment manifesting as total hepatectomy (i.e., complete removal of all macro- and micro- disease in the liver) with resection of primary and extrahepatic disease, and a liver allograft appears to be a legitimate option when one considers the excellent results reported by meticulously selected criteria.

Compared to the systematic review of Moris and colleagues [17] on liver transplantation for metastatic NEN, we identified updated multicentre experience from the United States (UNOS database) [42] reinforcing evidence that post-OLT outcomes have improved over time, and included the first study to report in detail on outcomes in cases of recurrence post-OLT [43], which suggests that even in these cases, multimodal therapy may be useful to prolong survival. Nevertheless, our overall conclusions align—documentation of selection criteria is generally poor to non-existent in most studies, there is inconsistent recording of indication, immunosuppression regimens, and patients tend to be heavily pre-treated [24,28]. While studies report 1-, 3-, 5-, and 10-year survival data, here is a lack of studies reporting outcomes after 10 years. Such data are essential in order to understand the true value of liver transplantation on disease outcomes.

There is some evidence that outcomes with OLT for NELM in the recent era in selected patients are not inferior to other, arguably more established "transplant oncology" indications [27,42]. However, the primary evidence gap limiting promotion of OLT pertains to selection—strict selection of patients is necessary, but the optimal approach is not discernible from current evidence.

In order to rectify this, robust derivation and evaluation of multivariable selection models in pooled, specialist centre data may offer utility. Movement away from reductionist, retrospectively derived "scores" with no solid validation towards a more nuanced risk estimation tool that is prospectively evaluated would represent a significant advance. As aforementioned, merely stratifying patients into two vague risk groups only demonstrates broad "average" expectancies, and the use of arbitrary cut-offs in such tools poses ethical questions that become rapidly apparent [24]. External implementation of the Milan NET criteria in other centres should be considered, with meticulous data collection, prospective validation and transparent reporting. Indeed, a novel transplant programme for NELM has been recently initiated for the United Kingdom and Ireland [45], which will be a valuable addition to the evidence base as they report later.

We conclude that, given the currently available evidence and limitations of this evidence, further single-centre or purely retrospective case series will not present any additional benefit to the literature. The relative rarity of NEN and uncommon consideration of

NELM for OLT presents logistical challenges, but with centralisation of care and through international interest groups, multi-centre collaboration for such prospective studies is not only possible, but something that is actively being arranged [45]. The literature for transplantation in the management of neuroendocrine liver metastases needs to mature, and relevant stakeholders must drive this. As part of these new studies, novel protocols could include the consideration of adjuvant therapies to reduce the risk of disease recurrence, should incorporate gold standard imaging protocols as part of follow-up and could consider novel omics-based biomarker technologies [46,47] to expedite the diagnosis of recurrent disease.

Supplementary Materials: The following supporting information can be downloaded at: https://www.mdpi.com/article/10.3390/jpm13101428/s1, Supplementary File S1: PRISMA checklist, Supplementary File S2: Search strategy; Supplementary File S3: Quality appraisal.

Author Contributions: Conceptualisation, V.P., C.H.L., A.F. and A.K.C.; Methodology, V.P., C.H.L., A.F. and A.K.C.; Formal analysis, V.P., C.H.L., A.F. and A.K.C.; Data curation, V.P., C.H.L. and A.K.C.; Writing—original draft preparation, V.P., C.H.L. and A.K.C.; Writing—review and editing, A.F.; Project administration, V.P., C.H.L. and A.K.C.; Supervision, A.F. and A.K.C. All authors have read and agreed to the published version of the manuscript.

Funding: This research received no external funding.

Conflicts of Interest: The authors declare no conflict of interest.

References

1. Dasari, A.; Shen, C.; Halperin, D.; Zhao, B.; Zhou, S.; Xu, Y.; Shih, T.; Yao, J.C. Trends in the Incidence, Prevalence, and Survival Outcomes in Patients With Neuroendocrine Tumors in the United States. *JAMA Oncol.* **2017**, *3*, 1335. [CrossRef]
2. Frilling, A.; Clift, A.K. Therapeutic strategies for neuroendocrine liver metastases. *Cancer* **2015**, *121*, 1172–1186. [CrossRef] [PubMed]
3. Miller, H.C.; Drymousis, P.; Flora, R.; Goldin, R.; Spalding, D.; Frilling, A. Role of Ki-67 proliferation index in the assessment of patients with neuroendocrine neoplasias regarding the stage of disease. *World J. Surg.* **2014**, *38*, 1353–1361. [CrossRef] [PubMed]
4. Riihimäki, M.; Hemminki, A.; Sundquist, K.; Sundquist, J.; Hemminki, K. The epidemiology of metastases in neuroendocrine tumors. *Int. J. Cancer* **2016**, *139*, 2679–2686. [CrossRef] [PubMed]
5. Panzuto, F.; Boninsegna, L.; Fazio, N.; Campana, D.; Brizzi, M.P.; Capurso, G.; Scarpa, A.; De Braud, F.; Dogliotti, L.; Tomassetti, P.; et al. Metastatic and locally advanced pancreatic endocrine carcinomas: Analysis of factors associated with disease progression. *J. Clin. Oncol.* **2011**, *29*, 2372–2377. [CrossRef] [PubMed]
6. Pavel, M.; Öberg, K.; Falconi, M.; Krenning, E.P.; Sundin, A.; Perren, A.; Berruti, A. Gastroenteropancreatic neuroendocrine neoplasms: ESMO Clinical Practice Guidelines for diagnosis, treatment and follow-up. *Ann. Oncol.* **2020**, *31*, 844–860. [CrossRef]
7. Elias, D.; Lefevre, J.H.; Duvillard, P.; Goéré, D.; Dromain, C.; Dumont, F.; Baudin, E. Hepatic metastases from neuroendocrine tumors with a 'thin slice' pathological examination: They are many more than you think. *Ann. Surg.* **2010**, *251*, 307–310. [CrossRef]
8. Frilling, A.; Modlin, I.M.; Kidd, M.; Russell, C.; Breitenstein, S.; Salem, R.; Kwekkeboom, D.; Lau, W.-Y.; Klersy, C.; Vilgrain, V.; et al. Recommendations for management of patients with neuroendocrine liver metastases. *Lancet Oncol.* **2014**, *15*, e8–e21. [CrossRef]
9. Caplin, M.E.; Pavel, M.; Ćwikła, J.B.; Phan, A.T.; Raderer, M.; Sedláčková, E.; Cadiot, G.; Wolin, E.M.; Capdevila, J.; Wall, L.; et al. Lanreotide in metastatic enteropancreatic neuroendocrine tumors. *N. Engl. J. Med.* **2014**, *371*, 224–233. [CrossRef]
10. Strosberg, J.; El-Haddad, G.; Wolin, E.; Hendifar, A.; Yao, J.; Chasen, B.; Mittra, E.; Kunz, P.L.; Kulke, M.H.; Jacene, H.; et al. Phase 3 Trial of ^{177}Lu-Dotatate for Midgut Neuroendocrine Tumors. *N. Engl. J. Med.* **2017**, *376*, 125–135. [CrossRef]
11. Raymond, E.; Dahan, L.; Raoul, J.-L.; Bang, Y.-J.; Borbath, I.; Lombard-Bohas, C.; Valle, J.; Metrakos, P.; Smith, D.; Vinik, A.; et al. Sunitinib malate for the treatment of pancreatic neuroendocrine tumors. *N. Engl. J. Med.* **2011**, *364*, 501–513. [CrossRef] [PubMed]
12. Yao, J.C.; Lombard-Bohas, C.; Baudin, E.; Kvols, L.K.; Rougier, P.; Ruszniewski, P.; Hoosen, S.; Peter, J.S.; Haas, T.; Lebwohl, D.; et al. Daily oral everolimus activity in patients with metastatic pancreatic neuroendocrine tumors after failure of cytotoxic chemotherapy: A phase II trial. *J. Clin. Oncol.* **2010**, *28*, 69–76. [CrossRef]
13. Braat, A.J.A.T.; Bruijnen, R.C.G.; van Rooij, R.; Braat, M.N.; Wessels, F.J.; van Leeuwaarde, R.S.; van Treijen, M.J.C.; de Herder, W.W.; Hofland, J.; Tesselaar, M.E.T.; et al. Additional holmium-166 radioembolisation after lutetium-177-dotatate in patients with neuroendocrine tumour liver metastases (HEPAR PLuS): A single-centre, single-arm, open-label, phase 2 study. *Lancet Oncol.* **2020**, *21*, 561–570. [CrossRef] [PubMed]

14. Strosberg, J.R.; Caplin, M.E.; Kunz, P.L.; Ruszniewski, P.B.; Bodei, L.; Hendifar, A.; Mittra, E.; Wolin, E.M.; Yao, J.C.; Pavel, M.E.; et al. ^{177}Lu-Dotatate plus long-acting octreotide versus high-dose long-acting octreotide in patients with midgut neuroendocrine tumours (NETTER-1): Final overall survival and long-term safety results from an open-label, randomised, controlled, phase 3 trial. *Lancet Oncol.* **2021**, *22*, 1752–1763. [CrossRef]
15. Frilling, A.; Clift, A.K. Surgical approaches to the management of neuroendocrine liver metastases. *Endocr. Metab. Clin. North. Am.* **2018**, *47*, 627–643. [CrossRef] [PubMed]
16. Saxena, A.; Chua, T.C.; Perera, M.; Chu, F.; Morris, D.L. Surgical resection of hepatic metastases from neuroendocrine neoplasms: A systematic review. *Surg. Oncol.* **2012**, *21*, e131–e141. [CrossRef]
17. Moris, D.; Tsilimigras, D.I.; Ntanasis-Stathopoulos, I.; Beal, E.W.; Felekouras, E.; Vernadakis, S.; Fung, J.J.; Pawlik, T.M. Liver transplantation in patients with liver metastases from neuroendocrine tumors: A systematic review. *Surgery* **2017**, *162*, 525–536. [CrossRef]
18. Mrzljak, A.; Kocman, B.; Skrtic, A.; Furac, I.; Popic, J.; Franusic, L.; Zunec, R.; Mayer, D.; Mikulic, D. Liver re-transplantation for donor-derived neuroendocrine tumor: A case report. *World J. Clin. Cases* **2019**, *7*, 2794–2801. [CrossRef]
19. Nikeghbalian, S.; Eshraghian, A.; Kazemi, K.; Shamsaeefar, A.; Hosseinzadeh, M.; Geramizadeh, B.; Malek-Hosseini, S.A. Liver Transplantation for High-Grade Primary Hepatic Neuroendocrine Tumor with Diffuse Liver Metastasis. *J. Gastrointest. Cancer* **2020**, *51*, 304–306. [CrossRef]
20. Sher, L.S.; Levi, D.M.; Wecsler, J.S.; Lo, M.; Petrovic, L.M.; Groshen, S.; Ji, L.; Uso, T.D.; Tector, A.J.; Hamilton, A.S.; et al. Liver transplantation for metastatic neuroendocrine tumors: Outcomes and prognostic variables. *J. Surg. Oncol.* **2015**, *112*, 125–132. [CrossRef]
21. Mangus, R.S.; Tector, A.J.; Kubal, C.A.; Fridell, J.A.; Vianna, R.M. Multivisceral transplantation: Expanding indications and improving outcomes. *J. Gastrointest. Surg.* **2013**, *17*, 179–186, discussion 186–187. [CrossRef] [PubMed]
22. Fan, S.T.; Le Treut, Y.P.; Mazzaferro, V.; Burroughs, A.K.; Olausson, M.; Breitenstein, S.; Frilling, A. Liver transplantation for neuroendocrine tumour liver metastases. *HPB* **2015**, *17*, 23–28. [CrossRef] [PubMed]
23. Olausson, M.; Friman, S.; Herlenius, G.; Cahlin, C.; Nilsson, O.; Jansson, S.; Wängberg, B.; Ahlman, H. Orthotopic liver or multivisceral transplantation as treatment of metastatic neuroendocrine tumors. *Liver Transpl.* **2007**, *13*, 327–333. [CrossRef]
24. Le Treut, Y.P.; Grégoire, E.; Klempnauer, J.; Belghiti, J.; Jouve, E.; Lerut, J.; Castaing, D.; Soubrane, O.; Boillot, O.; Mantion, G.; et al. Liver transplantation for neuroendocrine tumors in Europe-results and trends in patient selection: A 213-case European liver transplant registry study. *Ann. Surg.* **2013**, *257*, 807–815. [CrossRef]
25. Gedaly, R.; Daily, M.F.; Davenport, D.; McHugh, P.P.; Koch, A.; Angulo, P.; Hundley, J.C. Liver transplantation for the treatment of liver metastases from neuroendocrine tumors: An analysis of the UNOS database. *Arch. Surg.* **2011**, *146*, 953–958. [CrossRef]
26. Vilchez, V.; Gedaly, R. Liver transplantation for the treatment of neuroendocrine liver metastases. *Best. Pract. Res. Clin. Endocrinol. Metab.* **2016**, *30*, 141–147. [CrossRef] [PubMed]
27. Nguyen, N.T.T.; Harring, T.R.; Goss, J.A.; O'Mahony, C.A. Neuroendocrine Liver Metastases and Orthotopic Liver Transplantation: The US Experience. *Int. J. Hepatol.* **2011**, *2011*, 742890. [CrossRef]
28. Mazzaferro, V.; Sposito, C.; Coppa, J.; Miceli, R.; Bhoori, S.; Bongini, M.; Camerini, T.; Milione, M.; Regalia, E.; Spreafico, C.; et al. The Long-Term Benefit of Liver Transplantation for Hepatic Metastases From Neuroendocrine Tumors. *Am. J. Transplant.* **2016**, *16*, 2892–2902. [CrossRef]
29. Page, M.J.; McKenzie, J.E.; Bossuyt, P.M.; Boutron, I.; Hoffmann, T.C.; Mulrow, C.D.; Shamseer, L.; Tetzlaff, J.M.; Akl, E.A.; Brennan, S.E.; et al. The PRISMA 2020 statement: An updated guideline for reporting systematic reviews. *BMJ* **2021**, *372*, n71. [CrossRef]
30. Dousset, B.; Saint-Marc, O.; Pitre, J.; Soubrane, O.; Houssin, D.; Chapuis, Y. Metastatic endocrine tumors: Medical treatment, surgical resection, or liver transplantation. *World J. Surg.* **1996**, *20*, 908–915. [CrossRef]
31. Le Treut, Y.P.; Delpero, J.R.; Dousset, B.; Cherqui, D.; Segol, P.; Mantion, G.; Hannoun, L.; Benhamou, G.; Launois, B.; Boillot, O.; et al. Results of liver transplantation in the treatment of metastatic neuroendocrine tumors. A 31-case French multicentric report. *Ann. Surg.* **1997**, *225*, 355–364. [CrossRef] [PubMed]
32. Le Treut, Y.P.; Grégoire, E.; Belghiti, J.; Boillot, O.; Soubrane, O.; Mantion, G.; Cherqui, D.; Castaing, D.; Ruszniewski, P.; Wolf, P.; et al. Predictors of long-term survival after liver transplantation for metastatic endocrine tumors: An 85-case French multicentric report. *Am. J. Transplant.* **2008**, *8*, 1205–1213. [CrossRef] [PubMed]
33. Routley, D.; Ramage, J.K.; McPeake, J.; Tan, K.-C.; Williams, R. Orthotopic liver transplantation in the treatment of metastatic neuroendocrine tumors of the liver. *Liver Transpl. Surg.* **1995**, *1*, 118–121. [CrossRef] [PubMed]
34. Rosenau, J.; Bahr, M.J.; Wasielewski, V.; Mengel, M.; Schmidt, H.H.J.; Nashan, B.; Lang, H.; Klempnauer, J.; Manns, M.P.; Boeker, K.H.W. Ki67, E-cadherin, and p53 as prognostic indicators of long-term outcome after liver transplantation for metastatic neuroendocrine tumors. *Transplantation* **2002**, *73*, 386–394. [CrossRef]
35. Florman, S.; Toure, B.; Kim, L.; Gondolesi, G.; Roayaie, S.; Krieger, N.; Fishbein, T.; Emre, S.; Miller, C.; Schwartz, M. Liver transplantation for neuroendocrine tumors. *J. Gastrointest. Surg.* **2004**, *8*, 208–212. [CrossRef]
36. van Vilsteren, F.G.; Baskin-Bey, E.S.; Nagorney, D.M.; Sanderson, S.O.; Kremers, W.K.; Rosen, C.B.; Gores, G.J.; Hobday, T.J. Liver transplantation for gastroenteropancreatic neuroendocrine cancers: Defining selection criteria to improve survival. *Liver Transpl.* **2006**, *12*, 448–456. [CrossRef]

37. Marín, C.; Robles, R.; Fernández, J.; Bueno, F.; Ramírez, P.; Miras, M.; Parrilla, P. Role of liver transplantation in the management of unresectable neuroendocrine liver metastases. *Transplant. Proc.* **2007**, *39*, 2302–2303. [CrossRef]
38. Dhupar, R.; De Vera, M.E.; Marsh, J.W.; Fontes, P.A.; Steel, J.L.; Zeh, H.J.; Gamblin, T.C. Simultaneous pancreatectomy and liver transplantation: A single-institution experience. *HPB* **2009**, *11*, 242–246. [CrossRef]
39. Frilling, A.; Li, J.; Malamutmann, E.; Schmid, K.-W.K.-W.; Bockisch, A.; Broelsch, C.E. Treatment of liver metastases from neuroendocrine tumours in relation to the extent of hepatic disease. *Br. J. Surg.* **2009**, *96*, 175–184. [CrossRef]
40. Bonaccorsi-Riani, E.; Apestegui, C.; Jouret-Mourin, A.; Sempoux, C.; Goffette, P.; Ciccarelli, O.; Borbath, I.; Hubert, C.; Gigot, J.F.; Hassoun, Z.; et al. Liver transplantation and neuroendocrine tumors: Lessons from a single centre experience and from the literature review. *Transpl. Int.* **2010**, *23*, 668–678. [CrossRef]
41. Korda, D.; Doros, A.; Piros, L.; Gerlei, Z.; Haboub-Sandil, A.; Mándli, T.; Fazakas, J.; Deák, P.; Máthé, Z. Liver Transplant for Metastatic Neuroendocrine Tumors: A Single-Center Experience in Hungary. *Transplant. Proc.* **2019**, *51*, 1251–1253. [CrossRef] [PubMed]
42. Valvi, D.; Mei, X.; Gupta, M.; Shah, M.B.; Ancheta, A.; Marti, F.; Gedaly, R. Younger Age Is Associated with Improved Survival in Patients Undergoing Liver Transplantation Alone for Metastatic Neuroendocrine Tumors. *J. Gastrointest. Surg.* **2021**, *25*, 1487–1493. [CrossRef] [PubMed]
43. Sposito, C.; Rossi, R.E.; Monteleone, M.; Coppa, J.; Bongini, M.; Milione, M.; Bhoori, S.; Mazzaferro, V. Postrecurrence Survival After Liver Transplantation for Liver Metastases From Neuroendocrine Tumors. *Transplantation* **2021**, *105*, 2579–2586. [CrossRef] [PubMed]
44. Moons, K.G.M.; Altman, D.G.; Reitsma, J.B.; Ioannidis, J.P.A.; Macaskill, P.; Steyerberg, E.W.; Vickers, A.J.; Ransohoff, D.F.; Collins, G.S. Transparent Reporting of a multivariable prediction model for Individual Prognosis Or Diagnosis (TRIPOD): Explanation and Elaboration. *Ann. Intern. Med.* **2015**, *162*, W1. [CrossRef]
45. Shah, T.; Moore, J.; Venkataraman, H.; Caplin, M.; Smith, S.; O'Toole, D.; Hoti, E.; Dasari, B.; Ford, S.; Mazzaferro, V.; et al. Setting up of a national liver transplant programme for neuroendocrine tumour liver metastases in UK and Ireland: Opportunities for clinical study and research. *Endocr. Abstr.* **2021**, *80*, P1. [CrossRef]
46. Modlin, I.M.; Bodei, L.; Kidd, M. Neuroendocrine tumor biomarkers: From monoanalytes to transcripts and algorithms. *Best. Pract. Res. Clin. Endocrinol. Metab.* **2016**, *30*, 59–77. [CrossRef]
47. Oberg, K.; Modlin, I.M.; De Herder, W.; Pavel, M.; Klimstra, D.; Frilling, A.; Metz, D.C.; Heaney, A.; Kwekkeboom, D.J.; Strosberg, J.R.; et al. Consensus on biomarkers for neuroendocrine tumour disease. *Lancet Oncol.* **2015**, *16*, e435–e446. [CrossRef]

Journal of *Personalized Medicine*

MDPI

Article

Influence of the Bile Acid Transporter Genes *ABCB4, ABCB8,* and *ABCB11* and the Farnesoid X Receptor on the Response to Ursodeoxycholic Acid in Patients with Nonalcoholic Steatohepatitis

Henriette Kreimeyer [1], Katharina Vogt [2], Tobias Götze [3], Jan Best [1], Oliver Götze [1], Jochen Weigt [2], Alisan Kahraman [2], Mustafa Özçürümez [1], Julia Kälsch [2], Wing-Kin Syn [4,5], Svenja Sydor [1], Ali Canbay [1] and Paul Manka [1,*]

1 Department of Internal Medicine, University Hospital Knappschaftskrankenhaus, Ruhr-University Bochum, In der Schornau 23-25, 44892 Bochum, Germany; jan.best@kk-bochum.de (J.B.); oliver.goetze@kk-bochum.de (O.G.)
2 Department of Gastroenterology and Hepatology, University Hospital Essen, Hufelandstraße 55, 45147 Essen, Germany; jochen.weigt@med.ovgu.de (J.W.)
3 Department of Gastroenterology, Hepatology, and Infectious Diseases, Otto-von-Guericke-University Hospital Magdeburg, Leipziger Strasse 44, 39120 Magdeburg, Germany
4 Division of Gastroenterology and Hepatology, Department of Internal Medicine, Saint Louis University School of Medicine, St. Louis, MO 63104, USA
5 Department of Physiology, Faculty of Medicine and Nursing, University of Basque Country UPV/EHU, 48940 Leioa, Spain
* Correspondence: paul.manka@rub.de

Abstract: The prevalence of NAFLD and NASH is increasing worldwide, and there is no approved medical treatment until now. Evidence has emerged that interfering with bile acid metabolism may lead to improvement in NASH. In this study, 28 patients with elevated cholestatic liver function tests (especially GGT) were screened for bile acid gene polymorphisms and treated with UDCA. All patients had a bile acid gene polymorphism in *ABCB4* or *ABCB11*. Treatment with UDCA for 12 months significantly reduced GGT in all patients and ALT in homozygous patients. No difference in fibrosis was observed using FIb-4, NFS, and transient elastography (TE). *PNPLA3* and *TM6SF2* were the most common NASH-associated polymorphisms, and patients with *TM6SF2* showed a significant reduction in GGT and ALT with the administration of UDCA. In conclusion, NASH patients with elevated GGT should be screened for bile acid gene polymorphisms, as UDCA therapy may improve liver function tests. However, no difference in clinical outcomes, such as progression to cirrhosis, has been observed using non-invasive tests (NITs).

Keywords: NASH; UDCA; bile acid transporter; ALT; GGT; TE

check for updates

Citation: Kreimeyer, H.; Vogt, K.; Götze, T.; Best, J.; Götze, O.; Weigt, J.; Kahraman, A.; Özçürümez, M.; Kälsch, J.; Syn, W.-K.; et al. Influence of the Bile Acid Transporter Genes *ABCB4*, *ABCB8*, and *ABCB11* and the Farnesoid X Receptor on the Response to Ursodeoxycholic Acid in Patients with Nonalcoholic Steatohepatitis. *J. Pers. Med.* **2023**, *13*, 1180. https://doi.org/10.3390/jpm13071180

Academic Editor: Margarida Maria Fernandes Baptista e Silva

Received: 16 June 2023
Revised: 12 July 2023
Accepted: 18 July 2023
Published: 24 July 2023

1. Introduction

Non-alcoholic fatty liver disease (NAFLD) has a prevalence of nearly 25% worldwide [1–3] and eventually progresses to non-alcoholic steatohepatitis, fibrosis, cirrhosis, and finally hepatocellular carcinoma (HCC). Treatment focuses on weight loss, dietary modification, and the management of comorbidities [4]. Independent risk factors for non-alcoholic steatohepatitis (NASH) progression include type 2 diabetes (T2DM), insulin resistance, and fibrosis [5,6]. However, NAFLD and NASH can also occur in lean individuals, referred to as lean NASH. In patients with NAFLD, gamma-glutamyl transpeptidase (GGT) dynamics is a biomarker of advanced fibrosis [7].

Primary bile acids (BAs), such as cholic acid and chenodeoxycholic acid, are produced in hepatocytes and secreted against a gradient into the canaliculi by the bile salt export pump (BSEP; coding gene: ATP binding cassette subfamily B member 11; *ABCB11*) in the apical membrane of hepatocytes. Secretion depends on the amount of cholesterol and lipids in the membrane, which can be altered by the flippase familial intrahepatic cholestasis-1 (FIC, coding gene: ATPase phospholipid transporting 8B1; *ATP8B1*) and the flippase multidrug resistance protein (MDR3, coding gene: ATP binding cassette subfamily B member 4; *ABCB4*). The expression of BSEP is regulated by the farnesoid X receptor (FXR, coding gene: nuclear receptor subfamily 1 group H member 4; *NR1H4*) [8].

The synthesis and serum levels of BAs have been found to correlate with the severity of nonalcoholic NAFLD [9]. In recent years, there has been a growing emphasis on studying BA metabolism in the context of NASH [10]. One reason for this increased focus is the promising therapeutic response observed with obeticholic acid, an agonist of the bile acid receptor FXR, in NASH treatment. Notably, interim analysis of a phase 3 study has demonstrated histological improvement in fibrosis among patients, irrespective of polymorphisms in bile acid transporter genes. FXR plays a crucial role as a transcription factor in regulating the expression of various enzymes and transport proteins involved in bile acid synthesis and transport within the liver and intestine.

Mutations and polymorphisms in genes contributing to bile acid transportation can result in progressive familial intrahepatic cholestasis (PFIC), low phospholipid-associated cholestasis (LPAC), intrahepatic cholestasis of pregnancy (ICO), or benign recurrent intrahepatic cholestasis (BRIC) [11]. PFIC is a heterogeneous entity which leads to liver cirrhosis and HCC in newborns and adults. Presently, six different types have been described with slightly different clinical phenotypes. It is differentiated between high-GGT and low-GGT entities. Roughly, PFIC 3 (polymorphism in MDR3, *ABCB4*) leads to a disturbance in phospholipid secretion and consequently to an imbalance in micelle formation and an increase in free bile acids [12,13]. These lead to hepatocyte damage and an increase in GGT. PFIC 1,2,4,5,6 result in the reduced secretion of bile acids and therefore the accumulation of hepatocytes [14]. This leads to damage of hepatocytes and increase in alanine aminotransferase (ALT), but normally low GGT levels. BRIC seems to be a less harmful phenotype of all PFIC polymorphism. LPAC and ICP are described for MDR3 polymorphism or MDR and BSEP polymorphism and can be treated with ursodeoxycholic acid (UDCA) [15].

Several polymorphisms have been described in MDR3-associated diseases. *ABCB4* c. 711 A>T (p.I237=, *rs2109505*) is associated with elevated ALT and GGT and a higher risk of cirrhosis and HCC [16]. *ABCB4* c. 523 T>C (p.T175A, *rs58238559*) is correlated with liver stiffness [17].

UDCA is mainly used in cholestatic liver diseases. It activates a typical and alternative transport mechanism for bile acids. The accumulation of bile acids in hepatocytes contributes to an up-regulation of inflammatory cytokines and hepatic cell activation, resulting in fibrosis [18].

Steatotic liver disease is the most abundant liver disease, and patients can show a cholestatic condition. As therapy has emerged, bile acid metabolism seems to be one of the key players, and empirically it has been shown that patients improve with UDCA therapy. Following the human genome project, BA transporter polymorphism has been described and linked to rare genetic diseases. Empirically, it has been observed that NASH patients with cholestatic laboratory findings benefit from UDCA therapy [19–21]. However, a statistical evaluation and quantification of this effect, as well as an examination of the possible influence of gene polymorphisms of the above-mentioned transporters, are not yet available. Therefore, in this retrospective analysis, we investigated the course of various liver parameters in NASH patients who underwent "off-label" treatment with UDCA because of an existing cholestatic laboratory profile and in whom gene polymorphism determination was available.

2. Methods

2.1. Patient Information, Data Collections and Ethical Considerations

This retrospective study was performed in line with the "Declaration of Helsinki" (latest revision Fortaleza, Brasil; 2013). The study was approved by the ethics committee of the University of Essen (20-9137-BO, 25 May 2020).

Twenty-eight patients (53% male, 47% female) with NASH and elevated GGT level treated at Universal Hospital Essen and Universitätsklinikum Magdeburg between 2012 and 2019 were included. Relevant clinical data were extracted retrospectively from the electronic medical record. SNP analyses ("single nucleotide polymorphism") of the known polymorphisms of the relevant genes were performed in NASH patients with a cholestatic picture here in the context of clinical diagnostic.

NASH was diagnosed by ultrasound, elevated liver enzymes, and serologic exclusion of viral, autoimmune, and other metabolic diseases. Sequenced polymorphisms were analyzed for associations with GGT, alkaline phosphatase (ALP), transaminases, transient elastography and fibrosis scores (NFS and Fib-4). NFS and Fib-4 [22,23] were calculated as previously described. Data were compared for every polymorphism between wildtype, heterozygote, and homozygote patients.

2.2. Sequencing

Quantities of 5–10 mL EDTA blood was maintained from all patients and genotyping of ABCB4 (rs45575636, rs1202283, rs58238559, rs2109505), ABCB11 (rs72549402, rs497692, rs2287622), TM6SF2 (rs58542926), NR1H4 (rs56163822), ATP8B1 (rs146599962, rs34018205, rs121909100, rs765889649), and PNPLA3 (rs738409) (described in Table 1) was performed using qPCR (TaqMan fast 7500), as described previously [15,24].

Table 1. Description of polymorphisms.

Gen	SNP Cluster ID	c.	p.
ABCB4	rs45575636	c.1769G > A	p. R590Q
	rs1202283	c.504 C > T	p. N168=
	rs58238559	c.523 T> C	p. T175A
	rs2109505	c.711A>T	p. I237=
ABCB11	rs72549402	c.1445A>G	p. D482G
	rs497692	c. 3084 A>G	p. A1028=
	rs2287622	c.1331 T>C	p. A444V
ATP8B1	rs146599962	c.134A>C	p. N45T
	rs34018205	c.1286A>C	p. E429A
	rs121909100	c.1982T>C	p. I661T
	rs765889649	c.2855G>A	p. R952Q
PNPLA3	rs738409	C>G	p. I148M
TM6SF2	rs58542926	c.449 C>T	p. Glu167Lys;
FXR	rs56163822	c. −1 G>T	Intron

2.3. Statistics

Analyses were performed using Microsoft Office Excel 2013, Graph Pad 9.3.0 (GraphPad Software Inc., La Jolla, CA, USA) and R (4.2.3). Differences between and in groups were analyzed using the Friedman test (paired, non-normal distribution), Wilcoxon test (paired, non-normal distribution), p-adjustment using the Bonferroni method, and a paired *t*-test. If three timepoints were compared, the Friedman test was performed, if possible, followed by pairwise comparison using the Wilcoxon test. A value of $p < 0.05$ was considered to be statistically significant and adjusted for multiple comparisons (using Bonferroni) when performing multiple tests. All comparisons are displayed as medians with a minimum and maximum range; this is stated otherwise if not.

3. Results

3.1. Demographic Data

Due to elevated GGT levels, off-label therapy with UDCA was performed in 28 patients [42.29 yrs (17–75), 47% female]. In addition, these patients were screened for potential polymorphisms in bile acid transporter genes.

Mean BMI was 26.98 kg/m^2 (19–39), 19 (67.8%) patients were obese or suffered from adipositas. T2DM was only present in six (21.43) patients. Eight (28.6%) patients reported moderate alcohol consumption, and one (3.6%) rare alcohol consumption. Table 2 shows detailed clinical characteristics and co-morbidities.

Table 2. Clinical and demographic data. Data are displayed for all patients. In T2DM, vitamin D deficiency, anemia, iron deficiency, folate deficiency, osteopenia, and smoking data were only available for 27 patients, and high cholesterol only for 26 patients. Vitamin B 12 deficiency: serum vitamin B12 < 200 ng/mL, vitamin D deficiency: serum 25-OH vitamin D < 10 ng/mL, anemia: hemoglobin < 11.6 g/dL, high cholesterol: serum-LDL > 3 mmol/L, iron deficiency serum ferritin: <15 ng/mL, folate deficiency: serum folate: <5 ng/mL.

	Total Cohort	Min–Max (%)
Number	28	
Age (mean)	42.29	17–75
Sex (female)	13	
Height (cm)	171.68	132–192
Weight (kg)	80.39	35–115
BMI (kg/m^2)	26.98	19–39
FAST Score	0.68	0.3–0.97
Fib4 Score	1.56	0.0–5.82
NFS	−2.53	−8.08–1.53
Adipositas		
Normal weight	8	32
Obese	12	42.86
Adipostias Grad I	5	17.86
Adipositas Grad II	2	7.14
Alcohol		
None	19	67.86
Rarely	8	32
Sometimes	1	3.57
Comorbidities (no. of patients, %)		
T2DM (n = 27, 96)	6	21.42
Arterielle hypertonie (n = 28, 100)	8	32
Cardiovascular disease (n = 28, 100)	2	7.14
Chronic kidney disease(n = 28, 100)	2	7.14
Vitamin B12 deficiency (n = 27, 96)	2	7.14
Vitamin D deficiency (n = 27, 96)	5	17.86
Anemia (n = 27, 96)	2	7.14
Iron deficiency (n = 27, 96)	2	7.14
Folat deficiency (n = 27, 96)	0	0
Osteopenia (n = 27, 96)	3	10.71
High cholesterol (n = 26, 92.86)	14	0.5
Smoking (n = 27, 96)	5	17.86

Noninvasive testing for fibrosis and steatosis using transient elastography was performed in 21 and 14 patients, respectively. Steatosis was measured as the controlled attenuation parameter (CAP). When applying CAP values of >260 db/m as cut-off for NAFLD and NASH patients, advanced steatosis (S3) was present in five (35.7%) patients. One (7.1%) patient demonstrated mild steatosis (S1). Five (23.8%) patients presented

LSM values >9.7 kPa, indicating advanced fibrosis (F3), one (4.76) patient demonstrated moderate fibrosis (F2), and fourteen (66.7%) patients no or mild fibrosis (F0-F1).

3.2. *High Prevalence of Polymorphism in ABCB4 or ABCB11 in NASH Patients with Elevated GGT-Levels*

The genotyping of bile acid transporter gene polymorphism was performed as described above. The results are summarized in Table 3 and Figure 1. All patients were either carriers of a polymorphism in the bile acid transporter gene *ABCB4* (26 (92.9%)) or *ABCB11* (26 (92.9%)). None of the following polymorphism was detected in any patient: *ABCB4* p. R590Q (rs45575636), *ABCB4* c.523 T>C (p. T175A, rs58238559), *ABCB11* c. 1445 A>G (p. D482G, rs72549402), *ATP8B1* c. 134 A>C (p. N45T, rs146599962), *ATP8B1* c. 1286 A>C (p. E 429A, rs34018205), *ATP8B1* c. 1982 T>C (p. I661T, rs121909100).

Table 3. Frequency of polymorphisms.

Polymorphism	Heterozygote	Homozygote	Wildtype	No Data
ABCB4 c.504 C>T	11	11	6	0
ABCB4 c.771 A>T	5	21	2	0
ABCB11c. 3084 A>G	12	11	5	0
ABCB11 c. 1331 T>C	10	13	5	0
PNPLA3 rs738409 C>G	6	8	0	14
TM6SF2 c.449 C>T	2	0	10	16
FXR c. −1 G>T	3	0	25	0

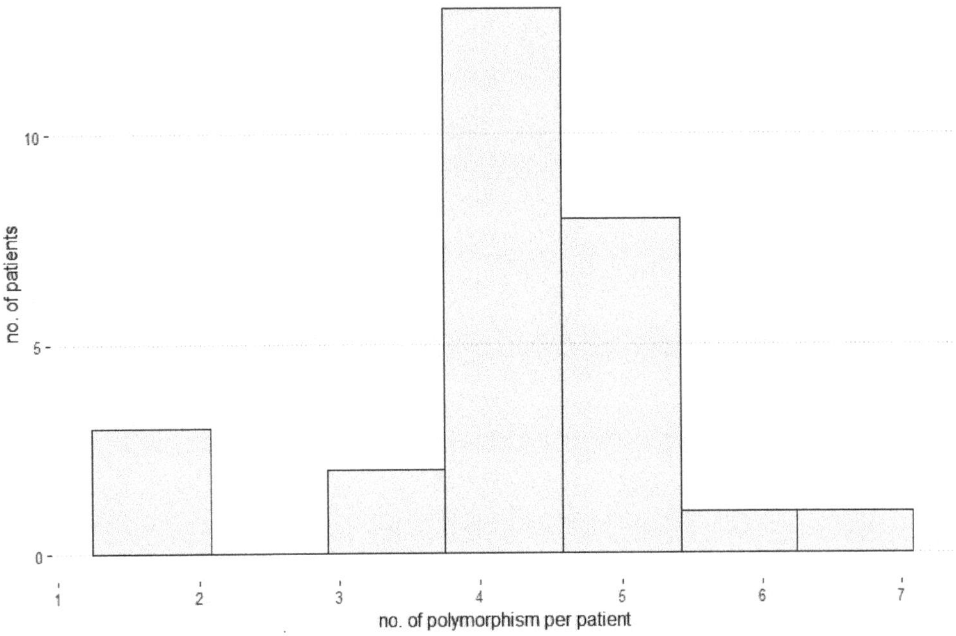

Figure 1. Number of polymorphisms per patient. Most patients showed 4 or 5 polymorphisms regarding all analyzed bile acid transporter genes and NAFLD-associated genes.

3.3. *Polymorphisms in PNPLA3 Were the Most Common among NASH-Associated Genes*

PNPLA3 (rs738409 C>G, p. I148M; patatin-like phospholipase domain containing 3) was present in 17 (43.5%) patients and in *TM6SF2* (rs58542926 c.449 C>T, p.Glu167Lys; transmembrane 6 superfamily member 2) in only 3 (7.7%) patients. All patients with one of these polymorphisms also carried a polymorphism in at least one other bile acid transporter gene. In patients with homozygote *PNPLA3* polymorphism, a significant decrease in ALT and GGT could be seen after 12 months. No statistical analysis was performed in *TM6SF2* polymorphism patients as only two had a heterozygous carrier status, but using descriptive analysis revealed a decrease in ALT and GGT (*ALT:* 45.5 (21–70) U/l vs. 24 (23–25) U/l; *GGT:* 299 (198–400) U/l vs. 186 (61–310) U/l).

3.4. UDCA Therapy Leads to Various Decreases in Liver Function Tests and Is Dependent on the Underlying Polymorphism

In all patients, GGT was elevated at baseline and decreased significantly over 12 months, except in patients with the *ATP8B1* c. 2855 G>A (p.R952Q, rs765889649) polymorphism, which was present in only three patients. ALP, a sensitive marker of bile duct epithelial damage, was significantly reduced in patients with ATP8B1 c. 2855 G>A and *TM6SF2*, but not in patients with *ATP8B1* c. 2855 G>A.

In addition, liver damage as measured by the ALT level was significantly reduced in all polymorphisms except *TM6SF2*. AST showed no difference in any group but in the whole cohort (Figure 2).

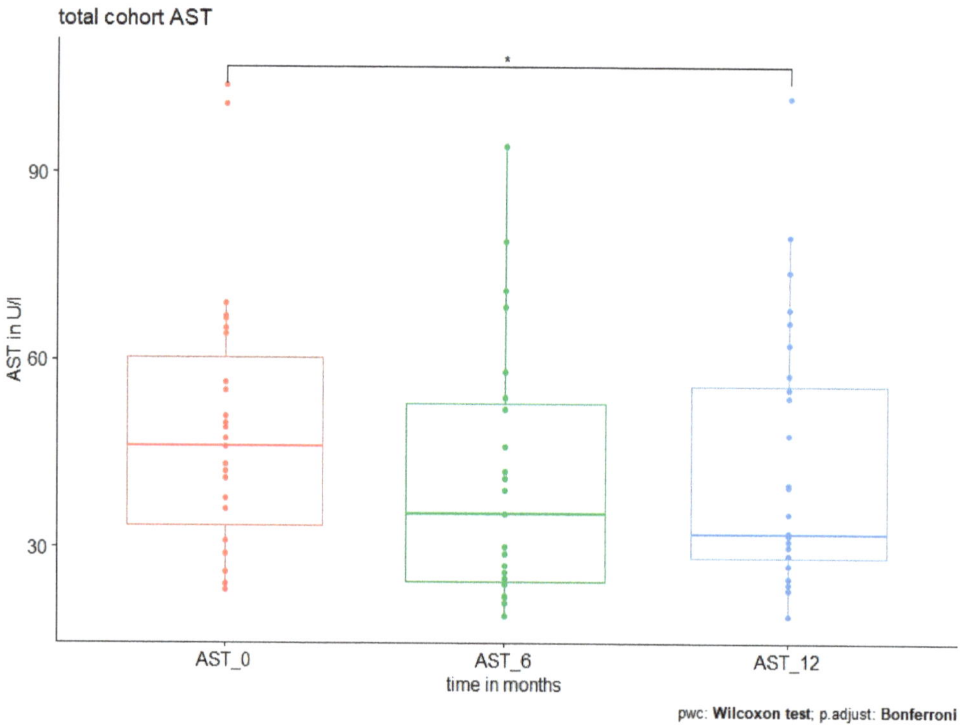

Figure 2. AST level at baseline and after 6 and 12 months in all patients. Median AST level was 46 (23–104) U/l at baseline vs. 35.4 (19–94.2) U/l at 6 months, p = 0.031. No further significant reduction was observed after 12 months (32.3 (19–102) U/l, p = 0.546 (0 vs. 12 months) and p = 1 (6 vs. 12 months).

In *ABCB4* c.504 C>T (p. N168=, rs1202283), GGT was reduced in patients with homozygous or heterozygous polymorphism, but in patients with polymorphism in ABCB4 c.A711 A>T, GGT was significantly reduced only in homozygous patients. In all patients with ABCB4 polymorphism, ALT was significantly reduced only in patients with homozygous polymorphisms (Figure 3).

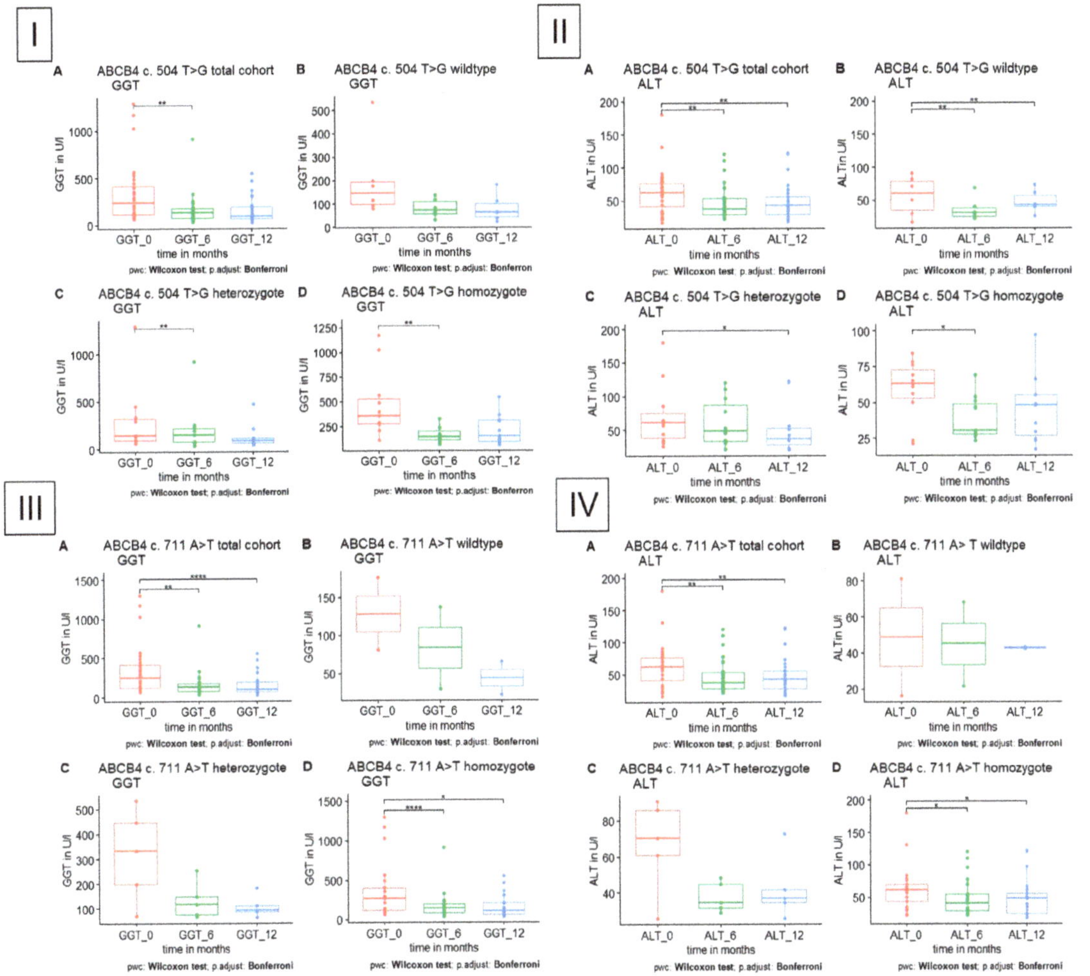

Figure 3. GGT and ALT level in patients with polymorphism in *ABCB4*, respectively, *A* total cohort, *B* wildtype, *C* heterozygote, *D* homozygote. (**I**) GGT is significantly reduced in heterozygote and homozygote polymorphism in *ABCB4* c.504 C>T after 6 months. (**II**) ALT is significantly reduced in heterozygote and homozygote polymorphism in *ABCB4* c.504 C>T after 6 and 12 months, respectively. (**III**,**IV**) GGT and ALT are significantly reduced in homozygote polymorphism in *ABCB4* c. 711 A>T after 6 and 12 months.

All patients with polymorphisms in *ABCB11*, homozygote or heterozygote, showed a significant decrease in GGT. ALT was significantly reduced in patients with homozygous polymorphism in *ABCB11* c. 3084 A>G (p. A1028=, rs497692), and homozygous polymorphism in *ABCB11* 1331T>C (p.A444V, rs2287622) (Figure 4). In patients with homozygous polymorphism in *ABCB11* c. 3084 A>G, there was a decrease in AST after 12 months ($p = 0.043$, 42.6 (24–104) U/l vs. 30 (19–66) U/l), which was not significant in patients with polymorphism in *ABCB11* c. 1331 T>C ($p = 0.13$, 43.2 (23–69) U/l vs. 32 (24–79.8) U/l).

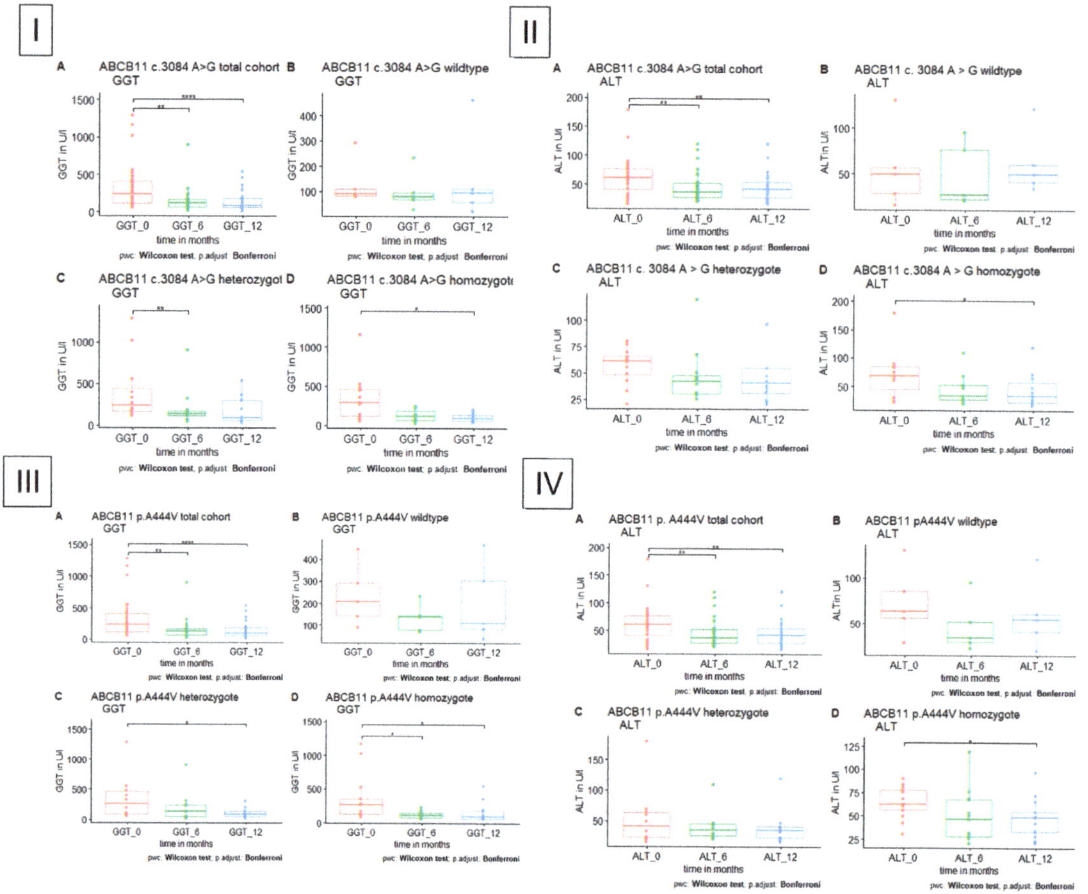

Figure 4. GGT and ALT level in patients with polymorphism in *ABCB11*, respectively, *A* total cohort, *B* wildtype, *C* heterozygote, *D* homozygote. (**I**) GGT is significantly reduced in heterozygote and homozygote polymorphism in *ABCB11* c. 3084 A>G after 6 months in heterozygote and after 12 months in homozygote patients. (**II**) ALT is significantly reduced in homozygote polymorphism in *ABCB11* c.3048 A>G after 12 months. (**III**) GGT is significantly reduced in heterozygote and homozygote polymorphism in *ABCB11* c. 1331 T>C after 12 months in heterozygote and after 6 and 12 months in homozygote patients. (**IV**) ALT is significantly reduced in homozygote polymorphism in *ABCB11* c. 1331 T>C after 12 months.

Patients with homozygote polymorphism in the *PNPLA3* showed a significant decrease in ALT and GGT under the administration of UDCA (Figure 5).

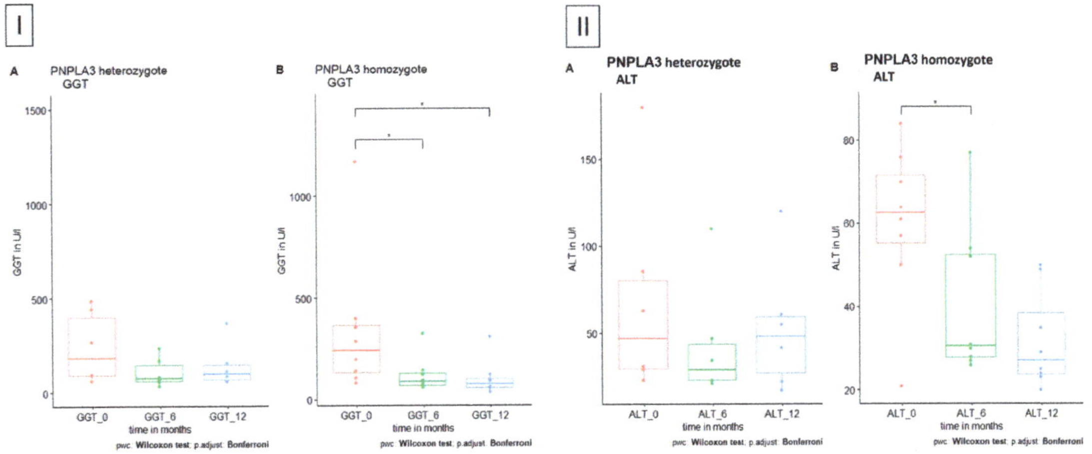

Figure 5. GGT and ALT level in patients with polymorphism in *PNPLA3*, respectively, *A* heterozygote and *B* homozygote. (**I**) GGT is significantly reduced in homozygote polymorphism in *PNPLA3* after 6 and 12 months. (**II**) ALT is significantly reduced in homozygote polymorphism in *PNPLA3* after 6 months.

When regarding all patients with a polymorphism in any bile acid transporter gene, *TM6SF2* and *PNLPA3*, GGT, ALT, ALP, and AST were reduced.

3.5. UDCA Therapy Has No Effect on the Progression of Fibrosis

No significant difference was observed in liver fibrosis or steatosis, as measured by transient elastography (TE) or with Fib-4 or NAFLD fibrosis score (NFS), when grouped by polymorphism.

3.6. The Combination of Different Polymorphisms Shows the Highest Effect on ALT: GGT and Fib-4

To determine whether there was a potential cumulative effect of multiple polymorphisms, we calculated the absolute and relative decrease in ALT, GGT, and Fib-4 for each group per number of polymorphisms between the baseline and 12 months. The greatest decrease in ALT and Fib-4 was observed in patients with three concurrent polymorphisms (mean decrease ALT 27.7 U/l (7–44.4); relative decrease ALT 38.3% (12–68.8%)). However, GGT levels decreased more, but not significantly, with each additional polymorphism (Table S1, Figure S1). Because of the small number of patients per group and the unequal distribution, we combined patients with 1–2 polymorphisms and patients with 3–5 polymorphisms. When comparing the relative and absolute reduction at 12 months, no significant results were found between the two groups (Table S2).

4. Discussion

In this retrospective study, we evaluated the effect of UDCA in patients with NASH and bile acid transporter gene polymorphisms after 6 and 12 months. In almost all patients, liver function tests improved after administration. However, there was no significant effect on fibrosis and steatosis measured by non-invasive testing.

In both homozygous and heterozygous patients, there was a significant reduction in GGT in most patients, as expected, regardless of the underlying specific polymorphism. Furthermore, UDCA therapy resulted in reduced ALT levels in homozygous patients with *ABCB4*, *ABCB11*, and *PNPLA3* polymorphisms. Therefore, we suggest that patients with a homozygous polymorphism in any bile acid transporter gene may benefit from UDCA therapy in terms of liver damage.

All patients studied (who initially presented with elevated GGT levels) had polymorphisms in *ABCB4* or *ABCB11*. In the population-based Icelandic Genome Study, 0.2% of participants had a polymorphism in this gene which was associated with cirrhosis and elevated serum levels of liver-related biomarkers, ALT, AST, and GGT [25]. Notably, in our cohort, we only evaluated data from preselected patients (NAFLD and a high level of cholestatic parameters), hence the data are not comparable to a population study.

Previously, it has been described that ABCB11 c.1331T>C is associated with biopsy-proven liver fibrosis and cirrhosis in patients with HCV but not in NAFLD [26]. In addition, a retrospective study observed a reduction in HCC in patients with HCV cirrhosis treated with UDCA for five years, probably due to the anti-inflammatory effect of UDCA [27]. In this cohort, a significant reduction in ALT levels was observed in patients with *ABCB11* c.1331 T>C. Another study evaluating liver fibrosis by TE did not show a significant association in patients with chronic hepatitis C [28]. In agreement with the results of Iwata et al. [26], patients with the *ABCB11* c.1331 T>C polymorphism did not show a significantly higher level of fibrosis, as measured by surrogate parameters, compared to the wild type in our cohort. However, they were not matched to healthy controls. In an underlying *ABCB11* c.1331 T>C polymorphism, UDCA therapy does not seem to affect fibrosis, but ultimately there is a reduction in inflammation.

In a large Icelandic population study, *ABCB4* polymorphisms were not only associated with rare monogenic liver diseases but also with chronic liver diseases [12,25]. *ABCB4* c.523T>C was found to be associated with increased liver stiffness, whereas in *ABCB4* c.711A>T, it was found to be associated with increased hepatic stiffening only in the presence of polymorphism in *PNLAP3* [17]. In another study including 227 patients, this was an independent risk factor for fibrosis (as measured by TE and Fib-4) [16]. Most of the analyzed patients in our cohort showed no or mild fibrosis (F0-F1). Hence, we cannot make any conclusion if BA transporter gene polymorphisms are an independent risk factor for fibrosis in NASH patients. In addition, there was no difference in fibrosis in our cohort after one year of UDCA therapy. However, the ALT level decreased and steatosis, as measured by CAP and Fib4, decreased but not significantly ($p = 0.175$ (CAP), 0.054 (Fib-4)) in the homozygote group. This may be due to the small cohort of patients and the short observation period. Probably there is a long-term effect of UDCA administration, as suggested in a case report by Frider et al. [29].

ABCB4 c. 504 A>T has only recently been discovered. It is mostly described in ICP. In one study, a patient with NAFLD who did not respond to therapy was genotyped and showed compound heterozygosity, including this polymorphism [30]. In this study, UDCA showed no difference in fibrosis or steatosis but a reduction in ALT and GGT level. As UDCA is well evaluated to decelerate the progress of MDR-3-associated diseases [13], it is possible that it could delay fibrosis in patients with NASH and *ABCB4* polymorphism.

Following the human genome project, polymorphisms associated with NAFLD have been described. Single-nucleotide polymorphism in *PNPLA3* (C>G, p. I148M) modifies NAFLD progression, is associated with the severity of steatohepatitis, severe fibrosis, and confers with an increased risk of developing HCC [31–33]. In Europeans, a heterozygote polymorphism has been described in 43.6–45% and a homozygote polymorphism in 12.1–25.5% of patients with NASH or NAFLD. Higher frequencies are described in Hispanic and lower frequencies in African American cohorts [31,34]. Lately, it has been shown that the incidence of liver-related events is higher in non-obese women in patients with *PNPLA3* polymorphism [35]. In this study cohort of 28 people, 21.4% of patients were heterozygous for the polymorphism in *PNPLA3* and 28.57% were homozygous. This can be explained by the fact that this is a small subset of patients who have already progressed to NASH and have high cholestatic parameters. This aspect must be considered in all the data presented here. In addition, *PNLPA3* sequencing was only available in 48.72% of patients. ALT and GGT decreased after treatment with UDCA. Given the high association of polymorphisms in *PNPLA3*, especially in non-obese women, with liver-related events, treatment with UDCA may be an affordable adjunct to more specific gene therapy, which is being evaluated in preclinical studies [35,36]. As each patient with a *PNPLA3* polymorphism had at least one additional polymorphism in one of the bile acid transporter genes, the implications of the findings need to be evaluated separately.

In a meta-analysis, UDCA treatment showed a significant effect on ALT and GGT in patients with liver disease, but in studies including only NAFLD or NASH patients, no significance was shown. Therefore, we believe that UDCA treatment may be beneficial only for patients with a bile acid transporter gene polymorphism [37]. In this study, we were able to show that 12 months of UDCA treatment resulted in a significant reduction in liver injury, as measured by ALT levels in patients with homozygote polymorphism in a BA transporter gene and NASH.

TM6SF2 c.449 C>T is associated with increased hepatic triglyceride accumulation and hepatic steatosis, fibrosis, cirrhosis, and HCC [38,39]. The frequency was higher in European individuals (7.2%) than in other ethnic groups. Comparably, in this study, heterozygote polymorphisms were found in 7.7% of patients, but less than half of the cohort was tested (43.59%). Data on decreases in GGT or ALT levels in this cohort are not representative as there are only two patients available. In the literature, *TM6SF2* polymorphism is associated with an increase in the ALT level [38].

Because of the high concordance of NASH-related and bile acid gene polymorphisms leading to increased fibrosis, as described in several case reports [29], it is possible that some polymorphisms only reach significance when they are a haplotype. However, due to the small number of patients, further studies are needed to investigate whether patients with certain haplotypes and NASH benefit from UDCA therapy.

T2DM is an independent risk factor for NASH in NAFLD and the prevalence of T2DM in NAFLD and NASH patients is estimated to be estimated to be 22.51% and 43.63% [1]. In this cohort, T2DM was present in only 17.9% of patients, which would be expected to be higher. It may be that the progression to NASH is also influenced by a pathophysiological mechanism other than T2DM, hence the BA transporter gene polymorphism.

As described above, the conclusions of this study are mainly limited by the small cohort of patients. In addition, the patients were not matched with NASH or healthy controls. In addition, the data could not be stratified for confounders such as diabetes or weight loss because these data were not available. Weight loss and antidiabetic drugs are part of the NAFLD therapy and can eventually lead to a resolution of NASH [40] and are therefore a confounding factor regarding ALT levels. Endpoints for drug approval by EMA and FDA are the histological resolution of NASH and/or improvement in fibrosis as fibrosis is the most predictive factor [41]. Histological data were not available in this cohort; moreover, no effect could be shown on fibrosis using non-invasive tests.

Due to the limitations of the study (retrospective, small number of patients), the results should be applied with caution in clinical practice. However, we believe that in patients with high GGT and NAFLD, sequencing for polymorphism may be performed in individual cases where standard of care does not provide an appropriate result. Prospective data should be collected for further application in clinical practice.

In conclusion, therapy with UDCA in NASH patients may lead to a reduction in GGT and ALT levels when there are polymorphisms in the genes encoding bile acid transporters and regulation. This study did not describe a difference in clinical outcomes such as progression to cirrhosis or death. Especially in patients with lean NAFLD and no response to therapy, a bile acid transporter gene polymorphism could be considered as the underlying pathophysiology.

We believe that our retrospective findings need to be evaluated in prospective studies enrolling a larger number of patients. Furthermore, the underlying mechanism of how UDCA may reduce liver injury in NAFLD patients with derivations in the bile acid transport system needs to be investigated.

Supplementary Materials: The supporting information can be downloaded at: https://www.mdpi.com/article/10.3390/jpm13071180/s1.

Author Contributions: Contributors H.K., T.G., K.V., S.S. and P.M. performed experiments and contributed intellectually to the study; J.B., O.G., J.W., A.K., M.Ö., J.K., W.-K.S., A.C. and P.M. provided samples and contributed to the study design, analysis, and interpretation of data and funding of the studies. H.K., K.V. and P.M. cowrote the manuscript. P.M. is the lead investigator, who designed and supervised the overall project and sub-studies, performed experiments, wrote the manuscript, and is the senior author and guarantor of the manuscript. All authors have read and agreed to the published version of the manuscript.

Funding: H.K. was supported by the Deutsche Forschungsgemeinschaft [KR 5843/1-1].

Institutional Review Board Statement: The study was conducted according to the guidelines of the Declaration of Helsinki (latest revision Fortaleza, Brasil; 2013) and approved by the Ethics Committee of the University of Essen (20-9137-BO, 25 May 2020).

Informed Consent Statement: Written informed consent was obtained from all subjects prospectivelyinvolved in the study. For patients retrospectively involved in the study, written informed consentwas waived by the local ethics committee.

Data Availability Statement: Data will be made available by the corresponding author upon reasonable request.

Acknowledgments: We dedicate this work to Professor Lars Bechmann, who has always dedicated his life to liver research.

Conflicts of Interest: The authors declare no conflict of interest.

References

1. Younossi, Z.M.; Koenig, A.B.; Abdelatif, D.; Fazel, Y.; Henry, L.; Wymer, M. Global epidemiology of nonalcoholic fatty liver disease—Meta-analytic assessment of prevalence, incidence, and outcomes. *Hepatology* **2016**, *64*, 73–84. [CrossRef] [PubMed]
2. Younossi, Z.M.; Golabi, P.; de Avila, L.; Paik, J.M.; Srishord, M.; Fukui, N.; Qiu, Y.; Burns, L.; Afendy, A.; Nader, F. The global epidemiology of NAFLD and NASH in patients with type 2 diabetes: A systematic review and meta-analysis. *J. Hepatol.* **2019**, *71*, 793–801. [CrossRef] [PubMed]
3. Estes, C.; Anstee, Q.M.; Arias-Loste, M.T.; Bantel, H.; Bellentani, S.; Caballeria, J.; Colombo, M.; Craxi, A.; Crespo, J.; Day, C.P.; et al. Modeling NAFLD disease burden in China, France, Germany, Italy, Japan, Spain, United Kingdom, and United States for the period 2016–2030. *J. Hepatol.* **2018**, *69*, 896–904. [CrossRef]
4. Marchesini, G.; Day, C.P.; Dufour, J.F.; Canbay, A.; Nobili, V.; Ratziu, V.; Tilg, H.; Roden, H.; Gastaldelli, A.; Yki-Järvinen, H.; et al. EASL-EASD-EASO Clinical Practice Guidelines for the management of non-alcoholic fatty liver disease. *J. Hepatol.* **2016**, *64*, 1388–1402. [CrossRef]
5. Canbay, A.; Kachru, N.; Haas, J.S.; Sowa, J.P.; Meise, D.; Ozbay, A.B. Patterns and predictors of mortality and disease progression among patients with non-alcoholic fatty liver disease. *Aliment. Pharmacol. Ther.* **2020**, *52*, 1185–1194. [CrossRef] [PubMed]
6. Angulo, P.; Kleiner, D.E.; Dam-Larsen, S.; Adams, L.A.; Björnsson, E.S.; Charatcharoenwitthaya, P.; Mills, P.R.; Keach, J.C.; Lafferty, H.D.; Stahler, A.; et al. Liver fibrosis, but no other histologic features, is associated with long-term outcomes of patients with nonalcoholic fatty liver disease. *Gastroenterology* **2015**, *149*, 389–397.e10. [CrossRef]
7. Ha, Y.; Chon, Y.E.; Kim, M.N.; Lee, J.H.; Hwang, S.G. Gamma-glutamyl transpeptidase dynamics as a biomarker for advanced fibrosis in non-alcoholic fatty liver disease. *J. Gastroenterol. Hepatol.* **2022**, *37*, 1624–1632. [CrossRef]
8. Gertzen, C.G.; Gohlke, H.; Häussinger, D.; Herebian, D.; Keitel, V.; Kubitz, R.; Mayatepek, E.; Schmitt, L. The many facets of bile acids in the physiology and pathophysiology of the human liver. *Biol. Chem.* **2021**, *402*, 1047–1062. [CrossRef]
9. Bechmann, L.P.; Kocabayoglu, P.; Sowa, J.-P.; Sydor, S.; Best, J.; Schlattjan, M.; Beilfuss, A.; Schmitt, J.; Hannivoort, R.A.; Kilicarslan, A.; et al. Free fatty acids repress small heterodimer partner (SHP) activation and adiponectin counteracts bile acid-induced liver injury in superobese patients with nonalcoholic steatohepatitis. *Hepatology* **2013**, *57*, 1394–1406. [CrossRef]
10. Gottlieb, A.; Canbay, A. Why Bile Acids Are So Important in Non-Alcoholic Fatty Liver Disease (NAFLD) Progression. *Cells* **2019**, *8*, 1358. [CrossRef]
11. Meier, Y.; Zodan, T.; Lang, C.; Zimmermann, R.; Kullak-Ublick, G.A.; Meier, P.J.; Stieger, B.; Pauli-Magnus, C. Increased susceptibility for intrahepatic cholestasis of pregnancy and contraceptive-induced cholestasis in carriers of the 1331T>C polymorphism in the bile salt export pump. *World J. Gastroenterol.* **2008**, *14*, 38–45. [CrossRef]
12. Sticova, E.; Jirsa, M. ABCB4 disease: Many faces of one gene deficiency. *Ann. Hepatol.* **2020**, *19*, 126–133. [CrossRef]
13. Gordo-Gilart, R.; Andueza, S.; Hierro, L.; Jara, P.; Alvarez, L. Functional Rescue of Trafficking-Impaired ABCB4 Mutants by Chemical Chaperones. *PLoS ONE* **2016**, *11*, e0150098. [CrossRef]
14. Eppens, E.F.; van Mil, S.W.; de Vree, J.L.; Mok, K.S.; Juijn, J.A.; Elferink, R.P.O.; Berger, R.; Houwen, R.H.; Klomp, L.W. FIC1, the protein affected in two forms of hereditary cholestasis, is localized in the cholangiocyte and the canalicular membrane of the hepatocyte. *J. Hepatol.* **2001**, *35*, 436–443. [CrossRef] [PubMed]

15. Jüngst, C.; Justinger, C.; Fischer, J.; Berg, T.; Lammert, F. Common ABCB4 and ABCB11 Genotypes Are Associated with Idiopathic Chronic Cholestasis in Adults. *Dig. Dis.* **2022**, *40*, 489–496. [CrossRef] [PubMed]
16. Smyk, W.; Weber, S.N.; Hall, R.; Gruenhage, F.; Lammert, F.; Krawczyk, M. Genetic variant c.711a>t in the hepatobiliary phospholipid transporter abcb4 is associated with significant liver fibrosis. *J. Physiol. Pharmacol.* **2020**, *71*, 321–327. [CrossRef]
17. Krawczyk, M.; Rau, M.; Grünhage, F.; Hepatology, J.S. 2017 Undefined. The ABCB4 p. T175A Variant as Potential Modulator of Hepatic Fibrosis in Patients with Chronic Liver Diseases: Looking beyond the Cholestatic Realm. Available online: https://journals.lww.com/hep/Fulltext/2017/08000/The_ABCB4_p_T175A_variant_as_potential_modulator.36.aspx?casa_token=xz8EzoBIEJUAAAAA:T6bIHHiUOmPNfbLBcNJ5MW21dmlBXENNpVXr85N_aUUGvLp4gew07sFM9D_uLGGzQP26UVW51gQpzJVTMTjYG3Mm (accessed on 18 March 2023).
18. Paumgartner, G.; Beuers, U. Mechanisms of action and therapeutic efficacy of ursodeoxycholic acid in cholestatic liver disease. *Clin. Liver Dis.* **2004**, *8*, 67–81. [CrossRef]
19. Elhini, S.H.; Wahsh, E.A.; Elberry, A.A.; El Ameen, N.F.; Saedii, A.A.; Refaie, S.M.; Elsayed, A.A.; Rabea, H.M. The Impact of an SGLT2 Inhibitor versus Ursodeoxycholic Acid on Liver Steatosis in Diabetic Patients. *Pharmaceuticals* **2022**, *15*, 1516. [CrossRef]
20. Lin, X.; Mai, M.; He, T.; Huang, H.; Zhang, P.; Xia, E.; Guo, H. Efficiency of ursodeoxycholic acid for the treatment of nonalcoholic steatohepatitis: A systematic review and meta-analysis. *Expert Rev. Gastroenterol. Hepatol.* **2022**, *16*, 537–545. [CrossRef]
21. Ratziu, V.; de Ledinghen, V.; Oberti, F.; Mathurin, P.; Wartelle-Bladou, C.; Renou, C.; Sogni, P.; Maynard, M.; Larrey, D.; Serfaty, L.; et al. A randomized controlled trial of high-dose ursodesoxycholic acid for nonalcoholic steatohepatitis. *J. Hepatol.* **2011**, *54*, 1011–1019. [CrossRef]
22. Sterling, R.K.; Lissen, E.; Clumeck, N.; Sola, R.; Correa, M.C.; Montaner, J.; Sulkowski, M.S.; Torriani, F.J.; Dieterich, D.T.; Thomas, D.L.; et al. Development of a simple noninvasive index to predict significant fibrosis in patients with HIV/HCV coinfection. *Hepatology* **2006**, *43*, 1317–1325. [CrossRef]
23. Angulo, P.; Hui, J.M.; Marchesini, G.; Bugianesi, E.; George, J.; Farrell, G.C.; Enders, F.; Saksena, S.; Burt, A.D.; Bida, J.P.; et al. The NAFLD fibrosis score: A noninvasive system that identifies liver fibrosis in patients with NAFLD. *Hepatology* **2007**, *45*, 846–854. [CrossRef]
24. Janik, M.K.; Smyk, W.; Kruk, B.; Szczepankiewicz, B.; Górnicka, B.; Lebiedzińska-Arciszewska, M.; Potes, Y.; Simões, I.C.M.; Weber, S.N.; Lammert, F.; et al. MARC1 p.A165T variant is associated with decreased markers of liver injury and enhanced antioxidant capacity in autoimmune hepatitis. *Sci. Rep.* **2021**, *11*, 24407. [CrossRef] [PubMed]
25. Gudbjartsson, D.F.; Helgason, H.; Gudjonsson, S.; Zink, F.; Oddson, A.; Gylfason, A.; Besenbacher, S.; Magnusson, G.; Halldorsson, B.V.; Hjartarson, E.; et al. Large-scale whole-genome sequencing of the Icelandic population. *Nat. Genet.* **2015**, *47*, 435–444. [CrossRef] [PubMed]
26. Iwata, R.; Baur, K.; Stieger, B.; Mertens, J.C.; Daly, A.K.; Frei, P.; Braun, J.; Vergopoulos, A.; Stickel, F.; Sabrane, K.; et al. A common polymorphism in the ABCB11 gene is associated with advanced fibrosis in hepatitis C but not in non-alcoholic fatty liver disease. *Clin. Sci.* **2011**, *120*, 287–296. [CrossRef] [PubMed]
27. Tarao, K.; Fujiyama, S.; Ohkawa, S.; Miyakawa, K.; Tamai, S.; Hirokawa, S.; Masaki, T.; Tanaka, K. Ursodiol use is possibly associated with lower incidence of hepatocellular carcinoma in hepatitis C virus-associated liver cirrhosis. *Cancer Epidemiol. Biomark. Prev.* **2005**, *14*, 164–169. [CrossRef]
28. Müllenbach, R.; Weber, S.N.; Krawczyk, M.; Zimmer, V.; Sarrazin, C.; Lammert, F.; Grünhage, F. A frequent variant in the human bile salt export pump gene ABCB11 is associated with hepatitis C virus infection, but not liver stiffness in a German population. *BMC Gastroenterol.* **2012**, *12*, 63. [CrossRef]
29. Frider, B.; Castillo, A.; Gordo-Gilart, R.; Bruno, A.; Amante, M.; Alvarez, L.; Mathet, V. Reversal of advanced fibrosis after long-term ursodeoxycholic acid therapy in a patient with residual expression of MdR3. *Ann. Hepatol.* **2015**, *14*, 745–751. [CrossRef]
30. Falcão, D.; Pedroto, I.; Moreira, T. The wide phenotypic and genetic spectrum of ABCB4 gene deficiency: A case series. *Dig. Liver Dis.* **2022**, *54*, 221–227. [CrossRef]
31. Romeo, S.; Kozlitina, J.; Xing, C.; Pertsemlidis, A.; Cox, D.; Pennacchio, L.A.; Boerwinkle, E.; Cohen, J.C.; Hobbs, H.H. Genetic variation in PNPLA3 confers susceptibility to nonalcoholic fatty liver disease. *Nat. Genet.* **2008**, *40*, 1461–1465. [CrossRef]
32. Anstee, Q.M.; Day, C.P. The genetics of NAFLD. *Nat. Rev. Gastroenterol. Hepatol.* **2013**, *10*, 645–655. [CrossRef]
33. Liu, Y.-L.; Patman, G.L.; Leathart, J.B.S.; Piguet, A.-C.; Burt, A.D.; Dufour, J.-F.; Day, C.P.; Daly, A.K.; Reeves, H.L.; Anstee, Q.M. Carriage of the PNPLA3 rs738409 C >g polymorphism confers an increased risk of non-alcoholic fatty liver disease associated hepatocellular carcinoma. *J. Hepatol.* **2014**, *61*, 75–81. [CrossRef]
34. Valenti, L.; Al-Serri, A.; Daly, A.K.; Galmozzi, E.; Rametta, R.; Dongiovanni, P.; Nobili, V.; Mozzi, E.; Roviaro, G.; Vanni, E.; et al. Homozygosity for the patatin-like phospholipase-3/adiponutrin I148M polymorphism influences liver fibrosis in patients with nonalcoholic fatty liver disease. *Hepatology* **2010**, *51*, 1209–1217. [CrossRef]
35. Rosso, C.; Caviglia, G.P.; Birolo, G.; Armandi, A.; Pennisi, G.; Pelusi, S.; Younes, R.; Liguori, A.; Perez-Diaz-Del-Campo, N.; Nicolosi, A.; et al. Impact of PNPLA3 rs738409 Polymorphism on the Development of Liver-Related Events in Patients With Nonalcoholic Fatty Liver Disease. *Clin. Gastroenterol. Hepatol.* **2023**, *5*, 100744. [CrossRef] [PubMed]
36. Lindén, D.; Romeo, S. Therapeutic opportunities for the treatment of NASH with genetically validated targets. *J. Hepatol.* 2023. [CrossRef]

37. Simental-Mendía MSánchez-García ASimental-Mendía, L.E. Effect of ursodeoxycholic acid on liver markers: A systematic review and meta-analysis of randomized placebo-controlled clinical trials. *Br. J. Clin. Pharmacol.* **2020**, *86*, 1476–1488. [CrossRef] [PubMed]
38. Kozlitina, J.; Smagris, E.; Stender, S.; Nordestgaard, B.G.; Zhou, H.H.; Tybjærg-Hansen, A.; Vogt, T.F.; Hobbs, H.H.; Cohen, J.C. Exome-wide association study identifies a TM6SF2 variant that confers susceptibility to nonalcoholic fatty liver disease. *Nat. Genet.* **2014**, *46*, 352–356. [CrossRef]
39. Liu, Y.-L.; Reeves, H.L.; Burt, A.D.; Tiniakos, D.; McPherson, S.; Leathart, J.B.S.; Allison, M.E.D.; Alexander, G.J.; Piguet, A.-C.; Anty, R.; et al. TM6SF2 rs58542926 influences hepatic fibrosis progression in patients with non-alcoholic fatty liver disease. *Nat. Commun.* **2014**, *5*, 4309. [CrossRef]
40. Promrat, K.; Kleiner, D.E.; Niemeier, H.M.; Jackvony, E.; Kearns, M.; Wands, J.R.; Fava, J.L.; Wing, R.R. Randomized controlled trial testing the effects of weight loss on nonalcoholic steatohepatitis. *Hepatology* **2010**, *51*, 121–129. [CrossRef] [PubMed]
41. Tacke, F.; Puengel, T.; Loomba, R.; Friedman, S.L. An integrated view of anti-inflammatory and antifibrotic targets for the treatment of NASH. *J. Hepatol.* **2023**, *79*, 552–566. [CrossRef]

Journal of *Personalized Medicine*

Article

Higher pNRF2, SOCS3, IRF3, and RIG1 Tissue Protein Expression in NASH Patients versus NAFL Patients: pNRF2 Expression Is Concomitantly Associated with Elevated Fasting Glucose Levels

Suzan Schwertheim [1,2], Malek Alhardan [1], Paul P. Manka [3], Jan-Peter Sowa [3], Ali Canbay [3], Hartmut H.-J. Schmidt [1], Hideo A. Baba [2] and Julia Kälsch [1,*]

1 Department of Gastroenterology, Hepatology and Transplant Medicine, University Hospital of Essen, University of Duisburg-Essen, 45147 Essen, Germany; suzan.schwertheim@gmail.com (S.S.); malek.alhardan@uk-essen.de (M.A.); hartmut.schmidt@uk-essen.de (H.H.-J.S.)
2 Institute of Pathology, University Hospital of Essen, University of Duisburg-Essen, 45147 Essen, Germany; hideo.baba@uk-essen.de
3 Department of Medicine, Ruhr University Bochum, University Hospital Knappschaftskrankenhaus Bochum, 44892 Bochum, Germany; paul.manka@rub.de (P.P.M.); jan.sowa@rub.de (J.-P.S.); ali.canbay@rub.de (A.C.)
* Correspondence: julia.kaelsch@uk-essen.de

Abstract: Non-alcoholic fatty liver disease (NAFLD) embraces simple steatosis in non-alcoholic fatty liver (NAFL) to advanced non-alcoholic steatohepatitis (NASH) associated with inflammation, fibrosis, and cirrhosis. NAFLD patients often have metabolic syndrome and high risks of cardiovascular and liver-related mortality. Our aim was to clarify which proteins play a role in the progression of NAFL to NASH. The study investigates paraffin-embedded samples of 22 NAFL and 33 NASH patients. To detect potential candidates, samples were analyzed by immunohistochemistry for the proteins involved in innate immune regulation, autophagy, apoptosis, and antioxidant defense: IRF3, RIG-1, SOCS3, pSTAT3, STX17, SGLT2, Ki67, M30, Caspase 3, and pNRF2. The expression of pNRF2 immunopositive nuclei and SOCS3 cytoplasmic staining were higher in NASH than in NAFL ($p = 0.001$); pNRF2 was associated with elevated fasting glucose levels. SOCS3 immunopositivity correlated positively with RIG1 ($r = 0.765$; $p = 0.001$). Further, in NASH bile ducts showed stronger IRF3 immunostaining than in NAFL ($p = 0.002$); immunopositive RIG1 tissue was higher in NASH than in NAFL ($p = 0.01$). Our results indicate that pNRF2, SOCS3, IRF3, and RIG1 are involved in hepatic lipid metabolism. We suggest that they may be suitable for further studies to assess their potential as therapeutics.

Keywords: NASH; pNRF2; SOCS3; immunohistochemistry; liver disease

Citation: Schwertheim, S.; Alhardan, M.; Manka, P.P.; Sowa, J.-P.; Canbay, A.; Schmidt, H.H.-J.; Baba, H.A.; Kälsch, J. Higher pNRF2, SOCS3, IRF3, and RIG1 Tissue Protein Expression in NASH Patients versus NAFL Patients: pNRF2 Expression Is Concomitantly Associated with Elevated Fasting Glucose Levels. *J. Pers. Med.* **2023**, *13*, 1152. https://doi.org/10.3390/jpm13071152

Academic Editor: Kenneth P.H. Pritzker

Received: 17 April 2023
Revised: 5 July 2023
Accepted: 14 July 2023
Published: 18 July 2023

1. Introduction

NAFLD is a worldwide increasing problem, and patients often have metabolic syndrome and high risks of cardiovascular and liver-related mortality. Recently, it was documented that constitutive active innate immune signaling can lead to excessive inflammatory cytokine release and consequently can promote the development of hepatic steatosis and fibrosis [1]. Nevertheless, the factors leading to the progression of NAFL to NASH are still unknown. Much of the liver research relates to outcomes in mouse models, but data on protein expression levels specifically in NAFLD patients on this topic are rare. Therefore, we asked if proteins of innate immunity are deregulated in our NAFLD group and if there is an association with the clinical data of our patients. Among others, pattern recognition receptors (PRRs) participate in the regulation of lipid metabolism. Extracellular pathogens or endogenous injury signals are initially detected by PRRs on cell membranes or endosomal membranes [1]. Retinoic acid-inducible gene 1 (RIG1/DDX58) belongs to a family of cytosolic pattern recognition receptors (PRR) and triggers an innate immune response. Briefly, the pathogenic association molecule pattern (PAMP) is recognized by RIG1 and activates mitochondrial antiviral signaling protein (MAVS)-dependent signals which lead to activation of IRF3 and NF-κB and subsequent production of type I/II IFN and inflammatory cytokines [1]. Among others, the proinflammatory cytokine IL-6 is activated by RIG1 and IL-6 activates the JAK-STAT signaling pathway. Based on our results, we assume signaling cascades activated by PRRs, FFA accumulation, and ROS formation (for more details see Figure 1). The signal transducer and activator of transcription 3 (STAT3) belong to the STAT family of cytoplasmic transcription factors [2]. STAT3 is to a large extent known as an oncogenic factor in various human cancers [2]. The suppressor of cytokine signaling 3 (SOCS3) proteins is also known as STAT-induced STAT inhibitor (SSI3) [3]. It has been documented that in obesity SOCS3 is upregulated in concert with increases in inflammation in the hypothalamus, adipose tissue, and liver [4]. The transcription factor nuclear factor-erythroid 2 related factor 2 (NRF2) is also known to participate in hepatic fatty acid metabolism [5] and to regulate the innate immune response [6]. NASH has been shown to elicit lipid peroxidation, accumulation of reactive oxygen species (ROS), and proinflammatory cytokines in the liver, which leads to liver injury and inflammation [7]. Exposure to oxidative stress and inflammatory conditions activates NRF2 inducing the expression of cytoprotective genes [8]. Furthermore, the involvement of autophagy in hepatocyte lipid metabolism has recently been demonstrated [9,10]. Impairment of autophagic flux is closely associated with NAFLD [9,11,12]. Therefore, we analyzed our patient group on the implication of autophagy by using Syntaxin (STX17), a SNARE protein, formerly successfully used in liver tissue for autophagy detection [13]. STX17 translocates to autophagosomes and mediates the fusion of autophagosomes with lysosomes which enables the degradation of autophagosome contents [14]. Briefly, we selected these proteins for our current studies based on previous study interests and staining availability. Further, a major risk factor for non-alcoholic steatohepatitis (NASH) is insulin resistance with elevated blood glucose [15]. Sodium–glucose cotransporter 2 (SGLT2) is the major cotransporter known to participate in glucose reabsorption in the kidney. In a mouse model with diabetes, NASH/cirrhosis/HCC SGLT2 expression was detected in liver tumors [16]. Recently, the SGLT2 inhibitor NGI001 inhibited diet-induced metabolic dysfunction and non-alcoholic fatty liver disease in mice [17]. As SGLT2 inhibitors dapagliflozin and empagliflozin improved liver enzymes and decreased liver fat [18] we analyzed also SGLT2 protein expression levels in our NAFLD group to elucidate the factors in the development of NASH.

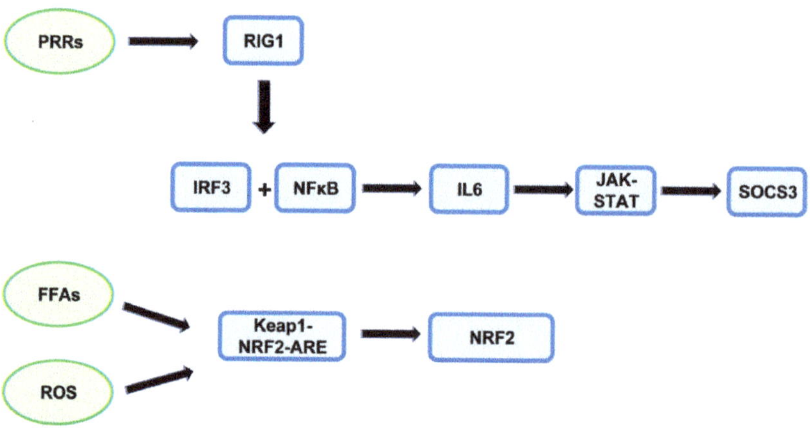

Figure 1. Summary figure representing the signaling cascades. We suppose that activation by PRRs triggers RIG1 expression. In addition, FFA-accumulation, and ROS formation leads to Keap1-NRF2-

ARE activation, whereby it is likely that many biological processes take place in the organism in a parallel manner (for example NFκB activation is known to induce ROS formation). Abbreviations: PRRs: pattern recognition receptors; FFAs: free fatty acids; ROS: reactive oxygen species; Keap1: Kelch-like ECH-associated protein 1; ARE: antioxidant response elements.

2. Materials and Methods

2.1. Patients

Our study was conducted with 55 morbidly obese patients (37 females, 18 males) who have undergone bariatric surgery at a bariatric surgery center. The study group was composed of 22 patients with steatosis (NAFL) and 33 with nonalcoholic steatohepatitis (NASH). Indication for bariatric surgery was based on National Institutes of Health (NIH) guidelines (BMI $\geq$ 40 kg/m^2 or $\geq$35 kg/m^2, plus co-morbidities) as described before [19]. In addition, the patient selection criteria were the same as described in our previous study [19]. Patients reporting excessive alcohol consumption (>20 g/day in males or >10 g/day in females) indicating alcoholic liver disease were excluded. The surgeon's choice—i.e., adjustable gastric band, Roux-Y, or gastric bypass surgery—was based on the current guidelines as adapted to the patient's clinical conditions and comorbidities as well as on clinical experience. Wedge liver biopsies were taken during the procedure.

There has been conflicting debate about the diagnostic criteria for NASH [20]. The degree of NAFLD can be quantified using the NAFLD activity score (NAS) according to Kleiner et al., where a NAS score $\geq$5 is defined as NASH [21], or according to the fatty liver inhibition and progression (FLIP) algorithm described by Bedossa et al. [22]. The FLIP algorithm was used by us to classify liver damage in morbid obesity because it allows a more accurate distinction between NAFL and NASH. By using the histologic features of "steatosis", "ballooning of hepatocytes", and "inflammation", the slides were classified as "NAFL" or "NASH". The HE-stained slides were assessed by two observers (HAB and JK) and the degree of NAFLD was quantified according to the FLIP algorithm of Bedossa et al. [22]. More detailed information on the characteristics of the patients is given in Table 1.

Table 1. Clinical and laboratory data of the study groups.

Characteristics	NAFL (*n* = 22)	NASH (*n* = 33)	*n* Valid Cases	*p* Value *
Age (years)	38 (24–67)	45 (20–67)	22/33	0.110
Gender (male/female)	4/18	14/19	22/33	0.082
BMI (kg/m^2)	49.9 (29.4–66.9)	53 (27.4–78.19)	21/31	0.208
Adiponectin (µg/mL)	3.35 (1.3–8.28)	2.87 (0.83–11.9)	20/24	0.759
CK18 M30 (IU/L)	174.7 (61.8–807.7)	366.6 (80.1–1573.9)	20/24	0.002
CK18 M65 (IU/L)	331.9 (87.9–960.1)	628.6 (255.8–5273.1)	19/24	<0.001
Fasting Glucose (mg/dL)	95.50 (73–150)	120 (72–385)	22/29	0.001
Total Cholesterol (mg/dL)	198 (120–261)	177.5 (116–247)	15/18	0.320
Triglyceride (mg/dL)	149 (34–218)	207 (55–421)	13/12	0.041
ALT (U/L)	20 (13–65)	39 (14–120)	22/31	0.001
AST (U/L)	23 (16–49)	32.5 (23–90)	14/20	<0.001
GGT (U/L)	20 (2–93)	43 (15–1213)	22/31	<0.001
Fibrosis Grade (0 + 1) (2 + 3)	 4 + 7 10 + 1	 6 + 7 18 + 2	22/33	0.58
Steatosis Grade (0 + 1) (2 + 3)	 0 + 19 3 + 0	 0 + 7 14 + 12	22/33	<0.001

Table 1. *Cont.*

Characteristics	NAFL (n = 22)	NASH (n = 33)	n Valid Cases	p Value *
Ballooning Grade				
(0 + 1)	17 + 3	0 + 16		
2	2	17	22/33	0.001
Lob. Inflam. Grade				
(0 + 1)	16 + 4	0 + 11		
(2 + 3)	2 + 0	20 + 2	22/33	<0.001

The presented values are medians; ranges are enclosed in parentheses. * p values correspond to the comparison of NAFL/NASH and n valid cases reports the number of valid NAFL/NASH cases used for the statistical analysis. Regarding fibrosis grade, steatosis grade, ballooning grade, and lobular inflammation grade, we placed the two grades 0 and 1 in a common group. For example, statistical evaluation was performed in the analysis of steatosis grade by comparing the number of NAFL and NASH patients in group 1 (grade 0 + 1) with the number of NAFL and NASH patients in group 2 (grade 2 + 3). We used the statistic test Mann–Whitney U test for continuous factors and two-sided Fisher's exact test for categorical parameters. $p \leq 0.05$ was defined as statistically significant. Abbreviations: n = number; NAFL = non-alcoholic fatty liver; NASH = non-alcoholic steatohepatitis; ALT = alanine aminotransferase; BMI = body mass index; AST = aspartate aminotransferase; GGT = gamma-glutamyl transferase; CK18 = Cytokeratin18; Lob. Inflam. Grade = lobular inflammation grade.

Individual patients' liver samples were obtained from the files of the Institute of Pathology of the University Hospital of Essen. For all cases, standardized prepared formalin-fixed and paraffin-embedded (FFPE) material was stained with HE, and immunohistochemical staining was performed according to institutional standards. Paraffin-embedded tissue was available in all cases and we reviewed all of them. Informed consent was obtained from every patient. The study was in accordance with the Helsinki Declaration of 1975 and approved by the Ethics Committee of the University Hospital Essen (reference number: 09-4252).

2.2. Histology and Immunohistochemistry

Tissue sections (1 to 2 µm thick) from formalin-fixed and paraffin-embedded (FFPE) tissue blocks were cut, dewaxed, and pretreated. The expression of selected candidate proteins was analyzed by immunohistochemistry, as described previously [23], with an automated staining device (Dako Autostainer, Dako, Glostrup, Denmark). The antibodies used were: anti-Vimentin (#M0725, Dako, Glostrup, Denmark; diluted 1:500 for 60 min at RT); anti-active Caspase 3 (#9661, Cell Signaling, Danvers, MA, USA; diluted 1:50 for 60 min at RT); ki67 (#5278384001, Roche Ventana, Tucson, AZ, USA; undiluted for 60 min at RT); anti-IRF3 (#712217, Invitrogen, San Diego, CA, USA; diluted 1:50 for 30 min at RT); anti-RIG1 (#PA5-110297, Invitrogen; diluted 1:200 for 30 min at RT); anti-M30 (#10700, TecoMedical, Sissach, Switzerland; diluted 1:4500 for 30 min at RT); anti-pSTAT3 (#9145, Cell Signaling; diluted 1:50 for 60 min at RT); anti-pNRF2 (#NBP2-67465, Novus, Centennial, CO, USA; diluted 1:25 for 60 min at RT); anti-syntaxin (#HPA001204, Sigma-Aldrich, Steinheim, Germany; diluted 1:200 for 60 min at RT); anti-SGLT2 (#NBP1-92384, Novus; diluted 1:50 for 30 min at RT); and anti-SOCS3 (#ab280884, Abcam, Cambridge, UK; diluted 1:100 for 30 min at RT). Detection of antigen–antibody binding was performed for vimentin, active Caspase 3, ki67, M30, pSTAT3, syntaxin, SGLT2 and SOCS3 with the ZytoChem Plus AP Polymer Kit (#POLAP-100, Zytomed, Berlin, Germany), and for IRF3, RIG1, and pNRF2 using POLYVIEW® PLUS AP anti-rabbit reagent Rb (#ENZ-ACC110-0150, Enzo Life Sciences, Lörrach, Germany) according to the manufacturer's protocols. Detailed information on staining protocols is given in Supplementary Material Table S1. Negative controls were included in every run and incubated with non-immune immunoglobulin in the same concentrations but instead of the primary antibody.

2.3. Sample and Immunohistochemistry Evaluation

Immunohistochemical staining was examined with manual IHC scoring and computer-assisted quantification with Aperio ImageScope depending on the protein analyzed. Vimentin, IRF3, M30, and Caspase 3 stainings were studied by visual IHC scoring. Briefly, two independent observers (M.A. and S.S.), blinded to the subgroups of the study, assessed the IHC stains using a semi-quantitative scoring system, analogous to the immunoreactivity scores (IRS) established by Remmele and Stegner [24], as described in our previous studies [25,26]. Vimentin IHC staining was carried out to assess the quality of the FFPE material. The use of vimentin is suitable to monitor the quality of antigen preservation and the uniformity of tissue fixation in FFPE tissues as the epitope of vimentin is partially susceptible to formaldehyde fixation [27].

In assessing IRF3 staining, we counted the number of bile ducts that showed IRF3 immunopositivity. The intensity of IRF3 staining was classified (weak: 1 point; moderate: 2 points; marked: 3 points) and in case of disagreement between the scores, a third observer was consulted. A study of apoptosis was performed by immunohistochemical staining analyses of the antibodies active Caspase 3 and M30. Active Caspase 3 detects endogenous levels of the large fragment of activated Caspase 3 deriving from cleavage adjacent to Asp175. We counted the number of hepatocytes showing positive cytoplasmic staining for cleaved Caspase 3 in each case in the whole slide. Additionally, we quantified apoptosis with the M30 antibody; M30 detects a neo-epitope on caspase-cleaved cytokeratin 18 (CK18) in apoptotic cells; uncleaved CK18 is not detected [28]. Analysis of M30 immunopositivity was performed by scoring the whole slide of each case on apoptotic cells showing intense red cytoplasmatic staining for M30 and counting them.

We examined the remaining immunohistochemical stains by computer-based automated quantitative IHC scoring. First, stained slides were digitized at 20× resolution using the Aperio AT2 all-slide scanner (Leica, Wetzlar, Germany). The quantification of nuclear stainings (Ki67, pSTAT3, and pNRF2) and cytoplasmic stainings (SOCS3, RIG1, syntaxin, and SGLT2) were performed by Leica image analysis software (Aperio ImageScope). Expression of Ki67, pSTAT3, and pNRF2 was quantified with Aperio nuclear algorithm and expression of SOCS3, RIG1, SGLT2, and syntaxin with Aperio's positive pixel count algorithm. The percentage of positive hepatocyte nuclei in relation to all hepatocyte nuclei and the percentage of positively stained cytoplasm in relation to the whole area was calculated.

2.4. Statistics

Analyses were carried out with the Statistical Package for Social Sciences (SPSS 28.0, Chicago, IL, USA). We used Mann–Whitney U test for continuous factors and two-sided Fisher's exact test categorical parameters to analyze associations between clinical/laboratory data and the study groups (NASH/NAFL). Using Mann–Whitney U test and Kruskal–Wallis tests we assessed the association between clinical/laboratory data and the immunopositivity of the proteins. We performed multivariable binary logistic regression analysis with NAFL/NASH as a dependent category and pNRF2, RIG1, and SOCS3 protein expression levels along with clinical and laboratory markers as independent covariates. This aimed to better assess the extent to which the presence of certain clinical and laboratory markers may influence the occurrence of certain protein expressions in NASH/NAFL. Additionally, Pearson's correlation coefficient was used to evaluate correlations between the protein expressions. All data are shown as medians with ranges presented in parentheses, if not stated otherwise; $p \leq 0.05$ was defined as statistically significant.

3. Results

3.1. Higher Expression of RIG1, pNRF2, and SOCS3 in NASH vs. NAFL

Our IHC studies demonstrated that several proteins involved in the innate immune system were differently expressed in NASH and NAFL, which suggests that they play a role in the progression of NAFL to NASH. We examined the percentage of RIG1 immunostained tissue compared to total tissue in all samples and detected cytoplasmic RIG1 expression (Figure 2A–C), whereby the percentage of RIG1 immunopositivity was in NASH (Figure 2C) significantly higher than in NAFL (Figure 2B), shown in boxplots ($p = 0.01$; Figure 2A). Further, to investigate NRF2 activity, we used the NRF2 [p Serine 40] antibody as phosphorylation at the serine 40 residue is required for the dissociation of NRF2 from KEAP1 and the transcriptional activation activity of NRF2 [29,30]. We found that the percentage of pNRF2 immunostained nuclei was in NASH (Figure 2F) significantly higher than in NAFL (Figure 2E), shown in boxplots ($p = 0.001$; Figure 2D). Additionally, the JAK-STAT pathway was activated as cytoplasmic SOCS3 immunostaining was significantly stronger in NASH than in NAFL ($p < 0.001$; Figure 2G–I). However, the comparison between the percentage of pSTAT3-immunostained nuclei in NASH and NAFL showed only a trend ($p = 0.059$; Supplementary Material Figure S1).

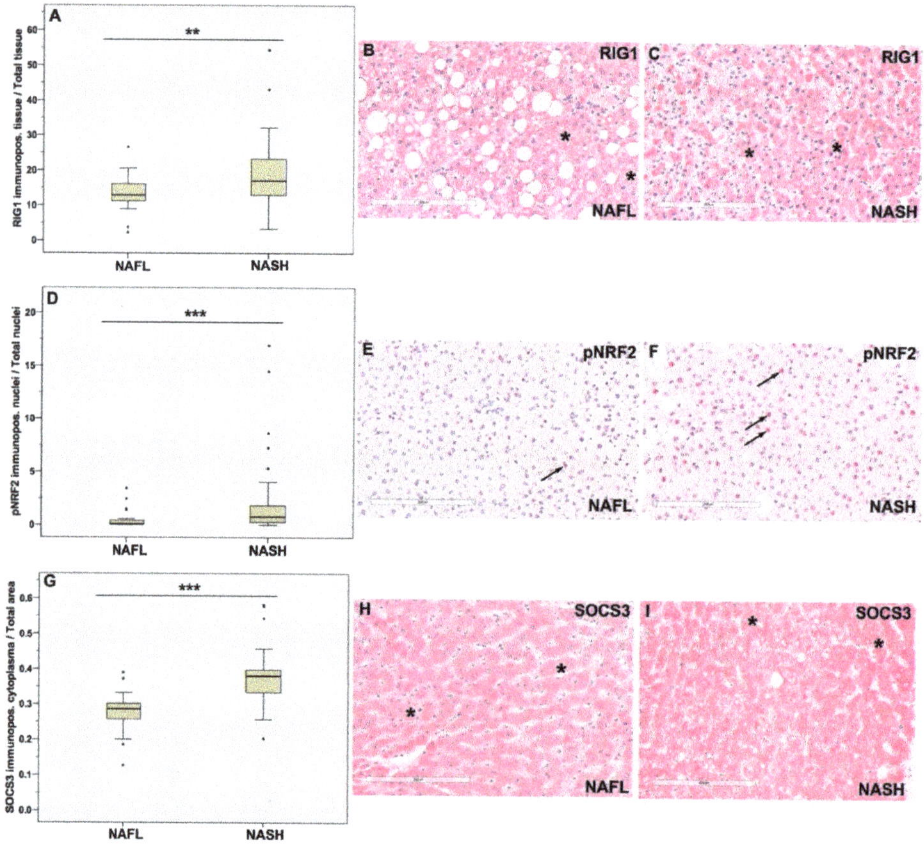

Figure 2. Semi-quantitative immunohistochemical analysis of RIG-1, pNRF2, and SOCS3 and protein expressions. Images show representative cases of our NAFLD cohort. (**A–C**) We detected stronger cytoplasmic RIG1 expression in NASH than in NAFL. (**A–C**) Boxplots depict that the percentage of RIG1 stained tissue to total tissue is higher in NASH than in NAFL; ** $p = 0.01$. (**A**) In the IHC images, representative cytoplasmic RIG1 staining in NAFL (**B**) and NASH (**C**) is shown (asterisk). Further, the ratio of pNRF2 stained nuclei to total nuclei was higher in NASH than in NAFL; *** $p = 0.001$ (**D**). Representative pNRF2 IHC images of NAFL (**E**) and NASH (**F**) demonstrate stronger nuclear staining (arrows) in NASH than in NAFL. Also, cytoplasmic SOCS3 expression was stronger in NASH than in NAFL (**G–I**). Boxplots depict the ratio of the area with cytoplasmic SOCS3 immunopositivity to the total area with increased SOCS3 expression in NASH compared with NAFL; *** $p < 0.001$ (**G**). Representative SOCS3 IHC images of NAFL (**H**) and NASH (**I**) show representative areas of cytoplasmic SOCS3 immunopositivity (asterisk). Differences between groups were analyzed using Mann–Whitney U tests; bold lines inside the box plot represent median levels. Results are significant at * $p \leq 0.05$, ** $p \leq 0.01$ and *** $p \leq 0.001$; bars = 200 μm.

3.2. Stronger IRF3 Immunostaining of Bile Ducts in NASH Than in NAFL

We investigated the bile ducts regarding their immunostainings for IRF3 and found out that the number of bile ducts with strong immunopositivity for IRF3, namely staining intensity of +3, was in the NASH patients significantly higher than in the NAFL group ($p = 0.027$; Figure 3A,B,D). Further, the average IRF3 staining intensity of the bile ducts was in NASH significantly higher than in NAFL ($p = 0.002$; Figure 3B–D), suggesting that the bile ducts play a role in the innate immune response.

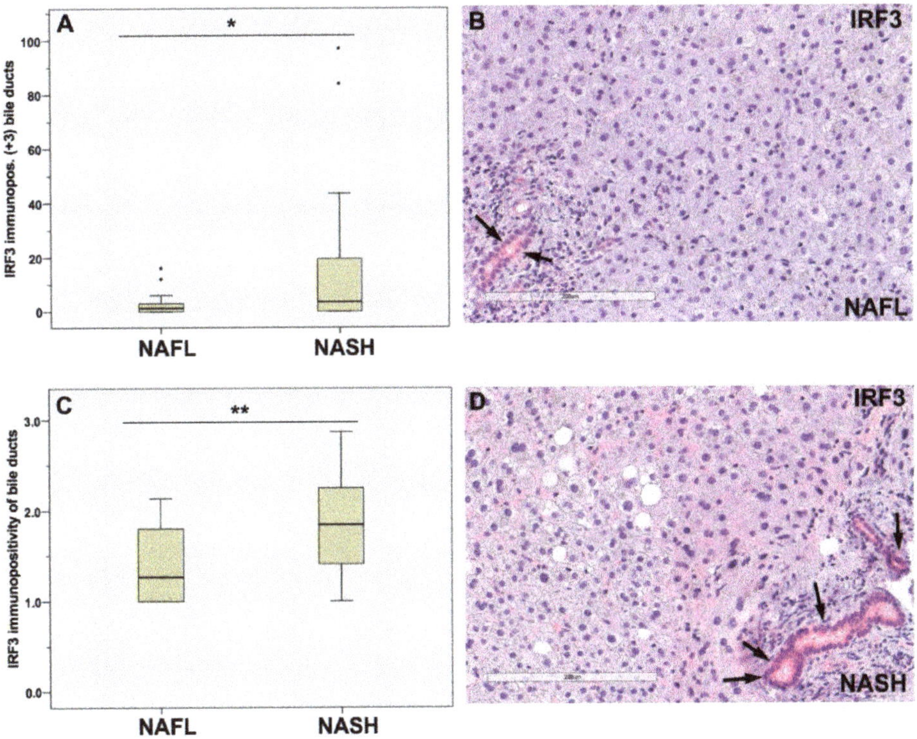

Figure 3. Associations between IRF3 immunostainings and disease progression. Study of the bile ducts in the cohort depicted that IRF3 staining intensity was in NASH higher than in NAFL (**A–D**). In NASH, there are more bile ducts having a strong IRF3 staining intensity of +3 than in NAFL; * $p = 0.027$ (**A**). The average IRF3 staining intensity of the bile ducts per case was in NASH higher than in NAFL; $p = 0.002$ (**C**). Representative IHC images of NAFL (**B**) and NASH (**D**) show representative areas of IRF3 immunopositive stained bile ducts (arrows) in NAFL (**B**) and NASH (**D**) with stronger IRF3 staining in NASH. Differences between groups were analyzed using Mann–Whitney U tests; bold lines inside the box plot represent median levels. Results are significant at * $p \leq 0.05$ and ** $p \leq 0.01$; bars = 200 μm.

3.3. Association between RIG1, pNRF2, SOCS3, IRF3 Immunopositivity and Histological/Laboratory Parameters

Fasting blood glucose values of patients were divided into three levels: normal range, 70–99 mg/dL; pre-diabetes, 100–125 mg/dL; and diabetes, ≥126 mg/dL according to the criteria of the American Diabetes Association (ADA) [31]. Then, the patients were classified into two groups: we combined the patients with pre-diabetes and diabetes into one group, and the second group aggregated all the patients with normal glucose levels. We detected a significantly higher NRF2 activity in the pre-diabetes and diabetes group than in the normal group ($p = 0.029$; Table 2). Further, we found that patients with steatosis grades 2 and 3 showed significantly higher pNRF2, SOCS3, and RIG1 expression levels than patients with steatosis grades 0 and 1 ($p = 0.011$; $p = 0.008$; $p = 0.032$; Table 2). Regarding ballooning grade, we observed that patients with grade 1 and grade 2 had significantly higher pNRF2 ($p < 0.001$) and SOCS3 ($p = 0.002$; $p < 0.001$; Table 2) protein expression levels than patients with ballooning grade 0. For RIG1 immunopositivity, only patients with ballooning grade 2 had significantly higher RIG1 expression levels than patients with ballooning grade 0 ($p = 0.010$; Table 2). We detected in patients with lobular inflammation 1 higher NRF2 activity than in patients with lobular inflammation 0 ($p = 0.022$; Table 2). Regarding SOCS3, patients with lobular inflammation 1, 2, and 3 had higher SOCS3 protein expressions than patients with lobular inflammation grade 0 ($p = 0.007$; $p = 0.007$; $p = 0.035$; Table 2). Association studies between IRF3 immunopositivity of the bile ducts and clinical/laboratory parameters showed as the only significant result that patients with lobular inflammation grade 2 had significantly stronger IRF3 immunopositivity of the bile ducts than patients with grade 0 ($p = 0.017$).

Table 2. Association of clinical/laboratory parameters with IHC studies.

		pNRF2		SOCS3		RIG1	
Parameters	***n* Valid Cases**	**Median Value [b]**	***p* Value ***	**MedianValue [b]**	***p* Value ***	**MedianValue [b]**	***p* Value ***
Fasting Glucose (mg/dL)							
Normal (70–99)	19	0.12 (0–7.25)		0.29 (0.12–0.45)		14.31 (2.33–25.25)	
Pre-diabetes and Diabetes [a]	32	0.69 (0.01–16.93)	0.029	0.33 (0.18–0.45)	0.090	15.98 (3.38–26.62)	0.205
Total Cholesterol (mg/dL)							
Normal (<200)	20	0.27 (0.01–4.06)		0.3 (0.18–0.45)		14.45 (3.38–26.28)	
Elevated (>200)	13	0.15 (0–7.25)	0.957	0.31 (0.2–0.45)	0.813	12.14 (3.84–26.62)	0.548
Triglyceride (mg/dL)							
Normal (<200)	18	0.06 (0–1.8)		0.3 (0.18–0.39)		12.82 (3.84–23.33)	
Elevated (>200)	7	0.2 (0.01–1.73)	0.297	0.27 (0.2–0.39)	0.701	14.6 (3.38–25.56)	0.495
ALT (U/L)							
Normal (<35/<50)	35	0.29 (0–7.25)		0.31 (0.12–0.45)		14.21 (2.33–26.62)	
Elevated (>35/>50)	18	0.67 (0.02–16.93)	0.195	0.34 (0.2–0.57)	0.276	16.29 (5.02–43.11)	0.244
AST (U/L)							
Normal (<35/<50)	24	0.1 (0–3.45)		0.3 (0.18–0.4)		14.16 (3.38–25.03)	
Elevated (>35/>50)	10	0.57 (0.1–16.93)	0.061	0.34(0.2–0.57)	0.223	14.53 (5.02–43.11)	0.539
GGT (U/L)							
Normal (<35/<55)	36	0.26 (0–16.93)		0.32 (0.12–0.57)		14.26 (2.33–43.11)	
Elevated (>35/>55)	17	0.93 (0.03–5.33)	0.054	0.33 (0.2–0.42)	0.607	16.02 (5.02–26.62)	0.261
Age Grade							
Young (20–40)	28	0.27 (0–14.48)		0.31 (0.2–0.57)		14.32 (3.38–43.11)	
Old (41–67)	27	0.74 (0–16.93)	0.055	0.33 (0.12–0.58)	0.245	16.21 (2.33- 54.32)	0.225
Fibrosis Grade							
0	10	1.4 (0.18–3.45)		0.34 (0.28–0.39)		16.69 (9.59–26.62)	
1	14	0.26 (0.04–7.25)	0.089	0.3 (0.12–0.45)	0.320	15.21 (2.33–25.25)	0.380
2	28	0.34 (0–16.93)	0.091	0.32 (0.2–0.58)	0.619	14.94 (3.38–54.32)	0.691
3	3	0.68 (0.21–2.45)	0.735	0.33 (0.33–0.37)	1.000	12.99 (12.3–17.32)	0.398
Steatosis Grade							
0 + 1	26	0.11 (0–7.25)		0.29 (0.12–0.45)		14.28 (2.33–25.25)	
2 + 3	29	0.77 (0–16.93)	0.011	0.35 (0.2–0.58)	0.008	17.03 (3.38–54.32)	0.032
Ballooning Grade							
0	17	0.05 (0–0.56)		0.28 (0.18–0.37)		12.14 (3.84–20.57)	
1	19	0.93 (0.1–16.93)	<0.001	0.34 (0.12–0.45)	0.002	15.93 (2.33–25.56)	0.068
2	19	1.42 (0–14.48)	<0.001	0.38 (0.2–0.58)	<0.001	17.03 (3.38–54.32)	0.010
Lobular Inflammation Grade							
0	16	0.09 (0–3.45)		0.28 (0.12–0.39)		12.91 (2.33–26.62)	
1	15	0.68 (0.05–7.25)	0.022	0.35 (0.18–0.45)	0.007	16.12 (5.02–26.28)	0.097
2	22	0.66 (0–16.93)	0.051	0.36 (0.2–0.58)	0.007	16.42 (3.38–54.32)	0.174
3	2	1.67 (0.1–3.26)	0.325	0.36 (0.33–0.39)	0.035	15.45 (15.1–15.82)	0.399

[a] We defined fasting blood glucose level of 100–125 mg/dL as pre-diabetes and glucose levels of ≥126 mg/dL as diabetes. [b] Values are the median of pNRF2, SOCS3, and RIG1 immunopositivity with ranges presented in parentheses. ALT and AST threshold for normal values were <35 U/L for females and <50 U/L for males; GGT threshold for normal values were <35 U/L for females and <55 U/L for males. * p values correspond to the analysis of the association between clinical/laboratory parameters with pNRF2, SOCS3, and RIG1 immunopositivity in the NAFLD cohort. Mann–Whitney U test and Kruskal–Wallis test were used for statistical analysis of the difference between two or more groups. Results are significant at $p \leq 0.05$.

3.4. Multivariable Binary Logistic Regression Analysis

We performed multivariable binary logistic regression analysis for the proteins pNRF2, SOCS3, and RIG1 (Table 3). We found that pNRF2 associated with NASH ($p = 0.046$) was independent of the presence of steatosis (Table 3(A); $p = 0.01$) and lobular inflammation ($p = 0.016$). Further, the multivariable binary logistic regression study showed that RIG1 associated with NASH (Table 3(B); $p = 0.045$) was independent of the presence of steatosis (Table 3(B); $p = 0.017$) and lobular inflammation (Table 3; $p = 0.009$). While RIG1 is significantly associated with NASH shown above by Mann–Whitney U test (Figure 2A; $p = 0.01$), the multivariable binary logistic regression analysis showed that this effect was not seen after adjusting for the variables fasting glucose and total cholesterol (Table 3(C); $p = 0.285$). We detected no associations for SOCS3 protein expression levels along with clinical and laboratory markers as independent covariates by logistic regression studies. Thus, the higher SOCS3 expression levels in NASH are not affected by other clinical and laboratory markers.

Table 3. Multivariable binary logistic regression studies.

	Exp (B)	95% CI		p Value
		Lower	Upper	
A				
pNRF2	2.456	1.017	5.93	0.046
Age (years)	1.006	0.926	1.092	0.893
BMI (kg/m^2)	1.102	0.947	1.283	0.208
Fibrosis Grade				
1 vs. 0	1.747	0.072	42.549	0.732
2 vs. 0	1.396	0.112	17.384	0.795
3 vs. 0	0.989	0.026	38.383	0.995
Steatosis Grade				
(2 + 3) vs. (0 + 1)	29.267	2.237	382.895	0.01
Ballooning Grade				
2 vs. (0 + 1)	0.628	0.041	9.674	0.739
Lob. Inflam. Grade				
(2 + 3) vs. (0 + 1)	19.197	1.738	211.979	0.016
B				
RIG1	1.21	1.005	1.457	0.045
Age (years)	1.085	0.997	1.18	0.058
BMI (kg/m^2)	1.018	0.898	1.153	0.782
Fibrosis Grade				
1 vs. 0	2.994	0.148	60.656	0.475
2 vs. 0	1.885	0.131	27.206	0.642
3 vs. 0	0.843	0.022	32.364	0.927
Steatosis Grade				
(2 + 3) vs. (0 + 1)	24.689	1.788	340.962	0.017
Ballooning Grade				
2 vs. (0 + 1)	0.384	0.021	7.164	0.521
Lob. Inflam. Grade				
(2 + 3) vs. (0 + 1)	35.992	2.454	527.784	0.009
C				
RIG1	1.080	0.938	1.243	0.285
Fasting glucose (mg/dL)	1.052	1.008	1.097	0.020
Total cholesterol (mg/dL)	0.995	0.972	1.018	0.682

Multivariable binary logistic regression analysis with NAFL/NASH as a dependent category and pNRF2 and RIG1 protein expression levels along with clinical and laboratory markers as independent covariates for (A) pNRF2 and (B) RIG1 together with age, BMI, fibrosis, steatosis, ballooning, and lobular inflammation grades as covariates. (C) RIG1 together with fasting glucose and total cholesterol levels as covariates. As the reference for the categorical covariate fibrosis "Grade 0" is chosen, the reference for the categorical covariates steatosis, ballooning, and lobular inflammation is "Grade (0 + 1)".

3.5. Analysis of Correlations between NRF2 Activation and Syntaxin, Ki67, M30 and SOCS3 Protein Levels

We detected a significant positive correlation of pNRF2 with syntaxin protein expression levels ($r = 0.604$; $p < 0.001$; Figure 4). Additionally, we found positive associations between pNRF2 and Ki67 ($r = 0.645$; $p < 0.001$; Figure 4) and M30 ($r = 0.431$; $p = 0.001$; Figure 4) and SOCS3 ($r = 0.560$; $p < 0.001$; Figure 4) protein expression levels.

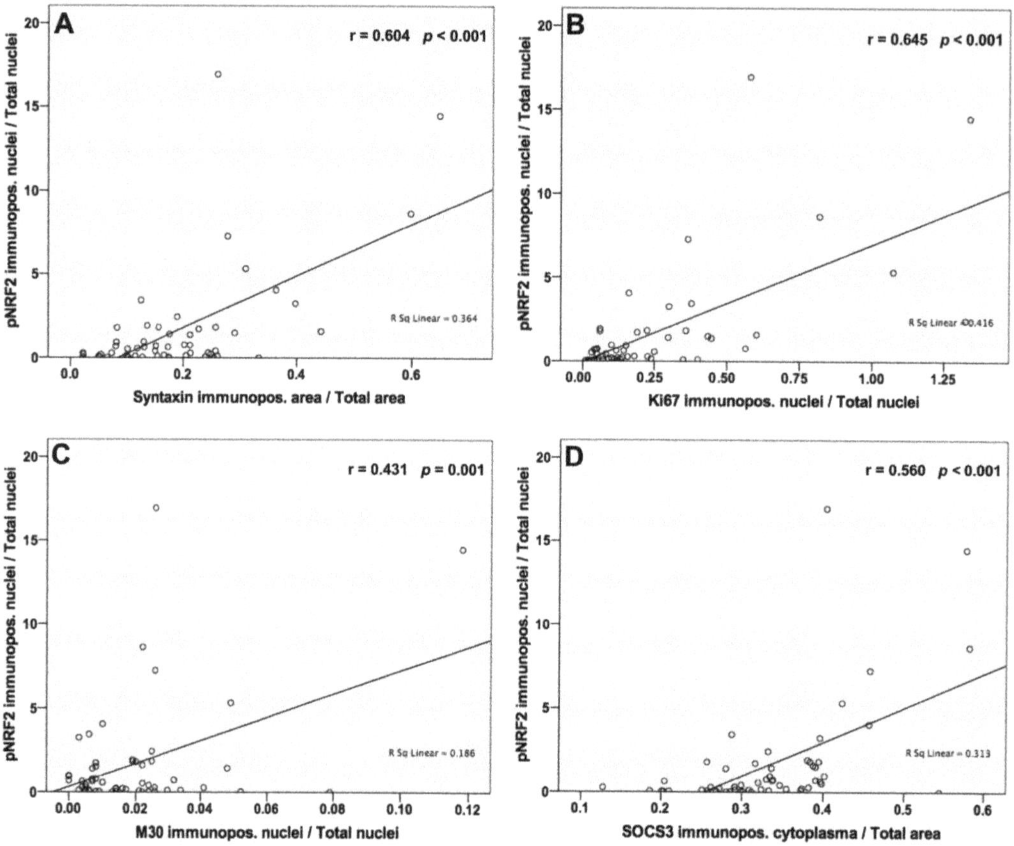

Figure 4. pNRF2 protein expressions are positively associated with RIG1, syntaxin, Ki67, M30, and SOCS3 expressions. (**A**,**B**) We observed a high degree of correlations between pNRF2 and syntaxin (r = 0.604; $p < 0.001$) and Ki67 (r = 0.645; $p < 0.001$). (**C**,**D**) The degree of correlation between pNRF2 and M30 was moderate (r = 0.431; $p = 0.001$) and between pNRF2 and SOCS3 was high (r = 0.560; $p < 0.001$). Pearson's correlation coefficient was used to evaluate protein expression results with significance set at $p \leq 0.05$. High degree of correlation: Pearson's correlation coefficient, also known as Pearson's r = 0.5–1.0; moderate degree of correlation: Pearson's r = 0.3–0.49; low degree of correlation: Pearson's r < 0.3.

3.6. Analysis of Correlations between SOCS3 and RIG1, Ki67, Syntaxin and SGLT2 Protein Levels

Pearson correlations showed significant positive associations between the protein levels of SOCS3 and RIG1 (r = 0.765; $p < 0.001$; Figure 5), Ki67 (r = 0.463; $p < 0.001$; Figure 5), syntaxin (r = 0.717; $p < 0.001$; Figure 5), and SGLT2 (r = 0.333; $p = 0.013$; Figure 5).

3.7. Analysis of Correlations between IRF3 and pNRF2 and SOCS3 Protein Levels

We found significant Pearson correlations between the IRF3 protein expression in the bile ducts of our patients and the pNRF2 immunopositivity in the nuclei (r = 0.364; $p = 0.006$; Figure 6). Additionally, IRF3 immunostaining of the bile ducts correlated positively with the SOCS3 immunopositivity of the cytoplasm (r = 0.368; $p = 0.006$; Figure 6).

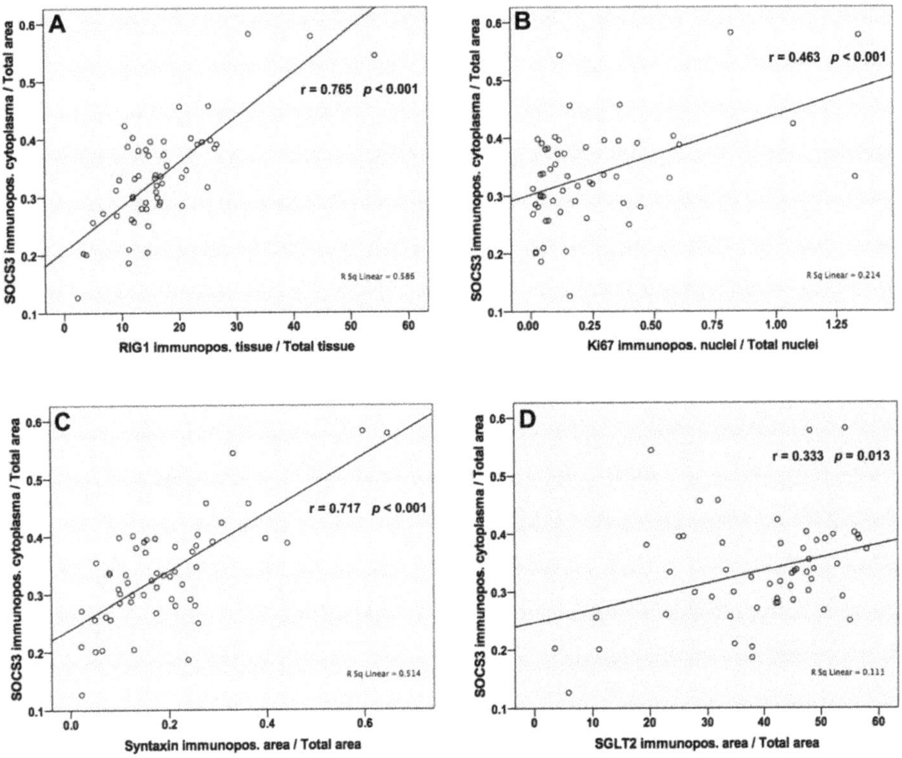

Figure 5. SOCS3 protein expressions are positively associated with RIG1, Ki67, syntaxin, and SGLT2 expressions. (**A**,**C**) High degree of correlations were detected to RIG1 (r = 0.765; $p < 0.001$) and syntaxin (r = 0.717; $p < 0.001$). (**B**,**D**) The degree of correlations to Ki67 and to SGLT2 were significant but moderate (r = 0.463; $p < 0.001$ and r = 0.333; $p = 0.010$). Pearson's correlation coefficient was used to evaluate protein expression results with significance set at $p \leq 0.05$.

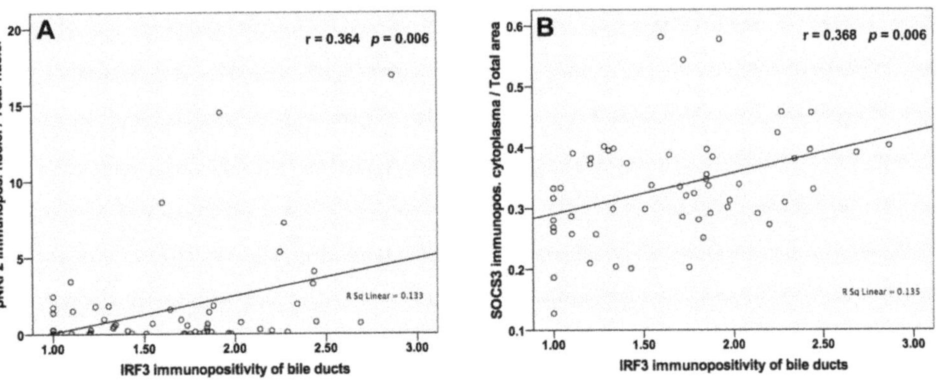

Figure 6. IRF3 immunostaining of the bile ducts is associated with NRF2 activity and with SOCS3 protein expressions. (**A**) We depicted positive significant correlations of moderate degree to pNRF2 immunopositivity of the nuclei (r = 0.364; $p = 0.006$) and (**B**) SOCS3 cytoplasmic staining (r = 0.368; $p = 0.006$). Pearson's correlation coefficient was performed with significance set at $p \leq 0.05$.

4. Discussion

In our present IHC study of NAFLD patients, we detected significantly higher RIG1, IRF3, pNRF2, and SOCS3 protein expressions in NASH patients compared with NAFL patients. These proteins are mainly involved in innate immune regulation [1]; dysregulation of innate immunity plays a role in the pathogenesis of NASH [32]. Further, these proteins were significantly associated with some of the clinical/laboratory parameters and with proteins involved in several biological processes such as autophagy, cell proliferation, and apoptosis. As fatty liver disease is a rising problem worldwide, it is essential to evaluate the mechanisms leading from simple steatosis to advanced non-alcoholic steatosis (NASH). In the following, we discuss how the differentially expressed proteins might be involved in the progression from NAFL to NASH. Based on our results, we assume signaling cascades activated by PRRs, FFA accumulation, and ROS formation (for more details see Figure 1).

RIG1 protein, belonging to the pattern recognition receptors (PRR) family, was significantly higher in NASH patients than in those with NAFL. It was reported that the levels of another PRR protein, namely STING (stimulator of interferon genes) 1, were increased in NASH patients [33,34]. It is documented that recognition of extracellular pathogens or endogenous injury signaling by PRRs leads to signal transmission to downstream MAPK signal cascades, resulting in the activation of ERK/JNK/p38MAPK/NF-κB signals and the transcription of proinflammatory cytokine genes and type I/II IFNs. Intriguingly, PRRs participate in the regulation of lipid metabolism [1]. Fatty acid toxicity is known to trigger the initiation and continuous activation of an inflammatory response in the liver of NAFLD patients [35]. Thus, our detection of higher RIG1 (a PRR) expression in NASH vs. NAFL is in line with the literature. There are only a few study groups analyzing RIG-1 expression immunohistochemically in patients; Frietze et al. analyzed 12 NASH patients and 5 normal controls and found a reduction in RIG-1 protein expression in NASH compared to normal controls. Nevertheless, their patient cohort was very small, and they did not compare NASH expression with NAFL as we did but with normal controls.

Additionally, we found that bile ducts showed stronger IRF3 immunostaining in NASH than in NAFL. We also detected that patients with lobular inflammation grade 2 showed higher IRF3 immunopositivity regarding bile ducts than patients with grade 0 ($p = 0.017$). This is in line with the literature as RIG-1 interacts with dsRNA and initiates downstream signaling which leads to IRF3 and NF-κB activation [36]. Expression of IRF3 in bile ducts is also reported in the literature [37–39]. It is documented that biliary innate immunity is implicated in the pathogenesis of various cholangiopathies in biliary tract diseases and biliary tract defense systems [38]. Cholangiocytes have an innate immune system and the biliary epithelial cells express a variety of PRRs such as Toll-like receptors (TLRs) and they recognize both bacterial and viral PAMPs. Cultured biliary epithelial cells were able to recognize viral PAMPs such as double-stranded RNA (dsRNA). After stimulation with poly(I:C), (analog of viral dsRNA), cultured human biliary epithelial cells expressed NF-κB and IRF3 [37,38]. Shimada performed experiments with human biliary epithelial HuCCT1 cells. They observed that TLR3 signaling led to the expression of CCL5 via NF-κB and IRF3 in bile duct cells, and they suggested the involvement of this signaling pathway in biliary atresia pathogenesis. The involvement of bile duct cells in the development of nonalcoholic fatty liver disease was reported; however, they did not study IRF3 expression but cellular senescence markers and chemokines [40].

NRF2 is described in the literature as a "double-edged sword" [41]. Under normal conditions, NRF2 is mainly localized in the cytoplasm attached to Keap1, which causes inhibition of NRF2 activity because of proteasomal degradation of NRF2. Under stressed conditions, NRF2 dissociates from Keap1, translocates to the nucleus, and promotes the expression of antioxidant response element (ARE) genes, which control stress response, antioxidant defense, drug metabolism, proteasomal degradation, and cell proliferation [8,42]. On the other side, NRF2 can function as a proto-oncogene promoting the growth of cancer cells [43]; liver tumorigenesis by NRF2 activation was reported [44]. Autophagy impairment and NRF2 activation have induced chaperone-mediated autophagy activation and tumor cell survival [45]. We detected higher NRF2 activity in NASH versus NAFL; our result is consistent with aberrant NRF2 activation in various cancer cells and cancer tissues [46] driven by different causes of NRF2 activation. In hepatocellular Huh1 carcinoma cell lines phosphorylated p62 accumulation was documented as the reason for NRF2 activation. This is in accordance with another IHC study of patients with chronic liver disease [41]. Mohs et al. reported an association of NRF2 activity in patients with the grade of inflammation, which we also observed. In contrast, they did not detect an association with steatosis, as we did. It is reported that age can explain different results found in the literature regarding the effects of NRF2 activation. In young animals, a protective role of NRF2 in the development of hepatitis and steatosis has been observed [5]. On the contrary, in older mice, NRF2 activation has induced the activation of genes involved in lipid synthesis and uptake [5]. Since we found a trend toward higher NRF2 activation in older patients, this is consistent to some extent with our results. The reasons for NRF2 activation in tumor tissue can be mutations in KEAP-1 [47] causing Keap-1 not to be able to bind to NRF2, resulting in constitutive NRF2 activation, increased ROS formation because of increased metabolism of fatty acids [5], and increased influx of unfolded proteins into the endoplasmatic reticulum (ER) inducing ER-stress [48]. Also, a noncanonical mechanism of NRF2 activation by autophagy deficiency is documented [49]. We found higher glucose levels in NASH patients; this is consistent with the literature, as those in transgenic mice with enhanced NRF2 activity blood glucose levels were elevated [50]. Also, Islam et al. documented the association of NRF2 activity with glucose metabolism [51]. He et al. [44] reported that NRF2 activation alters glucose and lipid metabolism; hepatocyte-specific NRF2 activation in a mouse model, caused by accumulation of p62 or inhibition of KEAP1 binding, resulted in hepatomegaly associated with increased glycogenosis, steatosis, and G2/M cell cycle arrest, favoring hyperplasia without cell division. Additionally, we detected that patients with ballooning grades 2 and 1 had higher pNRF2 levels than patients with grade 0. It is known that in ferroptosis, an iron-dependent, lipid peroxidation-driven cell death cascade, the formation of a "ballooning" phenotype describes its final critical feature; many of the key anti-ferroptotic pathway components are under the transcriptional control of NRF2 [52]. Dodson et al. studied the association between NRF2 activity and ferroptosis and found that NRF2 plays a critical role in attenuating lipid peroxidation and ferroptosis [52]. Interestingly, the double-sword effect of NRF2 activation is also observed in clinical trials: a phase 3 clinical trial of bardoxolone methyl (activator of the NRF2 pathway) for the treatment of type 2 diabetes and stage 4 chronic kidney disease was discontinued because it failed to reduce the risk of end-stage renal disease/death [53].

Further, we detected significantly higher SOCS3 expression in the cytoplasm of NASH patients than in NAFL. Usually, transcription of SOCS genes is activated following stimulation with cytokines; in the SOCS3 promoters STAT-binding sites were detected; transfection with dominant-negative STAT3 inhibits cytokine-induced expression of SOCS3 [54]. STAT3 overexpression and constitutive activation have been commonly recorded in HCC and are associated with poor prognosis [2]. Kim et al. found in high-fat diet-fed rat livers higher SOCS3 protein expressions than in normal diet-fed rat livers, which is in line with our results [55]. Zhang et al. detected that fructose-treated mouse cells had higher STAT3 and SOCS3 levels [56]. Fructose treatment-induced inflammation activity which was associated with lipid accumulation. We also found a tendential upregulation of pSTAT3 in NASH versus NAFL ($p = 0.059$) which is in accordance with the literature as SOCS3 is a target gene of STAT3 [57]. On the other side, as the association regarding STAT3 is only tendential, we suggest that additional other factors than STAT3 contribute to the activation of SOCS3. It is documented that the expression of the SOCS proteins is increased by cytokine signaling through the activation of STAT- and NF-KB-mediated pathways [58]. Ueki demonstrated in a mouse model that in both obesity and lipopolysaccharide (LPS)-induced endotoxemia there is an increase in SOCS proteins, SOCS1 and SOCS3, in the liver. Senn et al. observed that IL-6 treatment of mice leads to SOCS3 expression in the livers and also inhibited hepatic insulin receptor signaling in these animals [59]. We did not find a significant association between fasting glucose levels and SOCS3 protein expression. Nevertheless, the upregulation of SOCS3 in our patients can be caused by other factors than IL6 levels. It should be noted that few animal models faithfully replicate human disease. Bi et al. showed that in hepatocytes steatosis was alleviated by reducing SOCS3 by inhibiting JAK2/STAT3 pathway [60]. This is also in accordance with our results as we detected significant positive associations between enhanced SOCS3 expression and steatosis grade. Handa et al. observed increased gene expression of STAT3 in NASH patients versus NAFL. However, SOCS3 expression was significantly reduced in patients with NASH which is contradictory to our results [61]. They discussed why they did not observe upregulation of SOCS3, a known negative regulator of the JAK2/STAT3 signaling pathway in their cohort, and suggested that other mechanisms distinct from SOCS3 were involved in the suppression of inflammatory response. Ogata et al. studied SOCS3 in human HCC and detected that SOCS3 expression levels were significantly higher in non-HCC regions of the liver in 20 HCV-infected patients than in 17 non–HCV-infected patients [62]. However, their results were related to HCC patients and not NAFLD, and the cohort of patients studied was smaller than ours. Sharma et al. reported that SOCS3 expression was increased by adiponectin [63]. Bechmann, Canbay et al. studied NASH patients and reported that serum FFA, BA, and M30 were increased in NASH compared with simple steatosis, while adiponectin was significantly decreased [64]. In our patient cohort, we found no significant differences regarding adiponectin levels between the NASH and NAFL patients. However, the literature shows that adiponectin should not be considered only as a "good cytokine" and that its role is much more complex. Guo et al. reported that adiponectin knockout causes a protective effect against high-fat diet-induced liver injury, possibly related to autophagy regulation, despite persistent liver steatosis [65]. However, the reasons why we did not detect lower adiponectin in NASH may be that data referring to adiponectin was not available for all cases and, secondly, in our cohort, the number of females was relatively high. It is known that adiponectin is higher in women than in men [66], so this may also have an effect. We observed in two NASH cases adiponectin serum levels of about 8 and 11; we suppose that these patients had a more favorable disease course. Unfortunately, we cannot verify this as follow-up data is lacking.

We performed correlation studies to understand how the proteins upregulated in NASH are in association with each other and with autophagy, cell proliferation, and apoptosis markers. Upregulation of pNRF2 was positively correlated with the autophagy marker protein Syntaxin. In NASH patients, free fatty acids (FFA) are known to be elevated and correlate positively with disease severity [67]. FFAs induce ER stress response [68] leading to NRF2 activation and upregulation of NRF2 target genes, several of these genes induce autophagy [69]. Also, other studies described that the NRF2–KEAP1 pathway provides a positive feedback loop for autophagy activation. Also, the unfolded protein response (UPR) induced by ER stress leads to the activation of autophagy [70].

We detected a positive correlation between NRF2 activation and the cell proliferation marker Ki67 and, simultaneously, a positive correlation with apoptotic cleavage of CK18 indicated by the apoptosis marker M30. These results appear contradictory at first sight. The correlation between pNRF2 and cell proliferation ($r = 0.645$; $p > 0.000$) was stronger than the correlation between pNRF2 and M30 ($r = 0.431$; $p = 0.001$). In mouse hepatoma (Hepa-1) and human hepatoblastoma (HepG2) cells, NRF2 activation upregulated the antiapoptotic protein Bcl-2, prevented apoptosis, and increased tumor cell survival and growth/proliferation [71]. On the other side in transgenic mice, activated NRF2 delayed cell proliferation and enhanced the apoptosis of damaged liver cells [72]; apoptotic response after NRF2 activation by ROS is reported [73]. It is known that NRF2 activation causes cancer prevention and or progression; this depends on the cellular context and environment [74]. We consider that different biological processes take place in the organism in parallel. In addition, even if the patients are classified into a specific entity regarding NAFLD according to Bedossa criteria, each patient case represents a unique profile.

Further, we found a positive correlation between NRF2 activation and SOCS3 protein expression. This is in accordance with the literature as Meng et al. demonstrated that the p-STAT3/SOCS3 pathway and the KEAP1/NRF2 pathways are linked. They demonstrated that SOCS3 can directly bind to KEAP1 preventing the degradation of NRF2 and resulting in NRF2 activation [75]. We detected a positive correlation between SOCS3 and RIG1 which is consistent with the literature as the pattern recognition receptor RIG1 can lead to an activation of JAK-STAT signaling [1]. In our patient cohort, SOCS3 was positively correlated with the autophagy marker syntaxin. This is in accordance with the literature as Wan et al. showed a positive association between autophagy activation and upregulation of SOCS3. Briefly, AMPK-autophagy activation suppressed neuroinflammation and improved morphine tolerance via the upregulation of SOCS3 by inhibiting miRNA-30a-5p [76]. They also underlined the dual role of autophagy in tumor development: autophagy prevents tumor initiation in early tumorigenesis but once the tumor progresses autophagy contributes to tumor survival [76]. Further, we performed IHC staining for SGLT2 and detected immunopositive staining in the cytoplasm of hepatocytes which is in line with the literature [77]. We did not find significant differences regarding SGLT2 protein expression between NASH and NAFL. Nakano et al. also observed no significant difference in hepatic expression of SGLT2 in the stratified analysis according to age, sex, BMI, and the severity of the liver disease. We detected a positive correlation between SGLT2 and SOCS3 expression levels concomitant with the literature, as SOCS3 is a mediator of insulin resistance in the liver [78].

5. Conclusions

Our results demonstrate significantly higher expression of NRF2 and SOCS3 proteins in our NASH patients than in those with NAFL. We suggest that they are most likely promising candidates for further studies to detect drugs for therapy. The relationship between fasting glucose levels and NRF2 is of interest regarding metabolic syndrome therapy. We recommend immunohistochemical staining of liver tissue for pNRF2 during liver biopsy to detect prediabetic abnormality in advance. The association of SOCS3 with autophagy and SGLT2 expression are most certainly interesting and potential areas for further studies, such as the drug combination of SGLT2 inhibitors and SOCS3 modulators.

Supplementary Materials: The following supporting information can be downloaded at: https://www.mdpi.com/article/10.3390/jpm13071152/s1, Table S1: Immunohistochemistry antibodies and staining protocols; Figure S1: Nuclear pSTAT3 immunostaining shows a trend toward higher values in NASH than in NAFL; $p = 0.059$.

Author Contributions: H.A.B., J.K., S.S. and M.A. conceived and designed the experiments; S.S. and M.A. performed the experiments; in case of non-conformity of the results, H.A.B. determined a final IRS; S.S., M.A., P.P.M., J.-P.S., A.C., H.H.-J.S., H.A.B. and J.K. analyzed and interpreted the data; P.P.M., J.-P.S., A.C. and H.A.B. provided human samples; H.A.B. and J.K. performed scoring of the patients; S.S. and J.K. wrote the paper. All authors critically revised the manuscript. J.K. is the guarantor of this work and, as such, had full access to all of the data in the study and takes responsibility for the integrity of the data and the accuracy of the data analysis. All authors have read and agreed to the published version of the manuscript.

Funding: This research was supported by University Duisburg Essen intramural gender equity measure program "Welcome Back", grant number: 2020 4699 124.

Institutional Review Board Statement: The study was conducted in accordance with the Helsinki Declaration of 1975 and approved by the Ethics Committee of the University Hospital Essen (reference number: 09-4252; date of approval: 22 February 2010).

Informed Consent Statement: Informed consent was obtained from every patient involved in the study.

Data Availability Statement: Data is provided in the manuscript and/or Supplementary Data. The source of the data came from the Institute of Pathology, University Hospital of Essen.

Acknowledgments: We thank Dorothe Möllmann, Martin Schlattjan (Institute of Pathology, University Hospital of Essen, University of Duisburg-Essen), and Gabriele Ladwig (German Biobank Essen) for their excellent technical assistance. J.K and S.S were supported by a "Welcome Back" grant from the Medical Faculty of the University Duisburg Essen, grant number: 2020 4699 124.

Conflicts of Interest: The authors declare no conflict of interest.

References

1. Bai, L.; Li, H. Innate immune regulatory networks in hepatic lipid metabolism. *J. Mol. Med.* **2019**, *97*, 593–604. [CrossRef] [PubMed]
2. Lee, C.; Cheung, S.T. STAT3: An emerging therapeutic target for hepatocellular carcinoma. *Cancers* **2019**, *11*, 1646. [CrossRef] [PubMed]
3. Naka, T.; Narazaki, M.; Hirata, M.; Matsumoto, T.; Minamoto, S.; Aono, A.; Nishimoto, N.; Kajita, T.; Taga, T.; Yoshizaki, K.; et al. Structure and function of a new STAT-induced STAT inhibitor. *Nature* **1997**, *387*, 924–929. [CrossRef]
4. Jorgensen, S.B.; O'Neill, H.M.; Sylow, L.; Honeyman, J.; Hewitt, K.A.; Palanivel, R.; Fullerton, M.D.; Oberg, L.; Balendran, A.; Galic, S.; et al. Deletion of skeletal muscle SOCS3 prevents insulin resistance in obesity. *Diabetes* **2013**, *62*, 56–64. [CrossRef]
5. Chambel, S.S.; Santos-Goncalves, A.; Duarte, T.L. The dual role of Nrf2 in nonalcoholic fatty liver disease: Regulation of antioxidant defenses and hepatic lipid metabolism. *BioMed Res. Int.* **2015**, *2015*, 597134. [CrossRef] [PubMed]
6. Van der Horst, D.; Carter-Timofte, M.E.; van Grevenynghe, J.; Laguette, N.; Dinkova-Kostova, A.T.; Olagnier, D. Regulation of innate immunity by Nrf2. *Curr. Opin. Immunol.* **2022**, *78*, 102247. [CrossRef] [PubMed]
7. Lee, S.M.; Koh, D.H.; Jun, D.W.; Roh, Y.J.; Kang, H.T.; Oh, J.H.; Kim, H. Auranofin attenuates hepatic steatosis and fibrosis in nonalcoholic fatty liver disease via NRF2 and NF-κB signaling pathways. *Clin. Mol. Hepatol.* **2022**, *28*, 827–840. [CrossRef]
8. Bataille, A.M.; Manautou, J.E. Nrf2: A potential target for new therapeutics in liver disease. *Clin. Pharmacol. Ther.* **2012**, *92*, 340–348. [CrossRef]
9. Gonzalez-Rodriguez, A.; Mayoral, R.; Agra, N.; Valdecantos, M.P.; Pardo, V.; Miquilena-Colina, M.E.; Vargas-Castrillon, J.; Lo Iacono, O.; Corazzari, M.; Fimia, G.M.; et al. Impaired autophagic flux is associated with increased endoplasmic reticulum stress during the development of NAFLD. *Cell Death Dis.* **2014**, *5*, e1179. [CrossRef]
10. Wu, W.K.K.; Zhang, L.; Chan, M.T.V. Autophagy, NAFLD and NAFLD-Related HCC. *Obes. Fat. Liver Liver Cancer* **2018**, *1061*, 127–138.
11. Delgado, M.; Singh, S.; De Haro, S.; Master, S.; Ponpuak, M.; Dinkins, C.; Ornatowski, W.; Vergne, I.; Deretic, V. Autophagy and pattern recognition receptors in innate immunity. *Immunol. Rev.* **2009**, *227*, 189–202. [CrossRef]
12. Wang, Y.; Singh, R.; Xiang, Y.; Czaja, M.J. Macroautophagy and chaperone-mediated autophagy are required for hepatocyte resistance to oxidant stress. *Hepatology* **2010**, *52*, 266–277. [CrossRef]
13. Martinet, W.; Roth, L.; De Meyer, G.R.Y. Standard immunohistochemical assays to assess autophagy in mammalian tissue. *Cells* **2017**, *6*, 17. [CrossRef] [PubMed]
14. Yoshii, S.R.; Mizushima, N. Monitoring and measuring autophagy. *Int. J. Mol. Sci.* **2017**, *18*, 1865. [CrossRef]
15. Qiang, S.; Nakatsu, Y.; Seno, Y.; Fujishiro, M.; Sakoda, H.; Kushiyama, A.; Mori, K.; Matsunaga, Y.; Yamamotoya, T.; Kamata, H.; et al. Treatment with the SGLT2 inhibitor luseogliflozin improves nonalcoholic steatohepatitis in a rodent model with diabetes mellitus. *Diabetol. Metab. Syndr.* **2015**, *7*, 104. [CrossRef]
16. Jojima, T.; Wakamatsu, S.; Kase, M.; Iijima, T.; Maejima, Y.; Shimomura, K.; Kogai, T.; Tomaru, T.; Usui, I.; Aso, Y. The SGLT2 inhibitor canagliflozin prevents carcinogenesis in a mouse model of diabetes and non-alcoholic steatohepatitis-related hepatocarcinogenesis: Association with SGLT2 expression in hepatocellular carcinoma. *Int. J. Mol. Sci.* **2019**, *20*, 5237. [CrossRef] [PubMed]
17. Chiang, H.; Lee, J.-C.; Huang, H.-C.; Huang, H.; Liu, H.-K.; Huang, C. Delayed intervention with a novel SGLT2 inhibitor NGI001 suppresses diet-induced metabolic dysfunction and non-alcoholic fatty liver disease in mice. *Br. J. Pharmacol.* **2020**, *177*, 239–253. [CrossRef]
18. Hsiang, J.C.; Wong, V.W.-S. SGLT2 inhibitors in liver patients. *Clin. Gastroenterol. Hepatol.* **2020**, *18*, 2168–2172. [CrossRef] [PubMed]
19. Kälsch, J.; Bechmann, L.P.; Kälsch, H.; Schlattjan, M.; Erhard, J.; Gerken, G.; Canbay, A. Evaluation of biomarkers of NAFLD in a cohort of morbidly obese patients. *J. Nutr. Metab.* **2011**, *2011*, 369168. [CrossRef]
20. Ikura, Y. Transitions of histopathologic criteria for diagnosis of nonalcoholic fatty liver disease during the last three decades. *World J. Hepatol.* **2014**, *6*, 894–900. [CrossRef]
21. Kleiner, D.E.; Brunt, E.M.; Van Natta, M.; Behling, C.; Contos, M.J.; Cummings, O.W.; Ferrell, L.D.; Liu, Y.-C.; Torbenson, M.S.; Unalp-Arida, A.; et al. Design and validation of a histological scoring system for nonalcoholic fatty liver disease. *Hepatology* **2005**, *41*, 1313–1321. [CrossRef] [PubMed]
22. Bedossa, P.; FLIP, P.C. Utility and appropriateness of the fatty liver inhibition of progression (FLIP) algorithm and steatosis, activity, and fibrosis (SAF) score in the evaluation of biopsies of nonalcoholic fatty liver disease. *Hepatology* **2014**, *60*, 565–575. [CrossRef] [PubMed]

23. Schwertheim, S.; Kälsch, J.; Jastrow, H.; Schaefer, C.M.; Theurer, S.; Ting, S.; Canbay, A.; Wedemeyer, H.; Schmid, K.W.; Baba, H.A. Characterization of two types of intranuclear hepatocellular inclusions in NAFLD. *Sci. Rep.* **2020**, *10*, 16533. [CrossRef]
24. Remmele, W.; Stegner, H.E. Recommendation for uniform definition of an immunoreactive score (IRS) for immunohistochemical estrogen receptor detection (ER-ICA) in breast cancer tissue. *Pathologe* **1987**, *8*, 138–140.
25. Bertram, S.; Padden, J.; Kälsch, J.; Ahrens, M.; Pott, L.; Canbay, A.; Weber, F.; Fingas, C.; Hoffmann, A.C.; Vietor, A.; et al. Novel immunohistochemical markers differentiate intrahepatic cholangiocarcinoma from benign bile duct lesions. *J. Clin. Pathol.* **2016**, *69*, 619–626. [CrossRef]
26. Kälsch, J.; Padden, J.; Bertram, S.; Pott, L.L.; Reis, H.; Westerwick, D.; Schaefer, C.M.; Sowa, J.-P.; Möllmann, D.; Fingas, C.; et al. Annexin A10 optimally differentiates between intrahepatic cholangiocarcinoma and hepatic metastases of pancreatic ductal adenocarcinoma: A comparative study of immunohistochemical markers and panels. *Virchows Arch.* **2017**, *470*, 537–543. [CrossRef]
27. Battifora, H. Assessment of antigen damage in immunohistochemistry: The vimentin internal control. *Am. J. Clin. Pathol.* **1991**, *96*, 669–671. [CrossRef] [PubMed]
28. Bertram, S.; Myland, C.; Swoboda, S.; Gallinat, A.; Minor, T.; Lehmann, N.; Thie, M.; Kälsch, J.; Pott, L.; Canbay, A.; et al. Liver steatosis in pre-transplant liver biopsies can be quantified rapidly and accurately by nuclear magnetic resonance analysis. *Virchows Arch.* **2017**, *470*, 197–204. [CrossRef]
29. Apopa, P.L.; He, X.; Ma, Q. Phosphorylation of Nrf2 in the transcription activation domain by casein kinase 2 (CK2) is critical for the nuclear translocation and transcription activation function of Nrf2 in IMR-32 neuroblastoma cells. *J. Biochem. Mol. Toxicol.* **2008**, *22*, 63–76. [CrossRef]
30. Huang, H.-C.; Nguyen, T.; Pickett, C.B. Phosphorylation of Nrf2 at Ser-40 by protein kinase C regulates antioxidant response element-mediated transcription. *J. Biol. Chem.* **2002**, *277*, 42769–42774. [CrossRef]
31. Li, Y.; Teng, D.; Shi, X.; Qin, G.; Qin, Y.; Quan, H.; Shi, B.; Sun, H.; Ba, J.; Chen, B.; et al. Prevalence of diabetes recorded in mainland China using 2018 diagnostic criteria from the American Diabetes Association: National cross sectional study. *BMJ* **2020**, *369*, m997. [CrossRef]
32. Parthasarathy, G.; Revelo, X.; Malhi, H. Pathogenesis of nonalcoholic steatohepatitis: An overview. *Hepatol. Commun.* **2020**, *4*, 478–492. [CrossRef] [PubMed]
33. Luo, X.; Li, H.; Ma, L.; Zhou, J.; Guo, X.; Woo, S.-L.; Pei, Y.; Knight, L.R.; Deveau, M.; Chen, Y.; et al. Expression of STING is increased in liver tissues from patients with NAFLD and promotes macrophage-mediated hepatic inflammation and fibrosis in mice. *Gastroenterology* **2018**, *155*, 1971–1984. [CrossRef] [PubMed]
34. Yu, Y.; Liu, Y.; An, W.; Song, J.; Zhang, Y.; Zhao, X. STING-mediated inflammation in Kupffer cells contributes to progression of nonalcoholic steatohepatitis. *J. Clin. Investig.* **2019**, *129*, 546–555. [CrossRef]
35. Frietze, K.K.; Brown, A.M.; Das, D.; Franks, R.G.; Cunningham, J.L.; Hayward, M.; Nickels, J.T., Jr. Lipotoxicity reduces DDX58/Rig-1 expression and activity leading to impaired autophagy and cell death. *Autophagy* **2022**, *18*, 142–160. [CrossRef] [PubMed]
36. Szabo, G.; Dolganiuc, A.; Mandrekar, P. Pattern recognition receptors: A contemporary view on liver diseases. *Hepatology* **2006**, *44*, 287–298. [CrossRef] [PubMed]
37. Harada, K.; Sato, Y.; Itatsu, K.; Isse, K.; Ikeda, H.; Yasoshima, M.; Zen, Y.; Matsui, A.; Nakanuma, Y. Innate immune response to double-stranded RNA in biliary epithelial cells is associated with the pathogenesis of biliary atresia. *Hepatology* **2007**, *46*, 1146–1154. [CrossRef]
38. Harada, K.; Nakanuma, Y. Cholangiopathy with respect to biliary innate immunity. *Int. J. Hepatol.* **2012**, *2012*, 793569. [CrossRef]
39. Shimada, T.; Imaizumi, T.; Shirai, K.; Tatsuta, T.; Kimura, T.; Hayakari, R.; Yoshida, H.; Matsumiya, T.; Kijima, H.; Mizukami, H.; et al. CCL5 is induced by TLR 3 signaling in HuCCT1 human biliary epithelial cells: Possible involvement in the pathogenesis of biliary atresia. *Biomed. Res.* **2017**, *38*, 269–276. [CrossRef]
40. Chiba, M.; Sasaki, M.; Kitamura, S.; Ikeda, H.; Sato, Y.; Nakanuma, Y. Participation of bile ductular cells in the pathological progression of non-alcoholic fatty liver disease. *J. Clin. Pathol.* **2011**, *64*, 564–570. [CrossRef]
41. Mohs, A.; Otto, T.; Schneider, K.M.; Peltzer, M.; Boekschoten, M.; Holland, C.H.; Hudert, C.A.; Kalveram, L.; Wiegand, S.; Saez-Rodriguez, J.; et al. Hepatocyte-specific NRF2 activation controls fibrogenesis and carcinogenesis in steatohepatitis. *J. Hepatol.* **2021**, *74*, 638–648. [CrossRef]
42. Baird, L.; Dinkova-Kostova, A.T. The cytoprotective role of the Keap1-Nrf2 pathway. *Arch. Toxicol.* **2011**, *85*, 241–272. [CrossRef] [PubMed]
43. Hohmann, M.S.; Zaninelli, T.H.; Staurengo-Ferrari, L.; Manchope, M.F.; Badaro-Garcia, S.; de Freitas, A.; Casagrande, R.; Verri, W.A. Nrf2 in Immune Responses During Inflammation. In *Nrf2 and Its Modulation in Inflammation. Progress in Inflammation Research*; Deng, H., Ed.; Springer: Cham, Switzerland, 2020; pp. 23–49.
44. He, F.; Antonucci, L.; Yamachika, S.; Zhang, Z.; Taniguchi, K.; Umemura, A.; Hatzivassiliou, G.; Roose-Girma, M.; Reina-Campos, M.; Duran, A.; et al. NRF2 activates growth factor genes and downstream AKT signaling to induce mouse and human hepatomegaly. *J. Hepatol.* **2020**, *72*, 1182–1195. [CrossRef]
45. Dash, S.; Aydin, Y.; Moroz, K. Chaperone-mediated autophagy in the liver: Good or bad? *Cells* **2019**, *8*, 1308. [CrossRef] [PubMed]
46. Taguchi, K.; Yamamoto, M. The KEAP1-NRF2 system in cancer. *Front. Oncol.* **2017**, *7*, 85. [CrossRef] [PubMed]

47. Schulze, K.; Imbeaud, S.; Letouze, E.; Alexandrov, L.B.; Calderaro, J.; Rebouissou, S.; Couchy, G.; Meiller, C.; Shinde, J.; Soysouvanh, F.; et al. Exome sequencing of hepatocellular carcinomas identifies new mutational signatures and potential therapeutic targets. *Nat. Genet.* **2015**, *47*, 505–511. [CrossRef]
48. Engin, A.B.; Engin, A. *Obesity and Lipotoxicity*; Springer: Cham, Switzerland, 2017; pp. 59–61.
49. Lau, A.; Wang, X.-J.; Zhao, F.; Villeneuve, N.F.; Wu, T.; Jiang, T.; Sun, Z.; White, E.; Zhang, D.D. A noncanonical mechanism of Nrf2 activation by autophagy deficiency: Direct interaction between Keap1 and p62. *Mol. Cell. Biol.* **2010**, *30*, 3275–3285. [CrossRef]
50. Zhang, Y.-K.J.; Wu, K.C.; Liu, J.; Klaassen, C.D. Nrf2 deficiency improves glucose tolerance in mice fed a high-fat diet. *Toxicol. Appl. Pharmacol.* **2012**, *264*, 305–314. [CrossRef]
51. Islam, M.A.; Sooro, M.A.; Zhang, P. Autophagic regulation of p62 is critical for cancer therapy. *Int. J. Mol. Sci.* **2018**, *19*, 1405. [CrossRef]
52. Dodson, M.; Castro-Portuguez, R.; Zhang, D.D. NRF2 plays a critical role in mitigating lipid peroxidation and ferroptosis. *Redox Biol.* **2019**, *23*, 101107. [CrossRef]
53. Barbaro, B.; Romito, I.; Alisi, A. Commentary: The histone demethylase Phf2 acts as a molecular checkpoint to prevent NAFLD progression during obesity. *Front. Genet.* **2018**, *9*, 443. [CrossRef] [PubMed]
54. Brender, C.; Nielsen, M.; Kaltoft, K.; Mikkelsen, G.; Zhang, Q.; Wasik, M.; Billestrup, N.; Odum, N. STAT3-mediated constitutive expression of SOCS-3 in cutaneous T-cell lymphoma. *Blood J. Am. Soc. Hematol.* **2001**, *97*, 1056–1062. [CrossRef] [PubMed]
55. Kim, J.; Kim, C.-S.; Jo, K.; Lee, I.S.; Kim, J.-H.; Kim, J.S. POCU1b, the n-butanol soluble fraction of polygoni cuspidati rhizoma et radix, attenuates obesity, non-alcoholic fatty liver, and insulin resistance Via inhibitions of pancreatic lipase, cAMP-dependent PDE activity, AMPK activation, and SOCS-3 suppression. *Nutrients* **2020**, *12*, 3612. [CrossRef]
56. Zhang, X.; Zhang, J.-H.; Chen, X.-Y.; Hu, Q.-H.; Wang, M.-X.; Jin, R.; Zhang, Q.-Y.; Wang, W.; Wang, R.; Kang, L.-L.; et al. Reactive oxygen species-induced TXNIP drives fructose-mediated hepatic inflammation and lipid accumulation through NLRP3 inflammasome activation. *Antioxid. Redox Signal.* **2015**, *22*, 848–870. [CrossRef]
57. Fu, K.; Wang, C.; Ma, C.; Zhou, H.; Li, Y. The potential application of Chinese medicine in liver diseases: A new opportunity. *Front. Pharmacol.* **2021**, *12*, 771459. [CrossRef] [PubMed]
58. Ueki, K.; Kondo, T.; Kahn, C.R. Suppressor of cytokine signaling 1 (SOCS-1) and SOCS-3 cause insulin resistance through inhibition of tyrosine phosphorylation of insulin receptor substrate proteins by discrete mechanisms. *Mol. Cell. Biol.* **2004**, *24*, 5434–5446. [CrossRef]
59. Senn, J.J.; Klover, P.J.; Nowak, I.A.; Zimmers, T.A.; Koniaris, L.G.; Furlanetto, R.W.; Mooney, R.A. Suppressor of cytokine signaling-3 (SOCS-3), a potential mediator of interleukin-6-dependent insulin resistance in hepatocytes. *J. Biol. Chem.* **2003**, *278*, 13740–13746. [CrossRef]
60. Bi, J.; Sun, K.; Wu, H.; Chen, X.; Tang, H.; Mao, J. PPARγ alleviated hepatocyte steatosis through reducing SOCS3 by inhibiting JAK2/STAT3 pathway. *Biochem. Biophys. Res. Commun.* **2018**, *498*, 1037–1044. [CrossRef]
61. Handa, P.; Maliken, B.D.; Nelson, J.E.; Hennessey, K.A.; Vemulakonda, L.A.; Morgan-Stevenson, V.; Dhillon, B.K.; Gupta, R.; Yeh, M.M.; Kowdley, K.V. Differences in hepatic expression of iron, inflammation and stress-related genes in patients with nonalcoholic steatohepatitis. *Ann. Hepatol.* **2017**, *16*, 77–85. [CrossRef]
62. Ogata, H.; Kobayashi, T.; Chinen, T.; Takaki, H.; Sanada, T.; Minoda, Y.; Koga, K.; Takaesu, G.; Maehara, Y.; Iida, M.; et al. Deletion of the SOCS3 gene in liver parenchymal cells promotes hepatitis--induced hepatocarcinogenesis. *Gastroenterology* **2006**, *131*, 179–193. [CrossRef]
63. Sharma, D.; Wang, J.; Fu, P.P.; Sharma, S.; Nagalingam, A.; Mells, J.; Handy, J.; Page, A.J.; Cohen, C.; Anania, F.A.; et al. Adiponectin antagonizes the oncogenic actions of leptin in hepatocellular carcinogenesis. *Hepatology* **2010**, *52*, 1713–1722. [CrossRef]
64. Bechmann, L.P.; Kocabayoglu, P.; Sowa, J.-P.; Sydor, S.; Best, J.; Schlattjan, M.; Beilfuss, A.; Schmitt, J.; Hannivoort, R.A.; Kilicarslan, A.; et al. Free fatty acids repress small heterodimer partner (SHP) activation and adiponectin counteracts bile acid-induced liver injury in superobese patients with nonalcoholic steatohepatitis. *Hepatology* **2013**, *57*, 1394–1406. [CrossRef]
65. Guo, R.; Nair, S.; Zhang, Y.; Ren, J. Adiponectin deficiency rescues high-fat diet-induced hepatic injury, apoptosis and autophagy loss despite persistent steatosis. *Int. J. Obes.* **2017**, *41*, 1403–1412. [CrossRef]
66. Schöndorf, T.; Maiworm, A.; Emmison, N.; Forst, T.; Pfützner, A. Biological background and role of adiponectin as marker for insulin resistance and cardiovascular risk. *Clin. Lab.* **2005**, *51*, 489–494.
67. Gentile, C.L.; Frye, M.A.; Pagliassotti, M.J. Fatty acids and the endoplasmic reticulum in nonalcoholic fatty liver disease. *Biofactors* **2011**, *37*, 8–16. [CrossRef]
68. Jiao, P.; Ma, J.; Feng, B.; Zhang, H.; Alan-Diehl, J.; Eugene-Chin, Y.; Yan, W.; Xu, H. FFA-induced adipocyte inflammation and insulin resistance: Involvement of ER stress and IKKβ pathways. *Obesity* **2011**, *19*, 483–491. [CrossRef]
69. Nam, T.; Han, J.H.; Devkota, S.; Lee, H.-W. Emerging paradigm of crosstalk between autophagy and the ubiquitin-proteasome system. *Mol. Cells* **2017**, *40*, 897–905. [PubMed]
70. Korolchuk, V.I.; Menzies, F.M.; Rubinsztein, D.C. Mechanisms of cross-talk between the ubiquitin-proteasome and autophagy-lysosome systems. *FEBS Lett.* **2010**, *584*, 1393–1398. [CrossRef]
71. Niture, S.K.; Jaiswal, A.K. Nrf2 protein up-regulates antiapoptotic protein Bcl-2 and prevents cellular apoptosis. *J. Biol. Chem.* **2012**, *287*, 9873–9886. [CrossRef] [PubMed]

72. Köhler, U.A.; Kurinna, S.; Schwitter, D.; Marti, A.; Schäfer, M.; Hellerbrand, C.; Speicher, T.; Werner, S. Activated Nrf2 impairs liver regeneration in mice by activation of genes involved in cell-cycle control and apoptosis. *Hepatology* **2014**, *60*, 670–678. [CrossRef] [PubMed]
73. Olagnier, D.; Peri, S.; Steel, C.; van Montfoort, N.; Chiang, C.; Beljanski, V.; Slifker, M.; He, Z.; Nichols, C.N.; Lin, R.; et al. Cellular oxidative stress response controls the antiviral and apoptotic programs in dengue virus-infected dendritic cells. *PLoS Pathog.* **2014**, *10*, e1004566. [CrossRef] [PubMed]
74. Moon, E.J.; Giaccia, A. Dual roles of NRF2 in tumor prevention and progression: Possible implications in cancer treatment. *Free Radic. Biol. Med.* **2015**, *79*, 292–299. [CrossRef] [PubMed]
75. Meng, Y.; Lin, W.; Wang, N.; Wei, X.; Huang, Q.; Liao, Y. Bazedoxifene-induced ROS promote mitochondrial dysfunction and enhance osimertinib sensitivity by inhibiting the p-STAT3/SOCS3 and KEAP1/NRF2 pathways in non-small cell lung cancer. *Free Radic. Biol. Med.* **2023**, *196*, 65–80. [CrossRef] [PubMed]
76. Wan, L.; Jia, R.-M.; Ji, L.-L.; Qin, X.-M.; Hu, L.; Hu, F.; Han, Y.; Pan, Y.-B.; Jiang, C.-Y.; Liu, W.-T. AMPK-autophagy-mediated inhibition of microRNA-30a-5p alleviates morphine tolerance via SOCS3-dependent neuroinflammation suppression. *J. Neuroinflamm.* **2022**, *19*, 25. [CrossRef] [PubMed]
77. Nakano, D.; Akiba, J.; Tsutsumi, T.; Kawaguchi, M.; Yoshida, T.; Koga, H.; Kawaguchi, T. Hepatic expression of sodium-glucose cotransporter 2 (SGLT2) in patients with chronic liver disease. *Med. Mol. Morphol.* **2022**, *55*, 304–315. [CrossRef] [PubMed]
78. Torisu, T.; Sato, N.; Yoshiga, D.; Kobayashi, T.; Yoshioka, T.; Mori, H.; Iida, M.; Yoshimura, A. The dual function of hepatic SOCS3 in insulin resistance in vivo. *Genes Cells* **2007**, *12*, 143–154. [CrossRef]

Journal of *Personalized Medicine*

MDPI

Article

No Differences in Rotational Thromboelastometry Measurements between Portal and Peripheral Circulation in Cirrhotic Patients Undergoing TIPS

Sotiria Bedreli [1,†], Paul Manka [2,3,†], Matthias Buechter [4], Michael Jahn [5], Jens M. Theysohn [6], Ali Canbay [2,3] and Antonios Katsounas [2,3,7,*]

1 Department of Internal Medicine and Gastroenterology, Marienhospital Gelsenkirchen, 45886 Gelsenkirchen, Germany
2 Department of Medicine, University Hospital Knappschaftskrankenhaus Bochum, 44892 Bochum, Germany
3 Faculty of Medicine, Ruhr University Bochum, 44901 Bochum, Germany
4 Department of Medicine and Gastroenterology, Katholische Kliniken im Märkischen Kreis gem, GmbH, 58638 Iserlohn, Germany
5 Department of Nephrology, Alfried-Krupp Krankenhaus Essen, 45131 Essen, Germany
6 Department of Diagnostic and Interventional Radiology and Neuroradiology, University Hospital Essen, University Duisburg-Essen, 45122 Essen, Germany
7 Division Infectious Diseases and Critical Care Medicine, Department of Medicine, University Hospital Knappschaftskrankenhaus, 44892 Bochum, Germany
* Correspondence: antonios.katsounas@rub.de; Tel.: +49-234-299-3401; Fax: +49-234-299-3468
† These authors contributed equally to this work.

Abstract: Background: In patients with liver cirrhosis, transjugular intrahepatic portosystemic shunt (TIPS) is considered a standardized treatment of refractory ascites or variceal bleeding. TIPS thrombosis (TT) and/or portal vein thrombosis (PVT) are possible complications during/after TIPS placement. Previous studies suggested increased clotting activity in portal circulation (PORC). This pilot study aimed to evaluate alterations and differences of coagulation function in PORC and in peripheral circulation (PERC) via rotational thromboelastometry during TIPS. Methods: Blood samples were collected from cirrhotic patients (n = 13; median Model of End Stage Liver Disease, MELD Score: 12; median age: 60 years) undergoing TIPS (10/13 TIPSs were elective procedures due to refractory ascites) as follows: median cubital vein (MCV; PERC)—confluence of the three hepatic veins to the inferior cava vein (HV/ICV; PORC)—portal vein (PV; PORC)—TIPS (PORC). This research utilized four variables of the extrinsic test EXTEM, i.e., clotting time (CT), clot formation time (CFT), maximum clot firmness (MCF), and maximum lysis (ML). Results: EXTEM results [mean, M (range) ± standard deviation, SD (range)] showed no significant differences for CT [M (70–73) ± SD (9–13); p = 0.93] or CFT [M (137–155) ± SD (75–112); p = 0.97] or MCF [M (51–54) ± SD (9–10); p = 0.90] or ML [M (9–10) ± SD (4–5); p = 0.89] between the compartments, i.e., MCV vs. HV/ICV vs. PV vs. TIPS. Overall, we detected no differences in coagulation function between PERC and PORC. Conclusion: These results are in contrast to previous reports suggesting increased clotting activity in PORC vs. PERC in association with liver cirrhosis. Rotational thromboelastometry-based evaluation of coagulation function in PERC appears to reliably reflect coagulation function in PORC with respect to risk estimation for TT and/or PVT in cirrhotic patients undergoing TIPS.

Keywords: ROTEM®; TIPS; liver disease; thrombosis risk assessment; hypercoagulable state; portal hypertension; hepatic decompensation

check for updates

Citation: Bedreli, S.; Manka, P.; Buechter, M.; Jahn, M.; Theysohn, J.M.; Canbay, A.; Katsounas, A. No Differences in Rotational Thromboelastometry Measurements between Portal and Peripheral Circulation in Cirrhotic Patients Undergoing TIPS. *J. Pers. Med.* **2023**, *13*, 424. https://doi.org/10.3390/jpm13030424

Academic Editor: Konstantinos Tziomalos

Received: 22 January 2023
Revised: 5 February 2023
Accepted: 22 February 2023
Published: 26 February 2023

1. Introduction

In patients with end-stage liver disease, transjugular intrahepatic portosystemic shunt (TIPS) implantation is considered a standardized intervention for treatment of complications due to portal hypertension, such as refractory ascites [1,2], variceal bleeding [1,3], or other severe conditions, e.g., Budd-Chiari syndrome [1,4,5], hepatorenal syndrome, and hepatic hydrothorax [1,6]. However, TIPS remains one of the most challenging angiographic techniques, requiring high expertise to limit procedure- and/or shunt-associated complications [7]. Thrombosis is a post-TIPS complication that occurs in up to 10% of cases [8]. TIPS thrombosis (TT), which usually emerges within days after TIPS intervention or occasionally also during deployment, is often attributable to graft misplacement and/or underlying hypercoagulable pathologies [9,10], which may remain undetectable in routine screening tests [11]. Although molecular pathomechanisms accounting for the recently recognized hypercoagulable state in liver cirrhosis—favoring TT and/or portal vein thrombosis (PVT)—have not been completely elucidated yet [12], there is evidence that bacterial endotoxins predispose to thrombotic complications in the portal circulation (PORC) via tissue factor up-regulation, thereby increasing generation of thrombin [13]. In fact, previous research reported significantly higher endotoxin concentrations in the PORC in comparison to the peripheral circulation (PERC) [14,15]. Moreover, earlier studies investigating the potential interplay between clotting activation and endotoxins in cirrhotic patients undergoing TIPS suggested that thrombin generation and D-dimers were increased in the PERC of cirrhotic patients compared to controls [15,16]. According to the same sources, the grade of thrombin generation and hyperfibrinolysis, which correlated with the grade of endotoxemia, was higher in the PORC versus the PERC [14,15]. These findings suggest that specific conditions ruling the gut-liver-axis, such as increased endotoxemia in the PORC, which aggravates with worsening liver disease, among other possibly yet unidentified factors, may represent key mechanisms of overstimulation of clotting processes that favor TT and/or PVT [17]. However, to the best of our knowledge, coagulation pathway activation in the PORC—versus PERC—has never been evaluated using rotational thromboelastometry or other viscoelastic tests in patients with liver cirrhosis.

During the last two decades, viscoelastic tests using whole blood specimens have revolutionized our understanding in hemostaseology as, in contrast to conventional tests, they can evaluate coagulation dynamics from clot formation to clot lysis [11,18]. Especially in the field of clinical hepatology, rotational thromboelastometry has achieved acceptance as a feasible and reliable point-of-care tool, not only for differential hemostatic management during hemorrhages, but also for bleeding or thrombosis risk assessment [19,20].

Against this background, this analysis aimed to investigate and characterize coagulation profiles in PORC—in comparison to PERC—in patients with liver cirrhosis undergoing TIPS implantation using rotational thromboelastometry and check for potential associations with occurring TT and/or PVT within 30 days post-intervention.

2. Materials and Methods

2.1. Patients

Patients with liver cirrhosis (n = 13) treated in the Intermediate Care Unit of the Department of Gastroenterology and Hepatology at the University Hospital Essen were subjected to this analysis within twelve months (2016–2017). Patients' demographics and laboratory values are summarized in Table 1.

Table 1. Demographic characteristics and clinical parameters of patients undergoing TIPS. Data are presented as median (min; max) or as absolute count "n" with percentage (%).

Characteristics	TIPS; $n = 13$	Normal Range
Underlying liver disease (n) Primary sclerosing cholangitis (PSC) Alcoholic steatohepatitis (ASH) Hepatitis C virus (HCV) Autoimmune hepatitis Non-alcoholic steatohepatitis (ASH) Cryptogenic cirrhosis	 1 (8%) 6 (46%) 1 (8%) 2 (15%) 2 (15%) 1 (8%)	-
Sex (n) male female	 11 (85%) 2 (15%)	-
Age (years)	60 (22; 74)	-
MELD § Score ≥15 points <15 points	12 (8; 22) 4 (31%) 9 (69%)	-
Child-Pugh Score [Classification] 5–6 points [stage A] 7–9 points [stage B] 10–15 points [stage C]	8 (6; 11) 1 (8%) 8 (61%) 4 (31%)	-
aPTT ‡ (s)	29 (26; 66)	24.4–32.4
Thrombocytes (cells/nL)	112 (29; 259)	140–320
INR †	1.12 (1.05; 1.41)	-
Fibrinogen (mg/dL)	269 (101; 627)	180–350
Hemoglobin (g/dL)	10.6 (6.6; 14.5)	13.7–17.2
Bilirubin total (mg/dL)	1.0 (0.6; 3.7)	0.3–1.2
Creatinin (mg/dL)	1.1 (0.6; 5.0)	0.9–1.3
Albumin serum (g/dL)	2.3 (2.1; 2.8)	3.4–4.8
CRP (mg/dL)	1.3 (0.5; 13.7)	<0.5
Ascites total protein (mg/dL)	2086 (461; 3062)	-
Indication for TIPS variceal bleeding [high urgent procedure] refractory ascites [elective procedure]	 3 (23%) 10 (77%)	-

§ Model of End Stage Liver Disease; ‡ activated partial thromboplastin time; † International Normalized Ratio.

2.2. Study Design

This project, which was conceived as a monocentric pilot cohort study, included patients with liver cirrhosis ($n = 13$) undergoing TIPS implantation. Patients were assigned into three categories (cirrhosis: mild vs. moderate vs. severe) according to the Child-Pugh (CP) Score or into two categories (cirrhosis: mild/moderate vs. severe) according to the Model of End Stage Liver Disease (MELD) Score [21]:

- mild/moderate liver cirrhosis: MELD < 15 or CP 5–6 points [stage A: mild cirrhosis] or CP 7–9 points [stage B: moderate cirrhosis]
- severe liver cirrhosis: MELD ≥ 15 or CP 10–15 points [stage C]

The majority of patients with MELD < 15 (i.e., 8 of 9 patients) had a moderate cirrhosis (CP stage B). Furthermore, a median MELD score of 12 and a median CP score of 8 (i.e., stage B) has been calculated for thirteen patients included in this pilot study.

According to research protocol, patients were followed up for 30 days from the day of TIPS implantation regarding potential development of TT and/or PVT. Endpoint was to investigate and characterize coagulation profiles in the PERC in comparison to the PORC in patients with liver cirrhosis using rotational thromboelastometry (ROTEM®, ©Tem Innovations GmbH, Martin-Kollar-Strasse 13–15, D-81829, Munich, Germany) and check for potential associations with TT and/or PVT.

2.3. Blood Sampling

Patients ($n = 13$), whose blood samples were included in final analysis, were selected as a "convenience subset" by means of on-site availability (i.e., available vs. not available) of the respective ROTEM® delta device during TIPS implantation. For additional information and further clarification, readers are directed to the Discussion section.

PERC blood samples were collected from the median cubital vein (MCV). PORC blood samples were withdrawn from the prehepatic venous compartment, i.e., the portal vein (PV), from inside the stent lumen during the TIPS placement (TIPS), and from the posthepatic venous system at the level of the confluence of three hepatic veins to the inferior cava vein (HV/ICV). Blood sampling occurred in the following order: MCV, HV/ICV, PV, TIPS.

Overall, blood sampling was performed using a 5.4 mL Coagulation SARSTEDT Monovette® (Sarstedt, Germany). All PORC samples were collected as rest materials following standard TIPS protocol that foresees X-ray control and aspiration of the catheter as two safeguards used for placing the catheter in an optimal position; the use of this TIPS protocol was approved by the local institutional review board. PERC blood samples were obtained from an 18-G cannula that was routinely placed in the left MCV to ensure safe circulation management. Finally, all samples were subjected to ROTEM® analysis.

2.4. Rotational Thromboelastometry

According to the manufacturer's instructions, we chose the ROTEM® delta device to perform the coagulation assay EXTEM [22]. In this extrinsic screening test, coagulation was triggered by 20 µL tissue factor and 20 µL CaCl2 0.2 mol/l. ROTEM® values included clotting time (CT), clot formation time (CFT), maximum clot firmness (MCF), and maximum lysis (ML). Reference values have been defined according to the reference range stated by the manufacturer (©Tem Innovations GmbH, Martin-Kollar-Strasse 13–15, D-81829, Munich, Germany). Evaluation of ML was performed after 60 min.

2.5. Statistical Analysis

ROTEM® measurements are presented as median (minimum to maximum, min; max; Table 2) or mean (±standard deviation, SD; Figure 1) values. Comparison of ROTEM® values (mean ± SD) between the sample groups, i.e., MCV vs. HV/ICV vs. PV vs. TIPS, was performed using the one-way analysis of variance (ANOVA). Overall, $p < 0.05$ was considered significant. Calculations and graphs were generated using GraphPad version 5.00 for Windows, GraphPad Software, San Diego, CA, USA.

Table 2. Rotational Thromboelastometry (ROTEM®) measurements shown as median (min; max).

EXTEM Value	MCV	HV/ICV	PV	TIPS	Normal Range
CT, s	72 (57; 104)	69 (57; 100)	70 (59; 93)	74 (59; 110)	35–80
CFT, s	98 (78; 486)	136 (78; 321)	106 (75; 316)	112 (79; 368)	35–160
MCF, mm	57 (33; 66)	53 (37; 64)	57 (39; 66)	56 (37; 65)	53–72
ML, %	10 (4; 17)	12 (3; 19)	9 (3; 15)	9 (4; 16)	<15

MCV: median cubital vein; HV/ICV: confluence of three hepatic veins "HV" to the inferior cava vein "ICV"; PV: portal vein; TIPS: transjugular intrahepatic portosystemic shunt; CT: clotting time; CFT: clot formation time; MCF: maximum clot firmness; ML: maximum lysis; s: seconds; mm: millimeters.

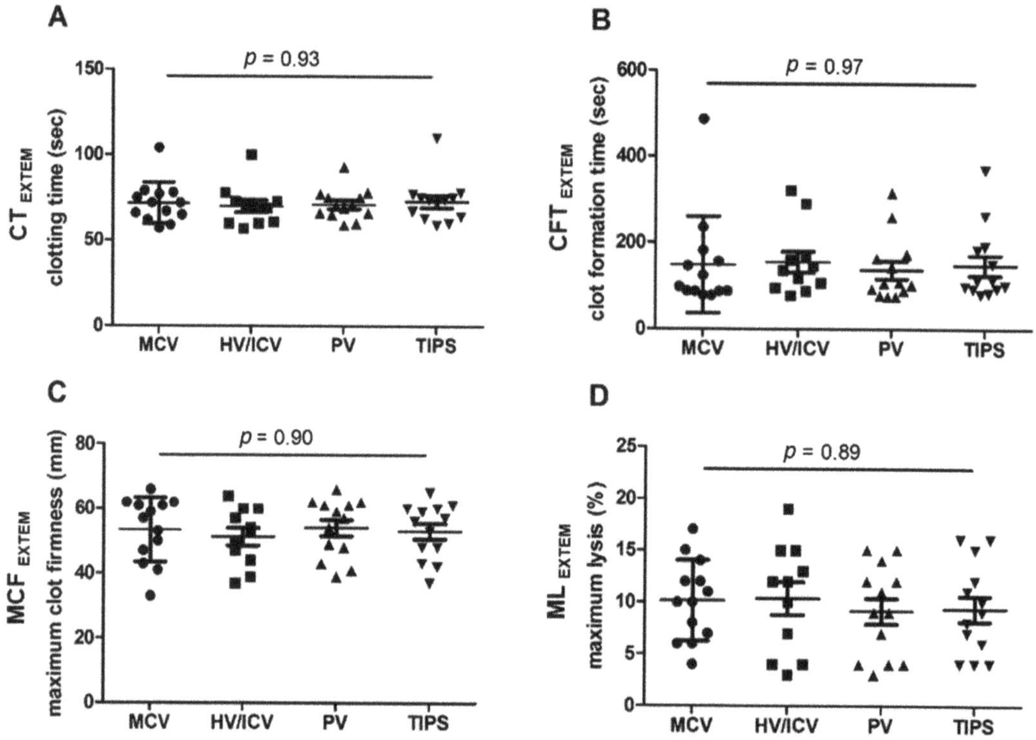

Figure 1. EXTEM/CT (**A**), EXTEM/CFT (**B**), EXTEM/MCF (**C**), and EXTEM/ML (**D**) measurements during TIPS implantation in patients with liver cirrhosis (n = 13) in peripheral (MCV) vs. pre-hepatic (PV, TIPS) vs. post-hepatic (HV/ICV) blood samples. Results are presented as mean ± standard deviation. Clotting time (CT); clot formation time (CFT); maximum clot firmness (MCF); maximum lysis (ML): reduction of the clot firmness (%) after MCF (in relation to MCF). Seconds (s); millimeters (mm). Median cubital vein (MCV); confluence of three hepatic veins to the inferior cava vein (HV/ICV); portal vein (PV); transjugular intrahepatic portosystemic shunt (TIPS).

3. Results

3.1. Demographic Data

This pilot study included thirteen patients with liver cirrhosis, who underwent TIPS (Table 1). The most common indication for TIPS was refractory ascites (n = 10, 76.9%). The median age was 60 (min.: 22; max.: 74) years. Liver disease severity assessment revealed a median MELD Score of 12 (min.: 8; max.: 22). Median INR was 1.12 (min.: 1.05; max.: 1.41) and median fibrinogen concentration was 269 mg/dL (min.: 101; max.: 627). Most of the patients had thrombocytopenia due to portal hypertension and splenomegaly with a median platelet count/nL of 112 (min. 29; max.: 259).

During the follow-up time of 30 days after TIPS, none of the patients developed TT and/or PVT.

3.2. TIPS Implantation and Procedure Related Data

Covered stents were placed in 13 (100%) patients. The stent diameters were 10 mm (n = 1, 7.8%), 8 mm (n = 7, 53.8%), and 6 mm (n = 5, 38.4%). The mean stent length was 7.8 ± 0.4 cm. The following lengths were used: 7 and 8 cm in 3 and 10 patients, respectively. Each stent was 2 cm longer in full length; this portion was uncovered. An adjunctive stent of 5 cm (+2 cm) was required to cover long track in one patient who initially received a stent of 7 cm (+2 cm).

The portal pressure gradient (PPG) was reduced from a mean of 22.1 ± 6.7 mm Hg to 8.9 ± 2.4 mm Hg. The mean decrease in post-TIPS PPG was 13.2 ± 4.2 mm Hg. Recurrent bleeding after TIPS placement did not occur during follow-up in any of the patients undergoing emergency TIPS due to variceal bleeding. One month after TIPS placement, complete or partial ascites response was seen in 100% of the cirrhotic patients (n = 10) undergoing TIPS due to refractory ascites. Gut decontamination was performed with oral lactulose (10–30 mL three times per day) after TIPS implantation for at least 7 days in all patients. Patients undergoing emergency TIPS due to variceal bleeding (n = 3) also received Ceftriaxon (2 g/24 h) intravenously for 7 days.

3.3. *EXTEM Analysis in Blood Samples from Different Vascular Compartments during TIPS*

EXTEM analysis showed no statistically significant differences for clotting time (CT, p = 0.93; Figure 1A), clot formation time (CFT, p = 0.97; Figure 1B), maximum clot firmness (MCF, p = 0.90; Figure 1C), or maximum lysis (ML, p = 0.89; Figure 1D) between the compartments, i.e., MCV vs. HV/ICV vs. PV vs. TIPS. Figure 1A–D shows mean ± standard deviation of CT, CFT, MCF, and ML:

- Mean CT was 72 ± 12 s in MCV blood samples, 70 ± 12 s in HV/ICV blood samples, 71 ± 9 s in PV blood samples, and 73 ± 13 s in TIPS blood samples.
- Mean CFT was 149 ± 112 s in MCV blood samples, 155 ± 80 s in HV/ICV blood samples, 137 ± 75 s in PV blood samples, and 148 ± 85 s in TIPS blood samples.
- Mean MCF was 53 ± 10 mm in MCV blood samples, 51 ± 9 mm in HV/ICV blood samples, 54 ± 9 mm in PV blood samples, and 53 ± 9 mm in TIPS blood samples.
- Mean ML was 10 ± 4% in MCV blood samples, 10 ± 5% in HV/ICV blood samples, 9 ± 4% in PV blood samples, and 9 ± 4% in TIPS blood samples.

In addition, medians were not significantly different between the compartments, i.e., MCV vs. HV/ICV vs. PV vs. TIPS. Median (min; max) as well as normal range values for CT, CFT, MCF, and ML are summarized in Table 2.

4. Discussion

This analysis, which has been based on four thromboelastometric variables in EXTEM, i.e., CT, CFT, MCF, and ML, detected no differences in coagulation function between blood samples withdrawn from the PERC versus the PORC in a convenience subset of patients with liver cirrhosis. These findings are in contrast to previous reports which suggest overall increased clotting activity in the PORC relative to the PERC in cirrhotic patients [14,15]. Thus, using rotational thromboelastometry, evaluation of coagulation function in peripheral blood—before TIPS implantation—appears to reliably depict coagulation function in portal blood among patients with liver cirrhosis.

As none of the thirteen TIPS recipients developed TT or PVT within 30 days post-intervention, this research failed to investigate the relation between potentially exaggerated coagulation in the PORC and thrombotic complications during and/or after TIPS in the setting of liver cirrhosis. In fact, based on consistent EXTEM measurements via CT and MCF in PERC and PORC blood samples, none of these thirteen patients was evaluated as being at risk for TT and/or PVT. Clearly, thirteen—by means of convenience sampling selected—patients with liver disease who have shown no prothrombotic alterations (and no significant differences) according to EXTEM within (or between) the PORC and the PERC, have indeed not developed TIPS thrombosis. Considering that none of these thirteen patients had received anticoagulant therapy before TIPS, these limited (but consistent) results may actually support causality. Again, based on rotational thromboelastometry, our data do not support the scenario of different coagulation states between the PERC and the PORC in patients with liver cirrhosis.

For the purposes of this study, we only applied EXTEM because this test delivers the most informative results of ROTEM®. EXTEM evaluates function of the extrinsic pathway, whereas EXTEM/CT provides information similar to the prothrombin time. EXTEM/MCF, which depends on platelets and/or fibrinogen concentration and/or function, evaluates clot stability. EXTEM/CFT, which also depends on platelets and fibrinogen, evaluates clot propagation. EXTEM/ML corresponds to the stability of the blood clot against fibrinolytic processes. In line with this approach, previous research also assessed coagulation activation, clot formation, and fibrinolysis by applying EXTEM as the main screening test [23].

As expected, application of oral lactulose for gut decontamination towards reduction of the risk of infection and/or hepatic encephalopathy after TIPS was effective in our cohort [24]. Intravenous antibiotic therapy with Ceftriaxon for 7 days was administered according to the guidelines for gastrointestinal bleeding in cirrhosis that can be found in the Baveno Statement [25]. Against this background no complications in terms of infection, recurrent bleeding and/or hepatic encephalopathy occurred among the study participants within 4 weeks after TIPS.

To the best of our knowledge, the present monocentric pilot study has been the first to evaluate coagulation activity in the PORC—versus PERC—using rotational thromboelastometry in patients with liver cirrhosis. However, it has various limitations that warrant special attention. First, these findings may be of limited statistical validity due to the small sample size of our cohort. Second, rotational thromboelastometry neglects the role of the vascular endothelium during the coagulation process [26]. Third, due to limited on-site availability of ROTEM® delta devices and the mandatory prioritization of critical emergencies, e.g., hemorrhages over measurements for research purposes, we have been able to timely perform rotational thromboelastometry tests (with blood samples collected as safeguards from the PORC and/or rest materials from the PERC) only in approximately 40% of all patients that underwent TIPS within 12 months (2016–2017). Most prominently, limitations of this study include that no patients developed TT or PVT, and thus, its validity in external populations of patients with liver cirrhosis undergoing TIPS remains to be further investigated. Moreover, all ROTEM measurements were performed by convenience sampling. Consequently, patients (n = 13) included in the final analysis should be considered a "convenience subset". Furthermore, as a single center research, a higher risk of bias may apply versus other multicenter trials. These issues should be addressed in larger cohorts with unbiased recruitment in the future.

Taken together, the herein presented methodic approach, once validated in larger studies, may provide a useful tool for individualized risk estimation for TT and/or PVT in patients with liver cirrhosis undergoing TIPS.

5. Conclusions

Further efforts to gain more differentiated insights into coagulation physiology in the PORC in comparison to the PERC—by means of viscoelastic tests such as ROTEM®—may lead to the establishment of personalized and perhaps more accurate thrombosis risk assessment and treatment strategies. Thus, we hope to stimulate the interest of the scientific community for more intense research on regulation of coagulation pathways within the gut-liver axis.

Ethical Approval

This research was performed in accordance with the Declaration of Helsinki of 1975, as revised in 2008, and the guidelines of the International Conference for Harmonization for Good Clinical Practice and was approved by the local institutional review board (IRB: Ethik Kommission am Universitätsklinikum Essen; IRB protocol no. 18-8184-BO). The IRB has found and documented that this research involves no more than minimal risk to the subjects; therefore, the IRB waived the requirements to obtain additional informed consents for collection of PORC blood samples since this step has been executed as a part of standard medical aid practices and/or as rest materials.

Author Contributions: P.M.: analysis and interpretation of data, drafting of the article, critical revision for important intellectual content. S.B.: analysis and interpretation of data, drafting of the article, critical revision for important intellectual content. M.B.: selection of patients and data acquisition, technical support for ROTEM® measurements. M.J.: interpretation and curation of data.

A.C.: investigation (data interpretation). J.M.T.: methodology (TIPS Interventions and technical support), data acquisition/validation. A.K.: study concept, study design, data acquisition, analysis and interpretation of data, drafting of the article. All authors have read and agreed to the published version of the manuscript.

Funding: The authors declare that no funds, grants, or other supports were received during preparation of this manuscript.

Institutional Review Board Statement: The study was conducted in accordance with the Declaration of Helsinki, and approved by the local institutional review board (IRB: Ethik Kommission am Universitätsklinikum Essen; IRB protocol no. 18-8184-BO; date of IRB approval 20 August 2018).

Informed Consent Statement: Standard written informed consent was obtained from all patients for being included in the study.

Data Availability Statement: The data that support the findings of this study are available from the corresponding author, upon reasonable request.

Conflicts of Interest: The authors declare no conflict of interest.

References

1. Rossle, M. TIPS: 25 years later. *J. Hepatol.* **2013**, *59*, 1081–1093. [CrossRef] [PubMed]
2. Kochar, N.; Tripathi, D.; McAvoy, N.C.; Ireland, H.; Redhead, D.N.; Hayes, P.C. Bleeding ectopic varices in cirrhosis: The role of transjugular intrahepatic portosystemic stent shunts. *Aliment. Pharmacol. Ther.* **2008**, *28*, 294–303. [CrossRef] [PubMed]
3. Kuiper, J.J.; de Man, R.A.; van Buuren, H.R. Review article: Management of ascites and associated complications in patients with cirrhosis. *Aliment. Pharmacol. Ther.* **2007**, *26* (Suppl. 2), 183–193. [CrossRef] [PubMed]
4. Attwell, A.; Ludkowski, M.; Nash, R.; Kugelmas, M. Treatment of Budd-Chiari syndrome in a liver transplant unit, the role of transjugular intrahepatic porto-systemic shunt and liver transplantation. *Aliment. Pharmacol. Ther.* **2004**, *20*, 867–873. [CrossRef] [PubMed]
5. Tripathi, D.; Macnicholas, R.; Kothari, C.; Sunderraj, L.; Al-Hilou, H.; Rangarajan, B.; Chen, F.; Mangat, K.; Elias, E.; Olliff, S. Good clinical outcomes following transjugular intrahepatic portosystemic stent-shunts in Budd-Chiari syndrome. *Aliment. Pharmacol. Ther.* **2014**, *39*, 864–872. [CrossRef]
6. Cardenas, A.; Kelleher, T.; Chopra, S. Review article: Hepatic hydrothorax. *Aliment. Pharmacol. Ther.* **2004**, *20*, 271–279. [CrossRef]
7. Buechter, M.; Manka, P.; Gerken, G.; Canbay, A.; Blomeyer, S.; Wetter, A.; Altenbernd, J.; Kahraman, A.; Theysohn, J.M. Transjugular Intrahepatic Portosystemic Shunt in Patients with Portal Hypertension: Patency Depends on Coverage and Interventionalist's Experience. *Dig. Dis.* **2018**, *36*, 218–227. [CrossRef]
8. Wan, Y.M.; Li, Y.H.; Wu, H.M.; Xu, Z.Y.; Xu, Y.; Yang, L.H.; Wu, X.N.; Yang, J.H. Portal vein thrombosis before and after transjugular intrahepatic portosystemic shunt placement: An observational study (STROBE compliant). *Medicine* **2017**, *96*, e8498. [CrossRef]
9. Jahangiri, Y.; Kerrigan, T.; Li, L.; Prosser, D.; Brar, A.; Righetti, J.; Schenning, R.C.; Kaufman, J.A.; Farsad, K. Risk factors for stent graft thrombosis after transjugular intrahepatic portosystemic shunt creation. *Cardiovasc. Diagn. Ther.* **2017**, *7*, S150–S158. [CrossRef]
10. Lv, Y.; He, C.; Wang, Z.; Guo, W.; Wang, J.; Bai, W.; Zhang, L.; Wang, Q.; Liu, H.; Luo, B.; et al. Association of Nonmalignant Portal Vein Thrombosis and Outcomes after Transjugular Intrahepatic Portosystemic Shunt in Patients with Cirrhosis. *Radiology* **2017**, *285*, 999–1010. [CrossRef]
11. Tripodi, A. Tests of coagulation in liver disease. *Clin. Liver Dis.* **2009**, *13*, 55–61. [CrossRef] [PubMed]
12. Tsochatzis, E.A.; Senzolo, M.; Germani, G.; Gatt, A.; Burroughs, A.K. Systematic review: Portal vein thrombosis in cirrhosis. *Aliment. Pharmacol. Ther.* **2010**, *31*, 366–374. [CrossRef]
13. Moore, K.L.; Andreoli, S.P.; Esmon, N.L.; Esmon, C.T.; Bang, N.U. Endotoxin enhances tissue factor and suppresses thrombomodulin expression of human vascular endothelium in vitro. *J. Clin. Investig.* **1987**, *79*, 124–130. [CrossRef] [PubMed]
14. Lumsden, A.B.; Henderson, J.M.; Kutner, M.H. Endotoxin levels measured by a chromogenic assay in portal, hepatic and peripheral venous blood in patients with cirrhosis. *Hepatology* **1988**, *8*, 232–236. [CrossRef] [PubMed]
15. Violi, F.; Ferro, D. Clotting activation and hyperfibrinolysis in cirrhosis: Implication for bleeding and thrombosis. *Semin. Thromb. Hemost.* **2013**, *39*, 426–433. [CrossRef] [PubMed]
16. Violi, F.; Ferro, D.; Basili, S.; Lionetti, R.; Rossi, E.; Merli, M.; Riggio, O.; Bezzi, M.; Capocaccia, L. Ongoing prothrombotic state in the portal circulation of cirrhotic patients. *Thromb. Haemost.* **1997**, *77*, 44–47. [CrossRef] [PubMed]
17. Lin, R.S.; Lee, F.Y.; Lee, S.D.; Tsai, Y.T.; Lin, H.C.; Lu, R.H.; Hsu, W.C.; Huang, C.C.; Wang, S.S.; Lo, K.J. Endotoxemia in patients with chronic liver diseases: Relationship to severity of liver diseases, presence of esophageal varices, and hyperdynamic circulation. *J. Hepatol.* **1995**, *22*, 165–172. [CrossRef]
18. Mallett, S.V. Clinical Utility of Viscoelastic Tests of Coagulation (TEG/ROTEM) in Patients with Liver Disease and during Liver Transplantation. *Semin. Thromb. Hemost.* **2015**, *41*, 527–537. [CrossRef]

19. Saner, F.H.; Gieseler, R.K.; Akiz, H.; Canbay, A.; Gorlinger, K. Delicate balance of bleeding and thrombosis in end-stage liver disease and liver transplantation. *Digestion* **2013**, *88*, 135–144. [CrossRef]
20. Bedreli, S.; Sowa, J.P.; Gerken, G.; Saner, F.H.; Canbay, A. Management of acute-on-chronic liver failure: Rotational thromboelastometry may reduce substitution of coagulation factors in liver cirrhosis. *Gut* **2016**, *65*, 357–358. [CrossRef]
21. Kamath, P.S.; Wiesner, R.H.; Malinchoc, M.; Kremers, W.; Therneau, T.M.; Kosberg, C.L.; D'Amico, G.; Dickson, E.R.; Kim, W.R. A model to predict survival in patients with end-stage liver disease. *Hepatology* **2001**, *33*, 464–470. [CrossRef] [PubMed]
22. Morrison-McKell, T.; Macik, B.G. 40-Point-of-Care Hemostasis Testing. In *Consultative Hemostasis and Thrombosis*, 3rd ed.; Kitchens, C.S., Kessler, C.M., Konkle, B.A., Eds.; W.B. Saunders: Philadelphia, PA, USA, 2013; pp. 717–729. [CrossRef]
23. Vorweg, M.; Hartmann, B.; Knuttgen, D.; Jahn, M.C.; Doehn, M. Management of fulminant fibrinolysis during abdominal aortic surgery. *J. Cardiothorac. Vasc. Anesth.* **2001**, *15*, 764–767. [CrossRef] [PubMed]
24. Patidar, K.R.; Sydnor, M.; Sanyal, A.J. Transjugular intrahepatic portosystemic shunt. *Clin. Liver Dis.* **2014**, *18*, 853–876. [CrossRef] [PubMed]
25. de Franchis, R. Baveno VIF. Expanding consensus in portal hypertension: Report of the Baveno VI Consensus Workshop: Stratifying risk and individualizing care for portal hypertension. *J. Hepatol.* **2015**, *63*, 743–752. [CrossRef]
26. Zipperle, J.; Schlimp, C.J.; Holnthoner, W.; Husa, A.M.; Nurnberger, S.; Redl, H.; Schochl, H. A novel coagulation assay incorporating adherent endothelial cells in thromboelastometry. *Thromb. Haemost.* **2013**, *109*, 869–877. [CrossRef] [PubMed]

MDPI
St. Alban-Anlage 66
4052 Basel
Switzerland
www.mdpi.com

Journal of Personalized Medicine Editorial Office
E-mail: jpm@mdpi.com
www.mdpi.com/journal/jpm

www.ingramcontent.com/pod-product-compliance
Lightning Source LLC
Chambersburg PA
CBHW061607210926
51850CB00009B/2248
9783725815289